A Short Course in General Relativity

Third Edition

James Foster J. David Nightingale

A Short Course in General Relativity

Third Edition

With 51 Illustrations

James Foster
formerly Senior Lecturer in Mathematics
University of Sussex
Brighton
UK

J. David Nightingale
Emeritus Professor of Physics
State University of New York
New Paltz, NY
USA

Library of Congress Control Number: 2005927384

ISBN-10: 0-387-26078-1 e-ISBN 0-387-27583-5
ISBN-13: 978-0387-26078-5

Printed on acid-free paper.

Printed in the United States of America. (EB)

9 8 7 6 5 4 3 2 1

springeronline.com

Preface

This book is a short introduction to general relativity, intended primarily as a one-semester course for first-year graduate students (or for seniors) in physics, or in related subjects such as astrophysics. While we expect such students to have been exposed to special relativity in their introductory modern physics courses (most likely in their sophomore year) it is unlikely that they have used the standard 4-vector methods, and so we supply such a review in Appendix A. We strongly advise reading Appendix A first.

Most students approaching general relativity require an introduction to tensors, and these are dealt with in Chapter 1 and the first half of Chapter 2, where geodesics, absolute and covariant differentiation, and parallel transport are discussed. This enables us to discuss the spacetime of general relativity in the latter half of the chapter and takes us on to a discussion of the field equations in Chapter 3. In Chapter 4 the results learned are applied to physics in the vicinity of a massive object, where we have tried to compare general relativistic results with their Newtonian counterparts. Chapters 5 and 6, on gravitational radiation and the elements of cosmology, respectively, give further applications of the theory, but students wanting a more detailed knowledge of these topics (and indeed all topics) would have to turn to the texts referred to in the body of the book.

Over the years, a version of this course has been offered variously (by JDN) at the University of Mississippi (Ole Miss), at Bard College, and at SUNY New Paltz, as well as (by JF) at the University of Sussex. It was often found that there was not enough time for Chapters 5 and 6, unless one made judicious cuts elsewhere. A few cuts may be made in the first two chapters, but it would probably be better to omit either Chapter 5, or Chapter 6 (or both) than to omit Appendix A, since a sound knowledge of the 4-vector formalism of special relativity is an essential prerequisite.

Exercises have been provided at the end of most sections and problems at the end of chapters. The former are often quite straightforward (but possibly tedious) verifications needed for a first reading of the book, while the latter are suitable for homework-type problems.

The original version of this book was published in 1979, with translations into other languages following in the 1980s and 1990s. That version placed mathematical demands on the reader which were not entirely appropriate for a physics student, requiring him or her to acquire mathematical skills beyond what is needed for a first course in general relativity. In the second edition, the mathematical sections were completely reorganized and rewritten, so as to make the text more accessible to the physics student who had the kind of background gained from following a course in vector calculus, with applications to field theories such as Newtonian gravitation and Maxwell's theory of electromagnetism. However, for the third edition, we have restored in Appendix C much of the original material on tensors and manifolds missing from the second edition. This would be appropriate reading for mathematics majors seeking a more formal approach to tensors than physics students might desire.

The third edition also includes some minor updating. In the chapter concerning physics in the vicinity of a massive object (Chapter 4) we have added a short section on the Kerr solution and its relevance to the *Gravity Probe B* experiment launched in 2004; and the chapter on cosmology (Chapter 6) has been supplemented by two sections concerning redshift and galaxy recession with speeds greater than that of light.

With gratitude, mention must be made of John Ray, Richard Halpern, Peter Skiff, Jeffrey Dunham, and Tarun Biswas, all of whom have been of assistance in one way or another. Marc Bensadoun kindly supplied the figure showing the measurements of cosmic background microwave radiation in Chapter 6 and gave us permission to reproduce it here. The first edition of the book was completed with the help and exemplary typing skills of Jill Foster, whose transcription of the original text to computer files served as a foundation for its revision and conversion to LaTeX format. We are also grateful to J. Snider, M.E. Horn, and N.B. Speyer for providing us with lists of errors from the first and second editions.

We have also included at the end of the book outline solutions (which are not model answers, and for which the student must supply all the details). Further, a beautifully written and detailed solution set for the exercises and problems is available from Professor J.S. Dunham, Department of Physics, Middlebury College, Middlebury, Vermont 05753.

David Nightingale, New Paltz
James Foster, Dumfries
January 2005

Contents

Appendices

Introduction

The originator of the general theory of relativity was Einstein, and in 1919 he wrote[1]: *The special theory, on which the general theory rests, applies to all physical phenomena with the exception of gravitation; the general theory provides the law of gravitation and its relation to the other forces of nature.* The claim that the general theory provides the law of gravitation does not mean that H.G. Wells' Mr Cavor could now introduce an antigravity material and glide up to the Moon, nor, for example, that we might produce intense permanent gravitational fields in the laboratory, as we can electric fields. It means only that all the properties of gravity of which we are aware are explicable by the theory, and that gravity is essentially a matter of geometry. Before saying how we get to the general from the special theory, we must first discuss the principle of equivalence.

In electrostatics, when a test particle of charge q and inertial mass m_i is placed in a static field $\mathbf{E}$, it experiences a force $q\mathbf{E}$ and undergoes an acceleration $\mathbf{a}$ given by

$$\mathbf{a} = (q/m_i)\mathbf{E}. \tag{0.1}$$

In contrast, a test particle of gravitational mass m_g and inertial mass m_i placed in a gravitational field $\mathbf{g}$ experiences a force $m_g\mathbf{g}$ and undergoes an acceleration $\mathbf{a}$ given by

$$\mathbf{a} = (m_g/m_i)\mathbf{g}. \tag{0.2}$$

It is an experimental fact (known since Galileo's time) that different particles placed in the same gravitational field acquire the same acceleration (see Fig. 0.1). This implies that the ratio m_g/m_i appearing in equation (0.2) is the same for all particles, and by an appropriate choice of units this ratio may be taken to be unity. This equivalence of gravitational and inertial mass (which allows us to drop the qualification, and simply refer to mass) has been checked experimentally by Eötvös (in 1889 and 1922), and more recently and more accurately (to one part in 10^{11}) by Dicke and his co-workers (in the

[1] *The Times*, London, 28 November 1919.

1960s). In contrast, the ratio q/m_i occurring in equation (0.1) is not the same for all particles (see Fig. 0.1).

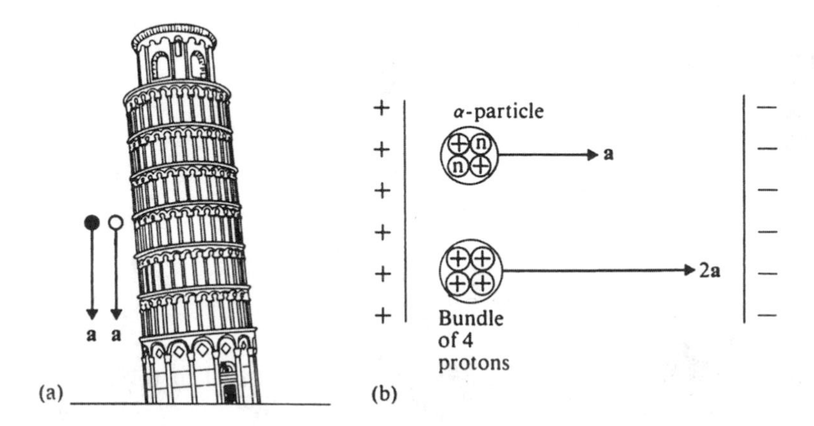

Fig. 0.1. Test particles in (a) a gravitational field, and (b) an electrostatic field.

Let us now consider the principle of equivalence, which it is instructive to do from the point of view of Einstein's freely falling elevator. If we consider a projectile shot from one side of the elevator cabin to the other, the projectile appears to go in a straight line (the elevator cable being cut) rather than in the usual curved trajectory. Projectiles that are released from rest relative to the cabin remain floating weightless in the cabin. Of course, if the cabin is left to fall for a long time, the particles gradually draw closer together, since they are falling down radial lines towards a common point which is the center of the Earth. However, if we make the proviso that the cabin is in this state for a short time, as well as being spatially small enough for the neglect of tidal forces in general, then the freely falling cabin (which may have X, Y, Z coordinates chalked on its walls, as well as a cabin clock measuring time T) looks remarkably like an inertial frame of reference, and therefore the laws of special relativity hold sway inside the cabin. (The cabin must not only occupy a small region of spacetime, it must also be nonrotating with respect to distant matter in the universe.[2]) All this follows from the fact that the acceleration of any particle relative to the cabin is zero because they both have the same acceleration relative to the Earth, and we see that the equivalence of inertial and gravitational mass is an essential feature of the discussion. We may incorporate these ideas into the *principle of equivalence*, which is this: *In*

[2]This statement is related to *Mach's principle*. For a discussion, see Weinberg, 1972, §3, and Anderson, 1967, Chap. 10.

a freely falling (nonrotating) laboratory occupying a small region of spacetime, the laws of physics are the laws of special relativity.[3]

As a result of the above discussion, the reader should not believe that we can actually transform gravity away by turning to a freely falling reference frame. It is absolutely impossible to transform away a permanent gravitational field of the type associated with a star (as we shall see in Chap. 3), but it is possible to get closer and closer to an ideal inertial reference frame if we make our laboratory occupy smaller and smaller regions of spacetime.

The way in which Einstein generalized the special theory so as to incorporate gravitation was extremely ingenious, and without precedent in the history of science. Gravity was no longer to be regarded as a force, but as a manifestation of the curvature of spacetime itself. The new theory, known as the *general theory of relativity* (or *general relativity* for short), yields the special theory as an approximation in exactly the way that the principle of equivalence requires. Because of the curvature of spacetime, it cannot be formulated in terms of coordinate systems based on inertial frames, as the special theory can, and we therefore use arbitrary coordinate systems. Indeed, global inertial frames can no longer be defined, the nearest we can get to them being freely falling nonrotating frames valid in limited regions of spacetime only. A full explanation of what is involved is given in Chapter 2, but we can give a limited preview here.

In special relativity, the invariant expression which defines the proper time τ is given by

$$c^2 d\tau^2 = \eta_{\mu\nu} dX^\mu dX^\nu, \tag{0.3}$$

where the four coordinates X^0, X^1, X^2, X^3 are given in terms of the usual coordinates T, X, Y, Z by

$$X^0 \equiv cT, \quad X^1 \equiv X, \quad X^2 \equiv Y, \quad X^3 \equiv Z. \tag{0.4}$$

(See Sec. A.0, but note the change to capital letters. See also Sec. 1.2 for an explanation of the summation convention.) If we change to arbitrary coordinates x^μ, which may be defined in terms of the X^μ in any way whatsoever (they may, e.g., be linked to an accelerating or rotating frame), then expression (0.3) takes the form

$$c^2 d\tau^2 = g_{\mu\nu} dx^\mu dx^\nu, \tag{0.5}$$

where

$$g_{\mu\nu} = \eta_{\rho\sigma} \frac{\partial X^\rho}{\partial x^\mu} \frac{\partial X^\sigma}{\partial x^\nu}.$$

This follows from the fact that $dX^\rho = (\partial X^\rho / \partial x^\mu) dx^\mu$. In terms of the coordinates X^μ, the equation of motion of a free particle is

$$d^2 X^\mu / d\tau^2 = 0, \tag{0.6}$$

[3]Some authors distinguish between weak and strong equivalence. Our statement is the strong statement; the weak one refers to freely falling particles only, and not to the whole of physics.

which, in terms of the arbitrary coordinates, becomes

$$\frac{d^2x^\mu}{d\tau^2} + \Gamma^\mu_{\nu\sigma}\frac{dx^\nu}{d\tau}\frac{dx^\sigma}{d\tau} = 0, \tag{0.7}$$

where

$$\Gamma^\mu_{\nu\sigma} = \frac{\partial x^\mu}{\partial X^\rho}\frac{\partial^2 X^\rho}{\partial x^\nu \partial x^\sigma},$$

as a short calculation (involving the chain rule) shows. Einstein's proposals for the general theory were that in any coordinate system the proper time should be given by an expression of the form (0.5) and that the equation of motion of a free particle (i.e., one moving under the influence of gravity alone, gravity no longer being a force) should be given by an expression of the form (0.7), but that (in contrast to the spacetime of special relativity) *there are no preferred coordinates* X^μ which will reduce these to the forms (0.3) and (0.6). This is the essential difference between the spacetimes of special and general relativity. The curvature of spacetime (and therefore gravity) is carried by the $g_{\mu\nu}$, and as we shall see, there is a sense in which these quantities may be regarded as gravitational potentials. We shall also see that the $\Gamma^\mu_{\nu\sigma}$ are determined by the $g_{\mu\nu}$, and that it is always possible to introduce *local* inertial coordinate systems of *limited extent* in which $g_{\mu\nu} \approx \eta_{\mu\nu}$ and $\Gamma^\mu_{\nu\sigma} \approx 0$, so that equations (0.3) and (0.6) hold as approximations. We thus recover special relativity as an approximation, and in a way which ties in with our discussion of the principle of equivalence.

Because the introduction of curvature forces us to use arbitrary coordinate systems, we need to formulate the theory in a way which is valid in all coordinate systems. This we do by using tensor fields, the mathematics of which is developed in Chapter 1; the way these fit into the theory is explained in Chapter 2. It might be thought that this arbitrariness causes problems, because the coordinates lose the simple physical meanings that the preferred coordinates X^μ of special relativity have. However, we still have contact with the special theory at the local level, and in this way problems of physical meaning and the correct formulation of equations may be overcome. The basic idea is contained in the principle of general covariance, which may be stated as follows: *A physical equation of general relativity is generally true in all coordinate systems if (a) the equation is a tensor equation (i.e., it preserves its form under general coordinate transformations), and (b) the equation is true in special relativity.* The way in which this principle works and the reason why it works are explained in Section 2.5.

General relativity should not only reduce to special relativity in the appropriate limit, it should also yield Newtonian gravitation as an approximation. Contacts and comparisons with Newtonian theory are made in Sections 2.6, 2.7, 2.8, and 2.9, and extensively in Chapter 4, where we discuss physics in the vicinity of a massive object. These reveal differences between the two theories which provide possible experimental tests of the general theory, and for convenience we list here the experimental and observational evidence concerning these tests.

1. *Perihelion advance.* General relativity predicts an anomalous advance of the perihelion of planetary orbits. The following (and many more) observations exist for the solar system[4]:
   Mercury $43.11 \pm 0.45''$ per century,
   Venus $8.4 \pm 4.8''$ per century,
   Earth $5.0 \pm 1.2''$ per century.
   The predicted values are $43.03''$, $8.6''$, and $3.8''$, respectively.
2. *Deflection of light.* General relativity predicts that light deviates from rectilinear motion near massive objects. The following (and many more) observed deflections exist for light passing the Sun at grazing incidence:
   1919 Greenwich Observatory $1.98 \pm 0.16''$,
   1922 Lick Observatory $1.82 \pm 0.20''$,
   1947 Yerkes Observatory $2.01 \pm 0.27''$,
   1972 Mullard Radio Observatory, Cambridge
   (using radio sources and interferometers) $1.82 \pm 0.14''$.
   The predicted value is $1.75''$.
3. *Spectral shift.* General relativity predicts that light emanating from near a massive object is red-shifted, while light falling towards a massive object is blue-shifted. Numerous observations of the spectra of white dwarfs, as well as the remarkable terrestrial experiments carried out at the Jefferson Laboratory[5] verify the general-relativistic prediction.
4. *Time delay in radar sounding.* General relativity predicts a time delay in radar sounding due to the gravitational field of a massive object. Experiments involving the radar sounding of Venus, Mercury, and the spacecrafts *Mariner 6* and *7*, performed in the 1960s and 1970s, have yielded agreement with the predicted values to well within the experimental uncertainties.[6]
5. *Geodesic effect.* General relativity predicts that the axis of a gyroscope which is freely orbiting a massive object should precess. For a gyroscope in a near-Earth orbit this precession amounts to about $8''$ per year, and an experiment involving a gyroscope in an orbiting satellite was launched in 2004.[7]
6. *Frame-dragging.* General relativity predicts that the axis of a gyroscope in polar orbit around a spinning massive body will also precess, being 'dragged around' in the same sense as the spin. The same gyroscopes orbiting in the above 2004 experiment are to measure this tiny effect as well.[8]

---

[4]The figures are taken from Duncombe, 1956.
[5]See Pound and Rebka, 1960.
[6]See Shapiro, 1968; Shapiro et al., 1971; and Anderson et al., 1975.
[7]See Sec. 4.7.
[8]See Sec. 4.10.

While all the above effects are small for our solar system, some larger, and presumably general-relativistic, effects have been observed since 1975 for the two mutually orbiting neutron stars PSR 1913+16. However, Einstein's theory of general relativity does not directly treat *two* such massive objects, and Chapter 4 (on physics in the vicinity of a massive object) looks only at the motions of a test particle in the field of *one* massive object.

Finally, let us say something about the notation used in this book. Wherever possible we have chosen it to coincide with that of the more recent and influential texts on general relativity. For working in spacetime, we use Greek suffixes (μ, ν, etc.) and these have the range 0, 1, 2, 3, while for three-dimensional space we use lower-case English suffixes from the middle of the alphabet (i, j, etc.) and these have the range 1, 2, 3. For working on a two-dimensional surface, we use upper-case English suffixes from the beginning of the alphabet (A, B, etc.) and these have the range 1, 2. The signature of the metric tensor is -2, which means that $\eta_{00} = 1$, $\eta_{11} = \eta_{22} = \eta_{33} = -1$. Rather than use gravitational units in which the gravitational constant G and the speed of light c are unity, we have retained G and c throughout, except in Chapter 6 where $c = 1$. In the sections dealing with general tensor fields and curvature, the underlying space or manifold is of arbitrary dimension, and we have used lower-case English suffixes from the beginning of the alphabet (a, b, etc.) to denote the arbitrary range $1, 2, \ldots, N$. Where an equation defines some quantity or operation, the symbol $\equiv$ is used on its first occurrence, and occasionally thereafter as a reminder. Important equations are displayed in boxes.

1

Vector and tensor fields

1.0 Introduction

In this first chapter we concentrate on the algebra of vector and tensor fields, while postponing ideas that are based on the calculus of fields to Chapter 2. Our starting point is a consideration of vector fields in the familiar setting of three-dimensional Euclidean space and how they can be handled using arbitrary curvilinear coordinate systems. We then go on to extend and generalize these ideas in two different ways, first by admitting tensor fields, and second by allowing the dimension of the space to be arbitrary and its geometry to be non-Euclidean.[1] The eventual goal is to present a model for the spacetime of general relativity as a four-dimensional space that is curved, rather than flat. While some aspects of this model emerge in this chapter, it is more fully developed in Chapters 2 and 3, where we introduce some more mathematical apparatus and relate it to the physics of gravitation.

1.1 Coordinate systems in Euclidean space

In this and the next five sections we shall be working in three-dimensional Euclidean space. We shall take it to be equipped with a Cartesian system of coordinates (x, y, z) and an associated set of unit vectors $\{\mathbf{i}, \mathbf{j}, \mathbf{k}\}$, each pointing in the direction of the corresponding coordinate axis. We shall regard this Cartesian setup as a fixed and permanent feature of our Euclidean space; its purpose is to serve as a basic reference system for the description of other (generally non-Cartesian) coordinate systems.

Suppose then that we have an alternate coordinate system (u, v, w) that is non-Cartesian, such as spherical coordinates (r, θ, ϕ), as in the example below. We can express the Cartesian coordinates x, y, z in terms of u, v, w,

[1] We use the term *non-Euclidean* simply to mean *not Euclidean*. Mathematicians sometimes restrict the term to describe the geometries that arise as a result of modifying Euclid's parallel postulate.

$$x = x(u, v, w), \quad y = y(u, v, w), \quad z = z(u, v, w), \tag{1.1}$$

and, in principle, invert these to get u, v, w in terms of x, y, z. Through any point P with coordinates (u_0, v_0, w_0) there pass three *coordinate surfaces*, given by $u = u_0$, $v = v_0$, and $w = w_0$, which meet in *coordinate curves*. The following example serves to illustrate these ideas.

Example 1.1.1

For spherical coordinates we have

$$x = r\sin\theta\,\cos\phi, \quad y = r\sin\theta\,\sin\phi, \quad z = r\cos\theta, \tag{1.2}$$

where the conventional ranges for the coordinates are[2]

$$r \geq 0, \quad 0 \leq \theta \leq \pi, \quad 0 \leq \phi < 2\pi.$$

The coordinate surface $r = r_0$ is a sphere of radius r_0 (because $x^2 + y^2 + z^2 = r_0^2$), the coordinate surface $\theta = \theta_0$ is an infinite cone with its vertex at the origin and its axis vertical, and the coordinate surface $\phi = \phi_0$ is a semi-infinite plane with the z axis as its edge. (See Fig. 1.1.)

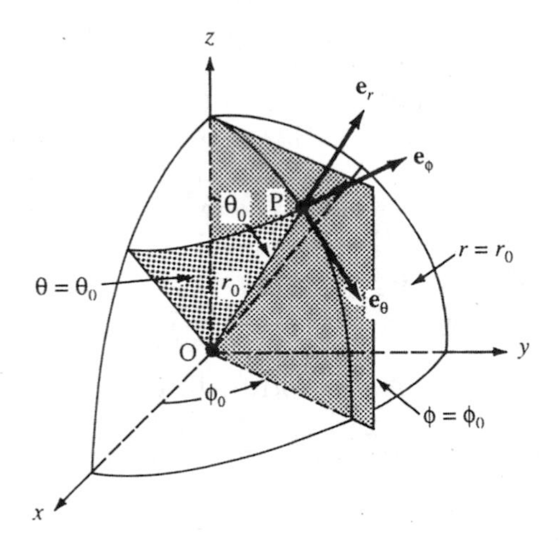

Fig. 1.1. The coordinate surfaces and coordinate curves of spherical coordinates.

The surfaces $\theta = \theta_0$ and $\phi = \phi_0$ intersect to give a coordinate curve which is a ray (part of a line) that emanates from O and passes through P; the surfaces $\phi = \phi_0$ and $r = r_0$ intersect to give a coordinate curve which is a semicircle having its endpoints on the z axis and passing through P; and the

[2]In practice, one usually lets ϕ wrap around and take all values, so that the ϕ coordinate of a point is unique only up to multiples of 2π.

surfaces $r = r_0$ and $\theta = \theta_0$ intersect to give a coordinate curve which is a horizontal circle passing through P with its center on the z axis.

The three equations (1.1) can be combined into a single vector equation that gives the position vector $\mathbf{r}$ of points in space as a function of the coordinates u, v, w that label the points:

$$\mathbf{r} = x(u, v, w)\,\mathbf{i} + y(u, v, w)\,\mathbf{j} + z(u, v, w)\,\mathbf{k}. \tag{1.3}$$

Setting w equal to the constant w_0, but leaving u, v to vary, gives

$$\mathbf{r} = x(u, v, w_0)\,\mathbf{i} + y(u, v, w_0)\,\mathbf{j} + z(u, v, w_0)\,\mathbf{k}, \tag{1.4}$$

which is a parametric equation for the coordinate surface $w = w_0$ in which the coordinates u, v play the rôle of parameters. Parametric equations for the other two coordinate surfaces arise similarly. If in equation (1.3) we set $v = v_0$ and $w = w_0$, but let u vary, then we get

$$\mathbf{r} = x(u, v_0, w_0)\,\mathbf{i} + y(u, v_0, w_0)\,\mathbf{j} + z(u, v_0, w_0)\,\mathbf{k}, \tag{1.5}$$

which is a parametric equation for the coordinate curve given by the intersection of $v = v_0$ and $w = w_0$, in which the coordinate u acts as a parameter along the curve. Parametric equations for the other two coordinate curves arise similarly.

If we differentiate equation (1.5) with respect to the parameter u, then we get a tangent vector to the coordinate curve. Since this differentiation is done holding v and w constant ($v = v_0$, $w = w_0$), it amounts to differentiating equation (1.3) *partially* with respect to u. Similarly, by differentiating equation (1.3) partially with respect to v and w, we get tangent vectors to the other two coordinate curves. Thus, the three partial derivatives

$$\boxed{\mathbf{e}_u \equiv \partial\mathbf{r}/\partial u, \quad \mathbf{e}_v \equiv \partial\mathbf{r}/\partial v, \quad \mathbf{e}_w \equiv \partial\mathbf{r}/\partial w,} \tag{1.6}$$

when evaluated at (u_0, v_0, w_0), give tangent vectors to the three coordinate curves that pass through P.

The usual way forward in vector calculus is made on the assumption that the coordinate system is *orthogonal* (which means that the coordinate surfaces intersect orthogonally, so that the three vectors (1.6) are mutually orthogonal) and involves normalizing the vectors by dividing them by their lengths to get unit vectors. Thus if

$$h_1 \equiv |\mathbf{e}_u|, \quad h_2 \equiv |\mathbf{e}_v|, \quad h_3 \equiv |\mathbf{e}_w|,$$

then the unit vectors are

$$\hat{\mathbf{e}}_u = \frac{1}{h_1}\mathbf{e}_u, \quad \hat{\mathbf{e}}_v = \frac{1}{h_2}\mathbf{e}_v, \quad \hat{\mathbf{e}}_w = \frac{1}{h_3}\mathbf{e}_w.$$

The set of vectors $\{\hat{\mathbf{e}}_u, \hat{\mathbf{e}}_v, \hat{\mathbf{e}}_w\}$ is then used as a *basis* for vectors at P, by writing any vector $\boldsymbol{\lambda}$ in the form

$$\boldsymbol{\lambda} = \alpha\,\hat{\mathbf{e}}_u + \beta\,\hat{\mathbf{e}}_v + \gamma\,\hat{\mathbf{e}}_w.$$

The triple (α, β, γ) comprises the *components* of $\boldsymbol{\lambda}$ relative to the basis $\{\hat{\mathbf{e}}_u, \hat{\mathbf{e}}_v, \hat{\mathbf{e}}_w\}$.

However, our way forward does not require the coordinate system to be orthogonal, nor do we bother normalizing the tangent vectors to make them unit vectors. So at each point P we have the *natural basis* $\{\mathbf{e}_u, \mathbf{e}_v, \mathbf{e}_w\}$ produced by the partial derivatives (as given by equation (1.6)) and, in general, these vectors are neither unit vectors nor mutually orthogonal. We continue with the example of spherical coordinates to illustrate the difference between the natural (or "hats-off") basis and the normalized (or "hats-on") basis.

Example 1.1.2

In terms of spherical coordinates (r, θ, ϕ) the position vector $\mathbf{r}$ is

$$\mathbf{r} = r\sin\theta\cos\phi\,\mathbf{i} + r\sin\theta\sin\phi\,\mathbf{j} + r\cos\theta\,\mathbf{k},$$

which gives the three natural basis vectors

$$\begin{aligned}
\mathbf{e}_r &= \partial\mathbf{r}/\partial r = \sin\theta\cos\phi\,\mathbf{i} + \sin\theta\sin\phi\,\mathbf{j} + \cos\theta\,\mathbf{k},\\
\mathbf{e}_\theta &= \partial\mathbf{r}/\partial\theta = r\cos\theta\cos\phi\,\mathbf{i} + r\cos\theta\sin\phi\,\mathbf{j} - r\sin\theta\,\mathbf{k},\\
\mathbf{e}_\phi &= \partial\mathbf{r}/\partial\phi = -r\sin\theta\sin\phi\,\mathbf{i} + r\sin\theta\cos\phi\,\mathbf{j}.
\end{aligned}$$

The directions of these vectors are shown in Figure 1.1. Their lengths are

$$h_1 = |\mathbf{e}_r| = 1, \quad h_2 = |\mathbf{e}_\theta| = r, \quad h_3 = |\mathbf{e}_\phi| = r\sin\theta,$$

so the normalized basis vectors are

$$\begin{aligned}
\hat{\mathbf{e}}_r &= \sin\theta\cos\phi\,\mathbf{i} + \sin\theta\sin\phi\,\mathbf{j} + \cos\theta\,\mathbf{k},\\
\hat{\mathbf{e}}_\theta &= \cos\theta\cos\phi\,\mathbf{i} + \cos\theta\sin\phi\,\mathbf{j} - \sin\theta\,\mathbf{k},\\
\hat{\mathbf{e}}_\phi &= -\sin\phi\,\mathbf{i} + \cos\phi\,\mathbf{j}.
\end{aligned}$$

The basis vectors satisfy

$$\mathbf{e}_r \cdot \mathbf{e}_\theta = \mathbf{e}_\theta \cdot \mathbf{e}_\phi = \mathbf{e}_\phi \cdot \mathbf{e}_r = 0,$$

so these coordinates are orthogonal.

There is, in fact, another way in which the coordinate system (u, v, w) can be used to construct a basis at P. This uses the normals to the coordinate surfaces rather than the tangents to the coordinate curves.

As remarked above, we can in principle invert equations (1.1) to obtain u, v, w in terms of x, y, z:

$$u = u(x, y, z), \quad v = v(x, y, z), \quad w = w(x, y, z). \tag{1.7}$$

This allows us to regard each coordinate as a scalar field and to calculate their gradients:

$$\begin{aligned} \nabla u &= \frac{\partial u}{\partial x}\mathbf{i} + \frac{\partial u}{\partial y}\mathbf{j} + \frac{\partial u}{\partial z}\mathbf{k}, \\ \nabla v &= \frac{\partial v}{\partial x}\mathbf{i} + \frac{\partial v}{\partial y}\mathbf{j} + \frac{\partial v}{\partial z}\mathbf{k}, \\ \nabla w &= \frac{\partial w}{\partial x}\mathbf{i} + \frac{\partial w}{\partial y}\mathbf{j} + \frac{\partial w}{\partial z}\mathbf{k}. \end{aligned} \tag{1.8}$$

At each point P, these gradient vectors are normal to the corresponding level surfaces through P, which are the coordinate surfaces $u = u_0$, $v = v_0$, $w = w_0$. We therefore obtain $\{\nabla u, \nabla v, \nabla w\}$ as an alternate basis at P. This basis is the *dual* of that obtained by using the tangent vectors to the coordinate curves and, to distinguish it from the previous one, we write its basis vectors with their suffixes as superscripts:

$$\boxed{\mathbf{e}^u \equiv \nabla u, \quad \mathbf{e}^v \equiv \nabla v, \quad \mathbf{e}^w \equiv \nabla w.} \tag{1.9}$$

Placing the suffixes in this position may seem odd at first (not least because of a possible confusion with powers), but it is part of a remarkably elegant and compact notation that will be developed more fully in later sections.

If the coordinate system is orthogonal, then the normals to the coordinate surfaces coincide with the tangents to the coordinate curves, making any distinction between $\{\mathbf{e}_u, \mathbf{e}_v, \mathbf{e}_w\}$ and its dual $\{\mathbf{e}^u, \mathbf{e}^v, \mathbf{e}^w\}$ just a matter of lengths, rather than the lengths and directions of the basis vectors. If the basis vectors are normalized, then the distinction disappears altogether. Consequently, to illustrate better the two bases that arise naturally from the coordinate system, we should use one that is not orthogonal, rather than continue using spherical coordinates for examples.

Example 1.1.3

Consider a coordinate system (u, v, w) defined by

$$x = u + v, \quad y = u - v, \quad z = 2uv + w, \tag{1.10}$$

where $-\infty < u < \infty$, $-\infty < v < \infty$, $-\infty < w < \infty$. Inverting these equations gives

$$u = \tfrac{1}{2}(x + y), \quad v = \tfrac{1}{2}(x - y), \quad w = z - \tfrac{1}{2}(x^2 - y^2), \tag{1.11}$$

from which we see that the coordinate surfaces $u = u_0$ form a family of planes, as do the surfaces $v = v_0$, while the surfaces $w = w_0$ form a family of hyperbolic paraboloids.

The position vector $\mathbf{r}$ is given by

$$\mathbf{r} = (u + v)\,\mathbf{i} + (u - v)\,\mathbf{j} + (2uv + w)\,\mathbf{k},$$

from which we get the basis vectors

$$\begin{aligned} \mathbf{e}_u &= \partial \mathbf{r}/\partial u = \mathbf{i} + \mathbf{j} + 2v\,\mathbf{k}, \\ \mathbf{e}_v &= \partial \mathbf{r}/\partial v = \mathbf{i} - \mathbf{j} + 2u\,\mathbf{k}, \\ \mathbf{e}_w &= \partial \mathbf{r}/\partial w = \mathbf{k}. \end{aligned} \tag{1.12}$$

Only the last of these is a unit vector.

None of the dot products

$$\mathbf{e}_u \cdot \mathbf{e}_v = 4uv, \quad \mathbf{e}_v \cdot \mathbf{e}_w = 2u, \quad \mathbf{e}_w \cdot \mathbf{e}_u = 2v$$

is in general zero, so the system is not orthogonal.

Using equations (1.11), we obtain the dual basis vectors

$$\begin{aligned} \mathbf{e}^u &= \nabla u = \tfrac{1}{2}\,\mathbf{i} + \tfrac{1}{2}\,\mathbf{j}, \\ \mathbf{e}^v &= \nabla v = \tfrac{1}{2}\,\mathbf{i} - \tfrac{1}{2}\,\mathbf{j}, \\ \mathbf{e}^w &= \nabla w = -x\,\mathbf{i} + y\,\mathbf{j} + \mathbf{k} = -(u+v)\,\mathbf{i} + (u-v)\,\mathbf{j} + \mathbf{k}. \end{aligned}$$

We see that in general $\mathbf{e}^u$ is not parallel to $\mathbf{e}_u$, $\mathbf{e}^v$ is not parallel to $\mathbf{e}_v$, and $\mathbf{e}^w$ is not parallel to $\mathbf{e}_w$.

Given a vector field $\boldsymbol{\lambda}$, we can at each point P refer $\boldsymbol{\lambda}$ to either the basis $\{\mathbf{e}_u, \mathbf{e}_v, \mathbf{e}_w\}$ or the dual basis $\{\mathbf{e}^u, \mathbf{e}^v, \mathbf{e}^w\}$:

$$\begin{aligned} \boldsymbol{\lambda} &= \lambda^u \mathbf{e}_u + \lambda^v \mathbf{e}_v + \lambda^w \mathbf{e}_w, \\ \boldsymbol{\lambda} &= \lambda_u \mathbf{e}^u + \lambda_v \mathbf{e}^v + \lambda_w \mathbf{e}^w. \end{aligned} \tag{1.13}$$

We have two sets of components, $(\lambda^u, \lambda^v, \lambda^w)$ and $(\lambda_u, \lambda_v, \lambda_w)$, the positioning of the suffixes serving to distinguish them. There are, in fact, connections between these two sets of components, as well as between the two bases involved. These connections will be established in the next section, after introducing the suffix notation, which allows easier handling of the equations involved.

Exercises 1.1

1. Cylindrical coordinates (ρ, ϕ, z) are defined by

$$\mathbf{r} = \rho \cos\phi\,\mathbf{i} + \rho \sin\phi\,\mathbf{j} + z\,\mathbf{k},$$

   where $0 \le \rho < \infty$, $0 \le \phi < 2\pi$, $-\infty < z < \infty$.
   Obtain expressions for the natural basis vectors $\mathbf{e}_\rho$, $\mathbf{e}_\phi$, $\mathbf{e}_z$ and the dual basis vectors $\mathbf{e}^\rho$, $\mathbf{e}^\phi$, $\mathbf{e}^z$ in terms of $\mathbf{i}$, $\mathbf{j}$, $\mathbf{k}$.

2. Show that, when referred to
   (a) the natural basis $\{\mathbf{e}_r, \mathbf{e}_\theta, \mathbf{e}_\phi\}$ of spherical coordinates,
   (b) the natural basis $\{\mathbf{e}_\rho, \mathbf{e}_\phi, \mathbf{e}_z\}$ of cylindrical coordinates,

(c) the natural basis $\{\mathbf{e}_u, \mathbf{e}_v, \mathbf{e}_w\}$ of the paraboloidal coordinates of Example 1.1.3,

the constant vector field $\mathbf{i}$ is given by

(a) $\mathbf{i} = \sin\theta\cos\phi\,\mathbf{e}_r + r^{-1}\cos\theta\cos\phi\,\mathbf{e}_\theta - r^{-1}\operatorname{cosec}\theta\sin\phi\,\mathbf{e}_\phi$,

(b) $\mathbf{i} = \cos\phi\,\mathbf{e}_\rho - \rho^{-1}\sin\phi\,\mathbf{e}_\phi$,

(c) $\mathbf{i} = \frac{1}{2}\,\mathbf{e}_u + \frac{1}{2}\,\mathbf{e}_v - (u+v)\,\mathbf{e}_w$.

1.2 Suffix notation

The suffix notation provides a way of handling collections of related quantities that otherwise might be represented by arrays. The coordinates of a point constitute such a collection, as do the components of a vector, and the vectors in a basis. The basic idea is to represent the members of such a collection by means of a kernel letter to which is attached a literal suffix (or literal suffixes) representing numbers that serve to label the quantities in the collection. A suffix can appear either as a subscript or a superscript, and there can be more than one suffix attached to a kernel letter. When used with the summation convention (explained below), the suffix notation gives an elegant and compact means of handling coordinates, components, basis vectors, and similar collections of related quantities. To see how it works, let us re-express the results of the last section using the suffix notation, and then extend them to learn a little more about natural bases, their duals, and the components of vectors.

We shall use u^i $(i = 1, 2, 3)$ in place of (u, v, w) for coordinates, $\{\mathbf{e}_i\}$ $(i = 1, 2, 3)$ in place of $\{\mathbf{e}_u, \mathbf{e}_v, \mathbf{e}_w\}$ for the natural basis, and $\{\mathbf{e}^i\}$ $(i = 1, 2, 3)$ in place of $\{\mathbf{e}^u, \mathbf{e}^v, \mathbf{e}^w\}$ for the dual basis. For a vector $\boldsymbol{\lambda}$, we denote its components relative to $\{\mathbf{e}_i\}$ by λ^i $(i = 1, 2, 3)$, and its components relative to $\{\mathbf{e}^i\}$ by λ_i $(i = 1, 2, 3)$. We can then re-express equations (1.13) as

$$\boldsymbol{\lambda} = \lambda^1\mathbf{e}_1 + \lambda^2\mathbf{e}_2 + \lambda^3\mathbf{e}_3 = \sum_{i=1}^{3} \lambda^i\mathbf{e}_i \tag{1.14}$$

and

$$\boldsymbol{\lambda} = \lambda_1\mathbf{e}^1 + \lambda_2\mathbf{e}^2 + \lambda_3\mathbf{e}^3 = \sum_{i=1}^{3} \lambda_i\mathbf{e}^i. \tag{1.15}$$

There are two ways that we can economize on our notation. The first is to agree that literal suffixes taken from around the middle of the alphabet (i.e., $i, j, k, \ldots$) always run through the values 1, 2, 3. That would allow us to drop the parenthetical comment $(i = 1, 2, 3)$ that occurs five times in the previous paragraph. The second is to agree that if a literal suffix occurs twice, *once as a subscript and once as a superscript*, then summation over the range indicated by the repeated suffix is implied *without the use of* $\sum_{i=1}^{3}$ *to indicate summation.* We can therefore shorten equations (1.14) and (1.15) to

$$\boldsymbol{\lambda} = \lambda^i \mathbf{e}_i \quad \text{and} \quad \boldsymbol{\lambda} = \lambda_i \mathbf{e}^i. \tag{1.16}$$

This agreement that a repeated suffix implies summation is known as the *summation convention* and is due to Einstein. Two suffixes used like this to indicate summation are called *dummy suffixes*, and they may be replaced by any other letter not already in use in the term involved. (Proper use of the convention requires that, in any term, a suffix should not occur more than twice, and that any repeated suffix should occur once as a superscript and once as a subscript.) Thus we could equally well write $\boldsymbol{\lambda} = \lambda^j \mathbf{e}_j$ and $\boldsymbol{\lambda} = \lambda_j \mathbf{e}^j$ to express $\boldsymbol{\lambda}$ in terms of basis vectors and components.

The components λ^i of a vector $\boldsymbol{\lambda}$ that arise from using the natural basis $\{\mathbf{e}_i\}$ are known as its *contravariant components*, while the components λ_i that arise from using the dual basis $\{\mathbf{e}^i\}$ are known as its *covariant components.* (A useful mnemonic is the rhyming of "co" with "below", which gives the position of the suffix.) It turns out that these components are given by[3] $\lambda^i = \boldsymbol{\lambda} \cdot \mathbf{e}^i$ and $\lambda_i = \boldsymbol{\lambda} \cdot \mathbf{e}_i$, but to establish this we need to look at the dot products $\mathbf{e}^i \cdot \mathbf{e}_j$.

Using the definitions of $\mathbf{e}^i$ and $\mathbf{e}_j$, we have

$$\mathbf{e}^i \cdot \mathbf{e}_j = \nabla u^i \cdot \frac{\partial \mathbf{r}}{\partial u^j} = \frac{\partial u^i}{\partial x}\frac{\partial x}{\partial u^j} + \frac{\partial u^i}{\partial y}\frac{\partial y}{\partial u^j} + \frac{\partial u^i}{\partial z}\frac{\partial z}{\partial u^j} = \frac{\partial u^i}{\partial u^j},$$

where we have used the chain rule of partial differentiation to simplify the three-term expression involving partial derivatives. If $i = j$, then $\partial u^i/\partial u^j = 1$, otherwise $\partial u^i/\partial u^j = 0$. So we can write

$$\mathbf{e}^i \cdot \mathbf{e}_j = \delta^i_j, \tag{1.17}$$

where δ^i_j is the *Kronecker delta* defined by

$$\delta^i_j = \begin{cases} 1, & \text{for } i = j \\ 0, & \text{for } i \neq j \end{cases}. \tag{1.18}$$

(Occasionally we need the forms δ_{ij} or δ^{ij} of the Kronecker delta, which are defined in a similar way.) We can now say that

$$\boldsymbol{\lambda} \cdot \mathbf{e}^j = \lambda^i \mathbf{e}_i \cdot \mathbf{e}^j = \lambda^i \delta^j_i = \lambda^j, \tag{1.19}$$

as asserted. (Note how in the sum $\lambda^i \delta^j_i$ the effect of the Kronecker delta is to substitute j for i in λ^i. This is because the only nonzero term in the sum occurs when $i = j$ and the Kronecker delta is then equal to one.) In a similar way we can establish that

$$\lambda_j = \boldsymbol{\lambda} \cdot \mathbf{e}_j \tag{1.20}$$

(see Exercise 1.2.1). Equations (1.20) and (1.19) show that the components of $\boldsymbol{\lambda}$ relative to one basis (the natural one or its dual) are given by taking dot

[3] Our use of the dot product here, and in the rest of this section, is not strictly correct: see Sec. 1.10.

products of $\boldsymbol{\lambda}$ with the vectors in the other basis (the dual or the natural one, respectively).

Using contravariant components, we can write the dot product of two vectors $\boldsymbol{\lambda}$ and $\boldsymbol{\mu}$ as

$$\boldsymbol{\lambda} \cdot \boldsymbol{\mu} = \lambda^i \mathbf{e}_i \cdot \mu^j \mathbf{e}_j = g_{ij} \lambda^i \mu^j, \tag{1.21}$$

where

$$\boxed{g_{ij} \equiv \mathbf{e}_i \cdot \mathbf{e}_j.} \tag{1.22}$$

Similarly, by using covariant components, we can write it as

$$\boldsymbol{\lambda} \cdot \boldsymbol{\mu} = \lambda_i \mathbf{e}^i \cdot \mu_j \mathbf{e}^j = g^{ij} \lambda_i \mu_j, \tag{1.23}$$

where

$$g^{ij} \equiv \mathbf{e}^i \cdot \mathbf{e}^j. \tag{1.24}$$

Or we could mix things and write

$$\boldsymbol{\lambda} \cdot \boldsymbol{\mu} = \lambda_i \mathbf{e}^i \cdot \mu^j \mathbf{e}_j = \lambda_i \mu^j \delta^i_j = \lambda_i \mu^i \tag{1.25}$$

(on using the substitutional effect of the Kronecker delta). We therefore have four different ways of writing the dot product:

$$\boxed{\boldsymbol{\lambda} \cdot \boldsymbol{\mu} = g_{ij} \lambda^i \mu^j = g^{ij} \lambda_i \mu_j = \lambda_i \mu^i = \lambda^i \mu_i.} \tag{1.26}$$

The fact that $g^{ij} \lambda_i \mu_j = \lambda_i \mu^i$ holds for *arbitrary* vector components λ_i implies that

$$\boxed{g^{ij} \mu_j = \mu^i,} \tag{1.27}$$

showing that the quantities g^{ij} can be used to raise the suffix and thereby obtain the contravariant components of μ from its covariant components. Similarly, because $g_{ij} \lambda^i \mu^j = \lambda^i \mu_i$ holds for *arbitrary* components λ^i, we have

$$g_{ij} \mu^j = \mu_i, \tag{1.28}$$

showing that the quantities g_{ij} give the reverse operation of lowering a suffix. Combining these two operations gives

$$\mu^i = g^{ij} \mu_j = g^{ij} g_{jk} \mu^k,$$

and because this holds for *arbitrary* components μ^i we can deduce that

$$\boxed{g^{ij} g_{jk} = \delta^i_k.} \tag{1.29}$$

From their definitions as dot products of basis vectors, it can be seen that g_{ij} and g^{ij} satisfy

$$g_{ij} = g_{ji}, \quad g^{ij} = g^{ji}. \tag{1.30}$$

Identities like this, that involve a transposition of suffixes, are referred to as *symmetry properties*. Using equation (1.30) we can write equation (1.29) in the form

$$g_{kj}g^{ji} = \delta^i_k. \tag{1.31}$$

We shall use the suffix notation in the development of any general theory, extending and developing it as the need arises. However, in particular examples it is often more convenient to revert to the nonsuffix notation, like that used in the previous section. It is also sometimes convenient to use matrix methods to handle the summations over repeated suffixes. These methods are restricted to quantities carrying either one or two suffixes, enabling them to be arranged as either one-dimensional arrays (row vectors or column vectors) or two-dimensional arrays (matrices). The notation $[\lambda^i]$ will be used to denote a column vector having λ^i as its ith entry (so the suffix labels the rows of the array), and $[\lambda_i]$ will be used in a similar way. If we need to think of the quantities as being arranged as a row matrix, then we shall use the notation $[\lambda^i]^{\mathrm{T}}$ or $[\lambda_i]^{\mathrm{T}}$ to indicate that we have transposed the column vector to obtain a row vector. For matrices, the notation $[A_{ij}]$ will be used to denote a matrix having A_{ij} as the entry in the ith row and jth column, and A^{ij} will be used in a similar way. This conforms to the normal conventions of matrix algebra, where the first suffix labels the row and the second the column. If we want to fit quantities A^i_j into an array, then we must specify which is the "first" suffix, to be used for labeling rows, and which is the "second", to be used for labelling columns. We shall regard the superscript as the 'first' suffix, so $[A^i_j]$ will be used to denote the matrix having A^i_j as the entry in its ith row and jth column. We are now in a position to give matrix versions of some of the equations established using the suffix notation.

Let $G \equiv [g_{ij}]$ and $\hat{G} \equiv [g^{ij}]$. Equations (1.30) then translate to

$$G = G^{\mathrm{T}}, \quad \hat{G} = \hat{G}^{\mathrm{T}}, \tag{1.32}$$

telling us that G and $\hat{G}$ are symmetric. Recognizing that $[\delta^i_j] = I$ (the 3×3 unit matrix), we see that equations (1.29) and (1.31) translate to

$$\hat{G}G = I, \quad G\hat{G} = I, \tag{1.33}$$

telling us that $\hat{G} = G^{-1}$. If we use $L \equiv [\lambda^i]$, $M \equiv [\mu^i]$ for the contravariant components of $\boldsymbol{\lambda}$, $\boldsymbol{\mu}$ and $L^* \equiv [\lambda_i]$, $M^* \equiv [\mu_i]$ for their covariant components, then equations (1.27) and (1.28) translate to

$$\hat{G}M^* = M, \quad GM = M^* \tag{1.34}$$

(which are consistent with $G\hat{G} = \hat{G}G = I$), while the four ways of writing the dot product in equation (1.26) translate to

$$\boldsymbol{\lambda} \cdot \boldsymbol{\mu} = L^{\mathrm{T}}GM = (L^*)^{\mathrm{T}}\hat{G}M^* = (L^*)^{\mathrm{T}}M = L^{\mathrm{T}}M^*. \tag{1.35}$$

The following example, which is a continuation of Example 1.1.3, illustrates some features of the suffix notation and matrix methods introduced in this section.

Example 1.2.1
For the coordinate system introduced in Example 1.1.3, the natural basis vectors are

$$\begin{aligned}\mathbf{e}_1 &= \mathbf{i} + \mathbf{j} + 2v\,\mathbf{k},\\ \mathbf{e}_2 &= \mathbf{i} - \mathbf{j} + 2u\,\mathbf{k},\\ \mathbf{e}_3 &= \mathbf{k},\end{aligned}$$

and the dual basis vectors are

$$\begin{aligned}\mathbf{e}^1 &= \tfrac{1}{2}\,\mathbf{i} + \tfrac{1}{2}\,\mathbf{j},\\ \mathbf{e}^2 &= \tfrac{1}{2}\,\mathbf{i} - \tfrac{1}{2}\,\mathbf{j},\\ \mathbf{e}^3 &= -(u+v)\,\mathbf{i} + (u-v)\,\mathbf{j} + \mathbf{k}.\end{aligned}$$

Calculating the dot products $g_{ij} \equiv \mathbf{e}_i \cdot \mathbf{e}_j$ and $g^{ij} \equiv \mathbf{e}^i \cdot \mathbf{e}^j$, we find that

$$G = [g_{ij}] = \begin{bmatrix} 2(1+2v^2) & 4uv & 2v \\ 4uv & 2(1+2u^2) & 2u \\ 2v & 2u & 1 \end{bmatrix},$$

$$\hat{G} = [g^{ij}] = \begin{bmatrix} 1/2 & 0 & -v \\ 0 & 1/2 & -u \\ -v & -u & 2u^2+2v^2+1 \end{bmatrix},$$

and it is straightforward to check that $G\hat{G} = \hat{G}G = I$.

If we take $\boldsymbol{\lambda}$ to be the unit vector $\mathbf{i}$, then the column vector L that holds the contravariant components of λ^i is (from Exercise 1.1.2, or equation (1.19))

$$L = [\lambda^i] = \begin{bmatrix} 1/2 \\ 1/2 \\ -(u+v) \end{bmatrix}.$$

The covariant components λ_i can then be obtained by saying

$$\begin{aligned}[\lambda_i] &= L^* = GL \\ &= \begin{bmatrix} 2(1+2v^2) & 4uv & 2v \\ 4uv & 2(1+2u^2) & 2u \\ 2v & 2u & 1 \end{bmatrix} \begin{bmatrix} 1/2 \\ 1/2 \\ -(u+v) \end{bmatrix} = \begin{bmatrix} 1 \\ 1 \\ 0 \end{bmatrix}.\end{aligned}$$

As a check, we note that

$$\boldsymbol{\lambda} \cdot \boldsymbol{\lambda} = L^{\mathrm{T}} G L = L^{\mathrm{T}} L^* = \tfrac{1}{2} + \tfrac{1}{2} + 0 = 1,$$

as it should be for the unit vector $\boldsymbol{\lambda}$.

The above example also illustrates how arrays can be used to display vector components in particular examples, as when writing

$$[\lambda^i] = \begin{bmatrix} 1/2 \\ 1/2 \\ -(u+v) \end{bmatrix}, \quad [\lambda_i] = \begin{bmatrix} 1 \\ 1 \\ 0 \end{bmatrix}.$$

We can also display the same information in suffix notation using Kronecker deltas:

$$\begin{aligned} \lambda^i &= \tfrac{1}{2}\delta^i_1 + \tfrac{1}{2}\delta^i_2 - (u+v)\delta^i_3, \\ \lambda_i &= \delta^1_i + \delta^2_i. \end{aligned}$$

A common practice, which achieves the same result, is simply to list the components in parentheses, separated by commas:

$$\lambda^i = \left(\tfrac{1}{2}, \tfrac{1}{2}, -(u+v)\right), \quad \lambda_i = (1, 1, 0).$$

We used this kind of notation in Appendix A, where special relativity was reviewed.

In the present section, the suffixes $i, j, k, \ldots$ take the values 1, 2, 3, and repeated literal suffixes imply summation over this range of values. When working in four-dimensional spacetime, we shall use Greek suffixes $\mu, \nu, \ldots$, and the understanding there will be that these take the values 0, 1, 2, 3; a repeated Greek suffix will then imply summation over these values. In Section 1.6, we discuss two-dimensional surfaces and there we shall use uppercase literal suffixes $A, B, \ldots$, taking the values 1, 2. At other times our underlying space may have arbitrary dimension N, and we shall then use lowercase letters $a, b, c, \ldots$ taken from the beginning of the alphabet to have the range $1, 2, \ldots, N$. The general idea is to avoid confusion by using different kinds of alphabets (or different parts of the same alphabet) to indicate the different ranges of values that our suffixes might take.

Exercises 1.2

1. Verify equation (1.20), which shows that the covariant components λ_j of a vector $\boldsymbol{\lambda}$ are given by taking dot products of $\boldsymbol{\lambda}$ with the natural basis vectors $\mathbf{e}_j$.

2. Show that $\mathbf{e}_i = g_{ij}\mathbf{e}^j$ and $\mathbf{e}^i = g^{ij}\mathbf{e}_j$.

3. Simplify the following expressions:

   (a) $\lambda^i \delta^j_i \lambda_j$, (b) $\mu_i g^{ij} g_{jk} \lambda^k$, (c) $g_{ij}\lambda^i \mu^j - \lambda^k \mu_k$.

4. If the coordinate system is orthogonal, what can you say about the matrices $G \equiv [g_{ij}]$ and $\hat{G} \equiv [g^{ij}]$?

5. Check that in Example 1.2.1 the matrices G and $\hat{G}$ satisfy $G\hat{G} = \hat{G}G = I$.

6. In the coordinate system of Example 1.2.1, a vector field $\boldsymbol{\mu}$ has covariant components given by

$$\mu_i = v\delta_i^1 - u\delta_i^2 + \delta_i^3.$$

What are its contravariant components μ^i?

7. A repeated suffix implies summation. What, then are the values of

(a) δ_i^i, (b) δ_A^A, (c) δ_a^a, (d) δ_μ^μ?

(Note the correspondence between alphabets and ranges of values explained in the last paragraph prior to these exercises.)

1.3 Tangents and gradients

By dropping the requirement that our coordinate systems be orthogonal, we have found ourselves in the position of having two different, but related, bases at each point of space. Is this one two many? To avoid confusion, should we reject one of them and retain the other? If so, which one? As we shall see, each has its uses, and there are situations where it is appropriate to use the natural basis $\{\mathbf{e}_i\}$ defined by the tangents to the coordinate curves, while in other situations it is appropriate to use the dual basis $\{\mathbf{e}^i\}$ defined by the normals to the coordinate surfaces. Let us start by looking at the tangent vector to a curve in space.

Suppose we put

$$u = u(t), \quad v = v(t), \quad w = w(t), \tag{1.36}$$

where $u(t)$, $v(t)$, $w(t)$ are differentiable functions of t for t belonging to some interval I. Then the points with coordinates given by equation (1.36) will lie on a curve γ parameterized by t. The position vector of these points is

$$\mathbf{r}(t) = x\,(u(t), v(t), w(t))\,\mathbf{i} + y\,(u(t), v(t), w(t))\,\mathbf{j} + z\,(u(t), v(t), w(t))\,\mathbf{k},$$

and for each t in I the derivative $\dot{\mathbf{r}}(t) \equiv d\mathbf{r}/dt$ gives a tangent vector to the curve (provided $\dot{\mathbf{r}}(t) \neq \mathbf{0}$). Using the chain rule we have

$$\frac{d\mathbf{r}}{dt} = \frac{\partial \mathbf{r}}{\partial u}\frac{du}{dt} + \frac{\partial \mathbf{r}}{\partial v}\frac{dv}{dt} + \frac{\partial \mathbf{r}}{\partial w}\frac{dw}{dt},$$

which can be written as

$$\dot{\mathbf{r}}(t) = \dot{u}(t)\mathbf{e}_u + \dot{v}(t)\mathbf{e}_v + \dot{w}(t)\mathbf{e}_w.$$

The suffix notation version of this last equation is

$$\dot{\mathbf{r}}(t) = \dot{u}^i(t)\mathbf{e}_i, \tag{1.37}$$

showing that the derivatives $\dot{u}^i(t)$ are the components of the tangent vector to the curve γ relative to the natural basis $\{\mathbf{e}_i\}$. So for tangents to curves, it is appropriate to use the natural basis $\{\mathbf{e}_i\}$.

The length of the curve γ is obtained by integrating $|\dot{\mathbf{r}}|$ with respect to t over the interval I. Now

$$|\dot{\mathbf{r}}|^2 = \dot{\mathbf{r}} \cdot \dot{\mathbf{r}} = \dot{u}^i\mathbf{e}_i \cdot \dot{u}^j\mathbf{e}_j = g_{ij}\dot{u}^i\dot{u}^j,$$

on using equation (1.22) that defines the quantities g_{ij}. So if I is given by $a \le t \le b$, then the length of γ is given by

$$\boxed{L = \int_a^b \left(g_{ij}\dot{u}^i\dot{u}^j\right)^{1/2} dt.} \tag{1.38}$$

The infinitesimal version of equation (1.37) is $d\mathbf{r} = du^i\mathbf{e}_i$, which gives

$$ds^2 = d\mathbf{r} \cdot d\mathbf{r} = du^i\mathbf{e}_i \cdot du^j\mathbf{e}_j$$

for the distance between points whose coordinates differ by du^i. We thus arrive at the formula

$$\boxed{ds^2 = g_{ij}du^i du^j,} \tag{1.39}$$

which is the generalization for arbitrary coordinates u^i of the Cartesian formula

$$ds^2 = dx^2 + dy^2 + dz^2.$$

Equation (1.39) can therefore be viewed as an expression of Pythagoras' Theorem. The expression $g_{ij}du^i du^j$ is often referred to as the *line element.* Equation (1.38) amounts to saying that the length of γ is given by the integral $\int ds$ taken along the curve.

Example 1.3.1

For spherical coordinates (r, θ, ϕ), the basis vectors $\mathbf{e}_1$, $\mathbf{e}_2$, $\mathbf{e}_3$ are the vectors $\mathbf{e}_r$, $\mathbf{e}_\theta$, $\mathbf{e}_\phi$, as given in Example 1.1.2. By working out the dot products $\mathbf{e}_i \cdot \mathbf{e}_j \equiv g_{ij}$ we get

$$[g_{ij}] = \begin{bmatrix} 1 & 0 & 0 \\ 0 & r^2 & 0 \\ 0 & 0 & r^2\sin^2\theta \end{bmatrix},$$

which gives

$$ds^2 = dr^2 + r^2 d\theta^2 + r^2\sin^2\theta\, d\phi^2 \tag{1.40}$$

for the line element of Euclidean space in spherical coordinates.

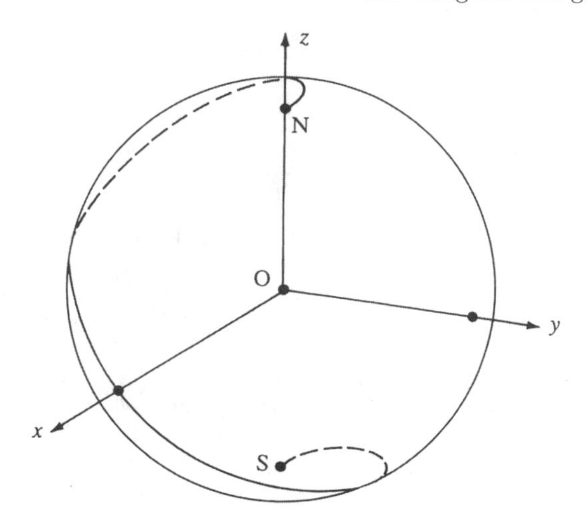

Fig. 1.2. The curve γ on a sphere.

Let us give a curve γ by putting

$$u^1 \equiv r = a, \quad u^2 \equiv \theta = t, \quad u^3 \equiv \phi = 2t - \pi,$$

where $0 \le t \le \pi$. This gives a curve on the sphere $r = a$ that winds down from the North Pole (where $\theta = 0$) to the South Pole (where $\theta = \pi$) (see Fig. 1.2). Its tangent vector has the components

$$\dot{u}^i = \dot{r}\delta^i_1 + \dot{\theta}\delta^i_2 + \dot{\phi}\delta^i_3 = \delta^i_2 + 2\delta^i_3,$$

so

$$g_{ij}\dot{u}^i\dot{u}^j = r^2 + 4r^2 \sin^2\theta = a^2 + 4a^2 \sin^2 t.$$

The length of the curve γ is therefore given by

$$\int_0^\pi \left(g_{ij}\dot{u}^i\dot{u}^j\right)^{1/2} dt = \int_0^\pi a\sqrt{1 + 4\sin^2 t}\, dt.$$

We shall not attempt to evaluate this integral.

Suppose now that we take a differentiable function $\phi(u, v, w)$ of the coordinates u, v, w. This will give us a function of position and therefore a scalar field. Its gradient is

$$\nabla\phi \equiv \frac{\partial\phi}{\partial x}\mathbf{i} + \frac{\partial\phi}{\partial y}\mathbf{j} + \frac{\partial\phi}{\partial z}\mathbf{k},$$

where, in calculating these partial derivatives, we are regarding ϕ as a function of x, y, z got by substituting the expressions for u, v, w in terms of x, y, z given by equation (1.7):

$$\phi = \phi\left(u(x,y,z), v(x,y,z), w(x,y,z)\right).$$

The chain rule gives

$$\frac{\partial \phi}{\partial x} = \frac{\partial \phi}{\partial u}\frac{\partial u}{\partial x} + \frac{\partial \phi}{\partial v}\frac{\partial v}{\partial x} + \frac{\partial \phi}{\partial w}\frac{\partial w}{\partial x},$$

with similar expressions for $\partial\phi/\partial y$ and $\partial\phi/\partial z$. Hence we can say that

$$\begin{aligned}\nabla\phi &= \frac{\partial \phi}{\partial u}\left(\frac{\partial u}{\partial x}\mathbf{i} + \frac{\partial u}{\partial y}\mathbf{j} + \frac{\partial u}{\partial z}\mathbf{k}\right) + \frac{\partial \phi}{\partial v}\left(\frac{\partial v}{\partial x}\mathbf{i} + \frac{\partial v}{\partial y}\mathbf{j} + \frac{\partial v}{\partial z}\mathbf{k}\right)\\ &\quad + \frac{\partial \phi}{\partial w}\left(\frac{\partial w}{\partial x}\mathbf{i} + \frac{\partial w}{\partial y}\mathbf{j} + \frac{\partial w}{\partial z}\mathbf{k}\right)\\ &= \frac{\partial \phi}{\partial u}\nabla u + \frac{\partial \phi}{\partial v}\nabla v + \frac{\partial \phi}{\partial w}\nabla w.\end{aligned}$$

That is,

$$\nabla\phi = \frac{\partial \phi}{\partial u}\mathbf{e}^u + \frac{\partial \phi}{\partial v}\mathbf{e}^v + \frac{\partial \phi}{\partial w}\mathbf{e}^w,$$

on using the defining equations (1.9). The suffix notation version of this is

$$\nabla\phi = \frac{\partial \phi}{\partial u^i}\mathbf{e}^i, \tag{1.41}$$

showing that the partial derivatives $\partial\phi/\partial u^i$ are the components of $\nabla\phi$ relative to the dual basis $\{\mathbf{e}^i\}$. Note that, in letting the repeated suffix imply summation in equation (1.41), we are regarding the suffix i on $\partial\phi/\partial u^i$ as a subscript. We can make this point more clearly by shortening the partial differential operator $\partial/\partial u^i$ to ∂_i, so that $\partial\phi/\partial u^i = \partial_i\phi$. The notation $\phi_{,i}$ is also used to mean the same thing. We can then rewrite equation (1.41) as

$$\nabla\phi = \partial_i\phi\,\mathbf{e}^i = \phi_{,i}\mathbf{e}^i, \tag{1.42}$$

with the suffix correctly occupying the subscript position.

Thus we see that when dealing with tangents to curves it is appropriate to use the natural basis $\{\mathbf{e}_i\}$ defined by the coordinate system, but when dealing with gradients of scalar fields it is appropriate to use the dual basis $\{\mathbf{e}^i\}$. This conclusion is not surprising, given the way in which the two bases are defined.

Exercises 1.3

1. What form does the line element $ds^2 = g_{ij}du^i du^j$ take for the paraboloidal coordinates of u, v, w of Example 1.1.3?

2. Describe the curve given in cylindrical coordinates by

$$\rho = a, \quad \phi = t, \quad z = t, \qquad -\pi \le t \le \pi$$

(where a is a positive constant) and find its length.

3. Show that if the arc-length s (measured along a curve from some base point) is used as a parameter, then at each point of the curve the tangent vector $\dot{\mathbf{r}}(s)$ has unit length.

1.4 Coordinate transformations in Euclidean space

Suppose we have two systems of curvilinear coordinates in Euclidean space, denoted by (u, v, w) and (u', v', w'). We can distinguish these by referring to them as the unprimed and primed coordinates, respectively. The purpose of this section is to explain how such things as the components of vectors relative to the bases defined by the coordinate systems transform, when we pass from the unprimed to the primed coordinate system (or *vice versa*). To this end, we shall use the suffix notation, with u^i representing the unprimed coordinates and $u^{i'}$ the primed coordinates. Placing the prime on the suffix, rather than the kernel letter u, may seem perverse, but it is part of the *kernel–index method* initiated by Schouten and his co-workers.[4] We shall use a similar notation for natural basis vectors, dual basis vectors, and components of vectors. So $\{\mathbf{e}_{i'}\}$ is the natural basis defined by the primed coordinate system and $\{\mathbf{e}^{i'}\}$ is its dual. The contravariant components of a vector $\boldsymbol{\lambda}$ relative to $\{\mathbf{e}_{i'}\}$ will be denoted by $\lambda^{i'}$, and similarly the covariant components relative to $\{\mathbf{e}^{i'}\}$ will be denoted by $\lambda_{i'}$. So we can write

$$\boldsymbol{\lambda} = \lambda^{i'} \mathbf{e}_{i'}, \tag{1.43}$$

with a similar expression giving $\boldsymbol{\lambda}$ relative to the dual basis $\{\mathbf{e}^{i'}\}$. An alternate (and much used) notation involves priming the kernel letter of the components and basis vectors, rather than the suffixes, but this has disadvantages in terms of later economy.

In the region of space covered by both coordinate systems, we have equations $u^{i'} = u^{i'}(u^j)$, giving the primed coordinates in terms of the unprimed, with inverses $u^i = u^i(u^{j'})$ giving the unprimed coordinates in terms of the primed. By definition, $\mathbf{e}_i \equiv \partial\mathbf{r}/\partial u^i$ and $\mathbf{e}_{i'} \equiv \partial\mathbf{r}/\partial u^{i'}$. The chain rule gives

$$\frac{\partial \mathbf{r}}{\partial u^j} = \frac{\partial \mathbf{r}}{\partial u^{i'}} \frac{\partial u^{i'}}{\partial u^j},$$

so we can write

$$\mathbf{e}_j = U^{i'}_j \mathbf{e}_{i'}, \tag{1.44}$$

where $U^{i'}_j$ is a short-hand for the partial derivative $\partial u^{i'}/\partial u^j$. We then have

$$\boldsymbol{\lambda} = \lambda^j \mathbf{e}_j = \lambda^j U^{i'}_j \mathbf{e}_{i'}.$$

Comparison with equation (1.43) gives

---

[4] Schouten, 1954, p.3, in particular footnote[1)].

$$\boxed{\lambda^{i'} = U^{i'}_j \lambda^j} \tag{1.45}$$

as the transformation formula for the contravariant components of a vector. We use a similar argument to deal with the covariant components.

By definition, $\mathbf{e}^i \equiv \nabla u^i$ and $\mathbf{e}^{i'} \equiv \nabla u^{i'}$. The chain rule gives

$$\frac{\partial u^j}{\partial x} = \frac{\partial u^j}{\partial u^{i'}} \frac{\partial u^{i'}}{\partial x},$$

with similar expressions for $\partial u^j/\partial y$ and $\partial u^j/\partial z$. So

$$\nabla u^j = \frac{\partial u^j}{\partial u^{i'}} \nabla u^{i'},$$

which we can write as

$$\mathbf{e}^j = U^j_{i'} \mathbf{e}^{i'}, \tag{1.46}$$

where $U^j_{i'}$ is a short-hand for the partial derivative $\partial u^j/\partial u^{i'}$. Then for covariant components we have

$$\boldsymbol{\mu} = \mu_j \mathbf{e}^j = \mu_j U^j_{i'} \mathbf{e}^{i'},$$

and comparison with $\boldsymbol{\mu} = \mu_{i'} \mathbf{e}^{i'}$ gives

$$\boxed{\mu_{i'} = U^j_{i'} \mu_j} \tag{1.47}$$

as the transformation formula for the covariant components.

There are two routes to the inverse transformations. The first is to note that primed and unprimed quantities are on an equal footing, so that primed and unprimed suffixes can be swapped. This gives

$$\mathbf{e}_{j'} = U^i_{j'} \mathbf{e}_i, \quad \mathbf{e}^{j'} = U^{j'}_i \mathbf{e}^i \tag{1.48}$$

for transforming basis vectors, and

$$\lambda^i = U^i_{j'} \lambda^{j'}, \quad \mu_i = U^{j'}_i \mu_{j'} \tag{1.49}$$

for transforming components. The second is to note that the chain rule gives (see Exercise 1.4.1)

$$\boxed{U^k_{i'} U^{i'}_j = \delta^k_j,} \tag{1.50}$$

so from equation (1.45) we have

$$U^k_{i'} \lambda^{i'} = U^k_{i'} U^{i'}_j \lambda^j = \delta^k_j \lambda^j = \lambda^k,$$

which reproduces the first of equations (1.49). A similar argument using equation (1.47) reproduces the second of equations (1.49) (see Exercise 1.4.2).

Matrix methods (as explained in Sec. 1.2) can be used to handle transformations of components of vectors. The matrices

$$U \equiv [U^{i'}_j], \quad \hat{U} \equiv [U^i_{j'}], \tag{1.51}$$

are the *Jacobian matrices* associated with the change of coordinates. Equation (1.50) translates to $\hat{U}U = I$, showing that $\hat{U} = U^{-1}$. If we put

$$L \equiv [\lambda^i], \quad L' \equiv [\lambda^{i'}], \quad M \equiv [\mu_i], \quad M' \equiv [\mu_{i'}],$$

then for transforming contravariant components we have

$$L' = UL, \quad L = \hat{U}L' \quad (\hat{U} = U^{-1})$$

and for transforming covariant components we have

$$M' = \hat{U}M, \quad M = UM' \quad (\hat{U} = U^{-1}).$$

The following example uses matrix methods.

Example 1.4.1

The equations connecting spherical and cylindrical coordinates are

$$\begin{aligned} x &= r\sin\theta\cos\phi = \rho\cos\phi, \\ y &= r\sin\theta\sin\phi = \rho\sin\phi, \\ z &= r\cos\theta. \end{aligned}$$

If we take cylindrical coordinates (ρ, ϕ, z) as the primed coordinates and spherical coordinates (r, θ, ϕ) as the unprimed ones, then

$$\begin{aligned} u^{1'} &= \rho = r\sin\theta = u^1 \sin u^2, \\ u^{2'} &= \phi = u^3, \\ u^{3'} &= z = r\cos\theta = u^1 \cos u^2, \end{aligned}$$

so that the matrix U is

$$U \equiv [U^{i'}_j] = \begin{bmatrix} \sin u^2 & u^1 \cos u^2 & 0 \\ 0 & 0 & 1 \\ \cos u^2 & -u^1 \sin u^2 & 0 \end{bmatrix} = \begin{bmatrix} \sin\theta & r\cos\theta & 0 \\ 0 & 0 & 1 \\ \cos\theta & -r\sin\theta & 0 \end{bmatrix}.$$

The inverse coordinate transformation equations are

$$\begin{aligned} u^1 &= r = \sqrt{\rho^2 + z^2} = \left((u^{1'})^2 + (u^{3'})^2\right)^{1/2}, \\ u^2 &= \theta = \arctan(\rho/z) = \arctan(u^{1'}/u^{3'}), \\ u^3 &= \phi = u^{2'}, \end{aligned}$$

which lead to

$$\hat{U} \equiv [U^i_{j'}] = \begin{bmatrix} \rho/\sqrt{\rho^2+z^2} & 0 & z/\sqrt{\rho^2+z^2} \\ z/(\rho^2+z^2) & 0 & -\rho/(\rho^2+z^2) \\ 0 & 1 & 0 \end{bmatrix}$$

$$= \begin{bmatrix} \sin\theta & 0 & \cos\theta \\ (\cos\theta)/r & 0 & -(\sin\theta)/r \\ 0 & 1 & 0 \end{bmatrix}.$$

We can check that $\hat{U}U = I$, confirming that $\hat{U} = U^{-1}$.

As we found in Exercise 1.1.2, the unit vector field $\boldsymbol{\lambda} = \mathbf{i}$ has contravariant components given by

$$L \equiv [\lambda^i] = \begin{bmatrix} \sin\theta\cos\phi \\ r^{-1}\cos\theta\cos\phi \\ -r^{-1}\operatorname{cosec}\theta\sin\phi \end{bmatrix},$$

while in cylindrical coordinates they are given by

$$L' \equiv [\lambda^{i'}] = \begin{bmatrix} \cos\phi \\ -\rho^{-1}\sin\phi \\ 0 \end{bmatrix} = \begin{bmatrix} \cos\phi \\ -r^{-1}\operatorname{cosec}\theta\sin\phi \\ 0 \end{bmatrix}.$$

Using these expressions, we can check that $L' = UL$ and $L = \hat{U}L'$.

In the primed coordinate system the quantities corresponding to g_{ij} in the unprimed coordinate system are defined by $g_{i'j'} \equiv \mathbf{e}_{i'} \cdot \mathbf{e}_{j'}$. Using the transformation equation for bases, we have

$$g_{i'j'} = \mathbf{e}_{i'} \cdot \mathbf{e}_{j'} = \left(U^k_{i'}\mathbf{e}_k\right) \cdot \left(U^l_{j'}\mathbf{e}_l\right) = U^k_{i'}U^l_{j'}\mathbf{e}_k \cdot \mathbf{e}_l = U^k_{i'}U^l_{j'}g_{kl}.$$

So the transformation formula for the quantities g_{ij} is

$$\boxed{g_{i'j'} = U^k_{i'}U^l_{j'}g_{kl},} \tag{1.52}$$

which is like that for the covariant components of a vector, but involves two sets of Jacobian matrix elements, $U^k_{i'}$ and $U^l_{j'}$. A similar argument gives

$$\boxed{g^{i'j'} = U^{i'}_k U^{j'}_l g^{kl}} \tag{1.53}$$

for the related quantities g^{ij}.

In equations (1.52) and (1.53), we have our first taste of how the components of a tensor transform. We look at tensors in Euclidean space in the next section, and in the more general setting of tensor fields on manifolds in

Section 1.8. We shall recognize the quantities g_{ij} as the components of the *metric tensor*, so called because it gives us access to metric properties such as the lengths of vectors and the angles between them (via the dot product $\boldsymbol{\lambda} \cdot \boldsymbol{\mu} = g_{ij}\lambda^i\mu^j$) and the distance between neighboring points (via the line element $ds^2 = g_{ij}du^i du^j$).

Exercises 1.4

1. Use the chain rule to show that

$$U^k_{i'}U^{i'}_j = \delta^k_j \quad \text{and} \quad U^{k'}_i U^i_{j'} = \delta^k_j.$$

   Obtain the same results by using the fact that

$$\delta^k_j = \mathbf{e}^k \cdot \mathbf{e}_j = \mathbf{e}^{k'} \cdot \mathbf{e}_{j'}.$$

2. Obtain the second of equations (1.49) using equation (1.47) and the result of Exercise 1.

3. Translate equation (1.52) into a matrix equation involving

$$\hat{U} \equiv [U^i_{j'}], \quad G \equiv [g_{ij}], \quad \text{and} \quad G' \equiv [g_{i'j'}].$$

   Hence, using G from Example 1.3.1 and $\hat{U}$ from Example 1.4.1, obtain the line element for Euclidean space in cylindrical coordinates.

1.5 Tensor fields in Euclidean space

While scalar and vector fields are sufficient to formulate Newton's theory of gravitation, tensor fields are an additional requirement for Einstein's theory. To introduce the idea of tensors and tensor fields, we shall look at an elastic body under stress in the classical theory of elasticity based on Newtonian mechanics.[5]

An elastic body is placed under stress by body forces (such as gravity) acting throughout its extent and by forces applied externally to its surface. If V is part of such a body, then it is postulated that the total force on V due to stresses in the body is given by a surface integral of the form

$$\iint_S \boldsymbol{\tau}(\mathbf{n})\, dS,$$

where S is the boundary of V. The vector $\boldsymbol{\tau}(\mathbf{n})$ is the force per unit area acting at a point of the bounding surface S of V, and the notation $\boldsymbol{\tau}(\mathbf{n})$ is used to indicate that it depends on the unit outward normal $\mathbf{n}$ to S. If any

[5]See, for example, Symon, 1971, Chap. 10, or Landau and Lifshitz, 1987, Chap. 1.

part of S coincides with the actual boundary of the elastic body, then on this part $\boldsymbol{\tau}(\mathbf{n})$ is the force per unit area due to the applied external force, whereas on any part of S that is inside the body, $\boldsymbol{\tau}(\mathbf{n})$ is the force per unit area due to the material outside V. In general, shearing forces are present, and $\boldsymbol{\tau}(\mathbf{n})$ is not parallel to $\mathbf{n}$ (see Fig. 1.3).

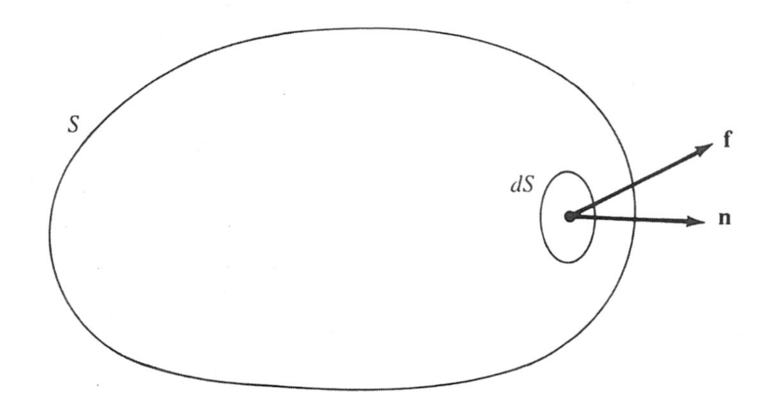

Fig. 1.3. Force per unit area $\mathbf{f}$ due to stress. In general, $\mathbf{f}$ is not in the same direction as the normal $\mathbf{n}$ to the element of area dS.

As the notation suggests, at each point of the body we can look on $\boldsymbol{\tau}$ as a function that acts on a vector $\mathbf{n}$ to produce another vector $\boldsymbol{\tau}(\mathbf{n})$ which (in general) is not parallel to $\mathbf{n}$. This function is assumed to act linearly, which means that for general vectors $\mathbf{u}, \mathbf{v}$ and general scalars α, β,

$$\boldsymbol{\tau}(\alpha\mathbf{u} + \beta\mathbf{v}) = \alpha\boldsymbol{\tau}(\mathbf{u}) + \beta\boldsymbol{\tau}(\mathbf{v}).$$

A consequence of this assumption is that $\boldsymbol{\tau}$ can act on general vectors $\mathbf{v}$ and not just unit vectors $\mathbf{n}$.

Suppose that we use curvilinear coordinates u^i to label points in the body and that at each point P of the surface S we express both the force per unit area $\mathbf{f}$, given by $\mathbf{f} = \boldsymbol{\tau}(\mathbf{n})$, and the unit normal $\mathbf{n}$ in terms of the natural basis vectors $\mathbf{e}_i$. So we have $\mathbf{f} = f^i\mathbf{e}_i$ and $\mathbf{n} = n^j\mathbf{e}_j$, which gives

$$f^i\mathbf{e}_i = \boldsymbol{\tau}(n^j\mathbf{e}_j) = n^j\boldsymbol{\tau}(\mathbf{e}_j), \tag{1.54}$$

on using the linearity of $\boldsymbol{\tau}$. For each j, the vector $\boldsymbol{\tau}(\mathbf{e}_j)$ can also be expressed in terms of the basis vectors $\mathbf{e}_i$. Thus we can write

$$\boldsymbol{\tau}(\mathbf{e}_j) = \tau^i_j\mathbf{e}_i, \tag{1.55}$$

where, for each j, the quantities τ^i_j $(i = 1, 2, 3)$ are the components of $\boldsymbol{\tau}(\mathbf{e}_j)$ relative to the basis $\{\mathbf{e}_i\}$. Equation (1.54) then gives

$$f^i \mathbf{e}_i = n^j \tau^i_j \mathbf{e}_i,$$

so

$$f^i = \tau^i_j n^j, \tag{1.56}$$

showing how the components f^i are obtained from the components n^i of the normal vector. (If we put $F \equiv [f^i]$, $T \equiv [\tau^i_j]$, and $N \equiv [n^i]$, then the matrix version of equation (1.56) is $F = TN$.) The linear function $\boldsymbol{\tau}$ is called the *stress tensor* and the quantities τ^i_j defined by equation (1.55) are its components.

If we worked using primed coordinates $u^{i'}$, then we would define primed components $\tau^{i'}_{j'}$ by means of

$$\boldsymbol{\tau}(\mathbf{e}_{j'}) = \tau^{i'}_{j'} \mathbf{e}_{i'}$$

and in place of equation (1.56) we would have

$$f^{i'} = \tau^{i'}_{j'} n^{j'}. \tag{1.57}$$

Now $f^{i'} = U^{i'}_k f^k$ and $n^{j'} = U^{j'}_l n^l$, so the above gives

$$U^{i'}_k f^k = \tau^{i'}_{j'} U^{j'}_l n^l.$$

But $f^k = \tau^k_l n^l$, so

$$U^{i'}_k \tau^k_l n^l = \tau^{i'}_{j'} U^{j'}_l n^l.$$

Since this holds for *all* unit vectors $\mathbf{n}$ at P, we conclude that

$$U^{i'}_k \tau^k_l = \tau^{i'}_{j'} U^{j'}_l.$$

Multiplying by $U^l_{m'}$ (and using $U^{j'}_l U^l_{m'} = \delta^j_m$) gives

$$\tau^{i'}_{m'} = U^{i'}_k U^l_{m'} \tau^k_l \tag{1.58}$$

as the transformation formula for the components of $\boldsymbol{\tau}$.

We now have three examples of how tensor components transform: equation (1.52) for the components g_{ij} of the metric tensor, equation (1.53) for the related quantities g^{ij}, and equation (1.58) for the components τ^i_j of the stress tensor. These transformation formulae are clearly similar in the way that each unprimed suffix is involved in a summation with Jacobian matrix elements of the kind $U^i_{j'}$ or $U^{i'}_j$; whichever kind is used is dictated by the requirement that a repeated suffix should occur once as a subscript and once as a superscript. The free suffixes (those not involved in summations) carry primes and those on the left of each transformation formula balance those on the right.

The three transformation formulae serve as prototypes for a general transformation formula for tensor components, where the components carry an arbitrary number of superscripts and an arbitrary number of subscripts. These are considered in Section 1.8 in the more general context of an N-dimensional manifold, which is the subject of Section 1.7. Before doing that, we shall look at surfaces in Euclidean space as examples of curved two-dimensional manifolds.

Exercise 1.5

1. Show that the components τ^i_j of the stress tensor $\boldsymbol{\tau}$ are given by

$$\tau^i_j = \mathbf{e}^i \cdot \boldsymbol{\tau}(\mathbf{e}_j)$$

and use this result to re-establish the transformation formula (1.58).

1.6 Surfaces in Euclidean space

A surface Σ in Euclidean space is given parametrically by expressing the Cartesian coordinates x, y, z as functions of two parameters u, v:

$$x = x(u, v), \quad y = y(u, v), \quad z = z(u, v). \tag{1.59}$$

In principle, it is possible to eliminate u, v from these equations to obtain an equation for Σ of the form $f(x, y, z) = 0$. However, our discussion here will be based on the use of parameters and our aim is to expose the similarities (and differences) between the use of parameters u, v to label points on a surface and the use of curvilinear coordinates u, v, w to label points in Euclidean space.

The position vector $\mathbf{r}$ of points on Σ can be given as a function of the parameters by combining equations (1.59) into a single vector equation:

$$\mathbf{r} = x(u, v)\mathbf{i} + y(u, v)\mathbf{j} + z(u, v)\mathbf{k}.$$

Analogous to the coordinate curves in space, we have at each point of Σ a pair of *parametric curves*. The first of these is obtained by keeping v constant and letting u act as a parameter along it, and the second is obtained in a similar way by interchanging the roles of u and v. So at a point P given by (u_0, v_0), the first coordinate curve through P is given parametrically by

$$\mathbf{r} = x(u, v_0)\mathbf{i} + y(u, v_0)\mathbf{j} + z(u, v_0)\mathbf{k}$$

and the partial derivative $\partial\mathbf{r}/\partial u$, when evaluated at (u_0, v_0), gives a tangent vector to the curve at P. Similarly, the second parametric curve through P is given by

$$\mathbf{r} = x(u_0, v)\mathbf{i} + y(u_0, v)\mathbf{j} + z(u_0, v)\mathbf{k}$$

and $\partial\mathbf{r}/\partial v$, when evaluated at (u_0, v_0), gives its tangent vector at P. So at each point P of Σ, both $\mathbf{e}_u \equiv \partial\mathbf{r}/\partial u$ and $\mathbf{e}_v \equiv \partial\mathbf{r}/\partial v$ are tangential to the surface, and together they define the *tangent plane* to Σ at P (see Fig. 1.4).

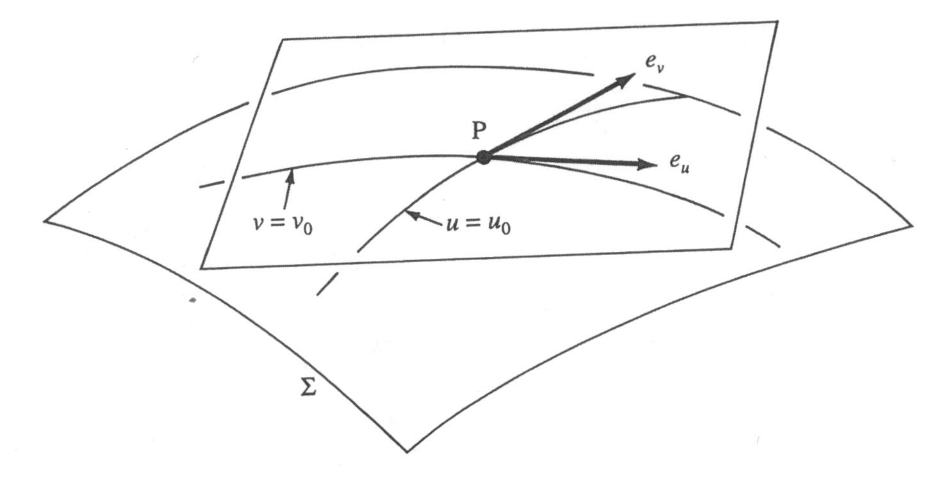

Fig. 1.4. The tangent plane to Σ at P.

A *vector field on a surface* Σ is an assignment to each point P of Σ of a vector $\boldsymbol{\lambda}$ that is *tangential* to Σ. So, for example, at any instant of time the horizontal wind-velocity $\mathbf{v}$ at each point of the Earth's surface E gives a vector field on E. It is this vector field that is represented by the arrows on a weather map. Returning to the general situation of a vector field $\boldsymbol{\lambda}$ on a surface Σ, we see that the vector at each point P lies in the tangent plane at P, so we can refer $\boldsymbol{\lambda}$ to the basis $\{\mathbf{e}_u, \mathbf{e}_v\}$ provided by the tangent vectors to the parametric curves:

$$\boldsymbol{\lambda} = \lambda^u \mathbf{e}_u + \lambda^v \mathbf{e}_v. \tag{1.60}$$

This is the *natural basis* for the tangent plane. It is induced by the system of parameters used to label points in exactly the same way as the natural basis associated with a curvilinear system of coordinates (u, v, w) in Euclidean space.

There is also a dual basis $\{\mathbf{e}^u, \mathbf{e}^v\}$, but it is not given in such a straightforward manner as its counterpart for curvilinear coordinates in space, where we used the gradient vectors ∇u, ∇v, ∇w to define $\mathbf{e}^u$, $\mathbf{e}^v$, $\mathbf{e}^w$. Each of the parameters u, v gives a scalar field on the surface Σ, and it is the gradients of these that provide the dual basis $\{\mathbf{e}^u, \mathbf{e}^v\}$. Off the surface, u and v have no meaning and it makes no sense to try to define $\mathbf{e}^u$, $\mathbf{e}^v$ as the gradients of scalar fields u, v defined throughout space, as we did in Section 1.1. However, we can fix the direction of the required basis vector $\mathbf{e}^u$ at P by noting that, as the gradient of the scalar field u on Σ, it is normal to the *level curve* of u that passes through P and points in the direction of increasing u. Since this level curve is given by $u = u_0$, it follows that $\mathbf{e}^u$ is orthogonal to the natural basis vector $\mathbf{e}_v$. The direction of the other dual basis vector $\mathbf{e}^v$ is given in a similar way by interchanging the roles of u and v. Figure 1.5 shows the relationship between the natural basis $\{\mathbf{e}_u, \mathbf{e}_v\}$ and its dual $\{\mathbf{e}^u, \mathbf{e}^v\}$ at a point P on Σ.

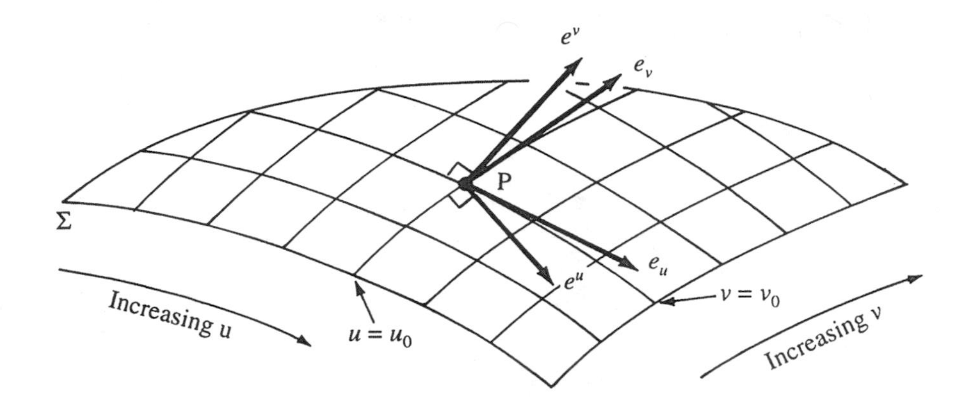

Fig. 1.5. The natural basis and its dual at a point P on a surface Σ.

From our observations on directions, we have

$$\mathbf{e}^u \cdot \mathbf{e}_v = 0, \qquad \mathbf{e}^v \cdot \mathbf{e}_u = 0. \tag{1.61}$$

We now fix the lengths of the dual basis vectors by requiring that

$$\mathbf{e}^u \cdot \mathbf{e}_u = 1, \qquad \mathbf{e}^v \cdot \mathbf{e}_v = 1. \tag{1.62}$$

If we use the suffix notation, with suffixes $A, B, \ldots$ taking the values $1, 2$, then the parameters u, v become u^A, the natural basis $\{\mathbf{e}_u, \mathbf{e}_v\}$ is denoted by $\{\mathbf{e}_A\}$ and its dual $\{\mathbf{e}^u, \mathbf{e}^v\}$ by $\{\mathbf{e}^A\}$. Equations (1.61) and (1.62) combine to give

$$\mathbf{e}^A \cdot \mathbf{e}_B = \delta^A_B, \tag{1.63}$$

which is the analog for surfaces of equation (1.17). Other equations and relationships developed in Section 1.2 have their counterparts here.

We can introduce a metric tensor with components g_{AB} defined by

$$g_{AB} \equiv \mathbf{e}_A \cdot \mathbf{e}_B$$

and a related tensor with components g^{AB} defined by

$$g^{AB} \equiv \mathbf{e}^A \cdot \mathbf{e}^B.$$

Equations involving g_{AB}, g^{AB}, and the components of vector fields on Σ, analogous to equations (1.25)–(1.31) of Section 1.2 then follow. We also have

$$\mathbf{e}_A = g_{AB}\mathbf{e}^B, \qquad \mathbf{e}^A = g^{AB}\mathbf{e}_B$$

(see Exercise 1.2.2), which provides a quick way of getting the dual basis $\{\mathbf{e}^A\}$ from the natural basis $\{\mathbf{e}_A\}$, as in the following example.

Example 1.6.1
The parametric equation

$$\mathbf{r} = (u+v)\,\mathbf{i} + (u-v)\,\mathbf{j} + 2uv\,\mathbf{k},$$

where $-\infty < u < \infty$, $-\infty < v < \infty$, gives a hyperbolic paraboloid. (It is the surface given by putting $w = 0$ in Example 1.1.3.) The two natural basis vectors are

$$\begin{aligned}
\mathbf{e}_1 &\equiv \mathbf{e}_u \equiv \partial\mathbf{r}/\partial u = \mathbf{i} + \mathbf{j} + 2v\,\mathbf{k}, \\
\mathbf{e}_2 &\equiv \mathbf{e}_v \equiv \partial\mathbf{r}/\partial v = \mathbf{i} - \mathbf{j} + 2u\,\mathbf{k}.
\end{aligned}$$

So the quantities $g_{AB} \equiv \mathbf{e}_A \cdot \mathbf{e}_B$ can be displayed as

$$[g_{AB}] = \begin{bmatrix} 2(1+2v^2) & 4uv \\ 4uv & 2(1+2u^2) \end{bmatrix}.$$

The inverse of this matrix is $[g^{AB}]$, which we calculate to be

$$[g^{AB}] = \begin{bmatrix} \frac{1+2u^2}{2(1+2u^2+2v^2)} & \frac{-uv}{1+2u^2+2v^2} \\ \frac{-uv}{1+2u^2+2v^2} & \frac{1+2v^2}{2(1+2u^2+2v^2)} \end{bmatrix}.$$

We can then give the dual basis vectors in terms of $\mathbf{i}, \mathbf{j}, \mathbf{k}$ by saying

$$\begin{aligned}
\mathbf{e}^u \equiv \mathbf{e}^1 &= g^{1A}\mathbf{e}_A = g^{11}\mathbf{e}_1 + g^{12}\mathbf{e}_2 \\
&= \frac{1+2u^2}{2(1+2u^2+2v^2)}(\mathbf{i}+\mathbf{j}+2v\,\mathbf{k}) + \frac{-uv}{1+2u^2+2v^2}(\mathbf{i}-\mathbf{j}+2u\,\mathbf{k}) \\
&= \frac{1+2u^2-2uv}{2(1+2u^2+2v^2)}\mathbf{i} + \frac{1+2u^2+2uv}{2(1+2u^2+2v^2)}\mathbf{j} + \frac{v}{1+2u^2+2v^2}\mathbf{k}
\end{aligned}$$

and

$$\begin{aligned}
\mathbf{e}^v \equiv \mathbf{e}^2 &= g^{2A}\mathbf{e}_A = g^{21}\mathbf{e}_1 + g^{22}\mathbf{e}_2 \\
&= \frac{-uv}{1+2u^2+2v^2}(\mathbf{i}+\mathbf{j}+2v\,\mathbf{k}) + \frac{1+2v^2}{2(1+2u^2+2v^2)}(\mathbf{i}-\mathbf{j}+2u\,\mathbf{k}) \\
&= \frac{1+2v^2-2uv}{2(1+2u^2+2v^2)}\mathbf{i} - \frac{1+2v^2+2uv}{2(1+2u^2+2v^2)}\mathbf{j} + \frac{u}{1+2u^2+2v^2}\mathbf{k}.
\end{aligned}$$

If we put $u^A = u^A(t)$, where each $u^A(t)$ is a differentiable function of t, for t belonging to some interval I, then (in a manner similar to that described in Sec. 1.3) we obtain a curve γ on the surface Σ whose tangent vector has components $\dot{u}^A$ relative to the natural basis $\{\mathbf{e}_A\}$. The length L of this curve is given by

$$L = \int_a^b \left(g_{AB}\dot{u}^A\dot{u}^B\right)^{1/2} dt, \tag{1.64}$$

where $a \leq t \leq b$ gives the interval I. For neighboring points on Σ, whose parameter differences are du^A, their distance apart ds is given by

$$ds^2 = g_{AB}du^A du^B, \tag{1.65}$$

as can be seen by working with

$$ds^2 = d\mathbf{r} \cdot d\mathbf{r} = du^A\mathbf{e}_A \cdot du^B\mathbf{e}_B.$$

Equation (1.65) specifies the *line element* for Σ and (as in Sec. 1.3) can be viewed as an expression of Pythagoras' Theorem.

We see that, in the main, much of the terminology and notation developed for use with arbitrary curvilinear coordinates in Euclidean space can be adapted for use on a surface Σ, where the parameters u^A play the role of coordinates, and we have presented the material in a way that brings out these similarities. However, there is one respect in which the situations are fundamentally different: the geometry of Euclidean space *is* Euclidean, whereas that of a surface is not, unless it happens to be flat. If the geometry is Euclidean then we can introduce a system of Cartesian coordinates (x, y, z) or (x, y), according to dimension. The line element then takes the form

$$ds^2 = dx^2 + dy^2 + dz^2 \quad \text{or} \quad ds^2 = dx^2 + dy^2,$$

and we have $g_{ij} = \delta_{ij}$ or $g_{AB} = \delta_{AB}$. Thus we can characterize the basic flatness of Euclidean geometry by the possibility of introducing a coordinate system in which the metric tensor components are given by a Kronecker delta. For a curved surface this is not possible. To fully appreciate why, we must wait for a discussion of curvature in Chapter 3.

Another difference between the two situations is that we can regard the vectors of a vector field in Euclidean space as being in the space itself, whereas for a vector field on a surface the vectors are not in the surface (unless it happens to be flat), but tangential to it. We shall have more to say on this matter in Section 1.10, after we have introduced manifolds and discussed vector and tensor fields on manifolds.

Exercises 1.6

1. Check that the natural basis $\{\mathbf{e}_A\}$ of Example 1.6.1 and its dual $\{\mathbf{e}^A\}$ satisfy

$$\mathbf{e}^A \cdot \mathbf{e}_B = \delta^A_B.$$

2. Write down the line element for
   (a) a sphere of radius a, using angles (θ, ϕ) borrowed from spherical coordinates as parameters;
   (b) a cylinder whose cross section is a circle of radius a, using (ϕ, z) borrowed from cylindrical coordinates as parameters;
   (c) the hyperbolic paraboloid of Example 1.6.1, using the parameters (u, v) of that example.

3. Is the cylinder of Exercise 2b curved or flat?

1.7 Manifolds

The model for the spacetime of general relativity makes use of a certain kind of four-dimensional manifold, so we need to explain what this involves. In doing this, we shall not give a precise mathematical definition, but rather explain and describe the properties of an N-dimensional manifold. We shall assume that this manifold is endowed with a metric tensor field (which is not a general requirement of manifolds) and explain how this is used to define and handle metric properties. We shall be guided by the notation and terminology developed when considering arbitrary curvilinear coordinates in Euclidean space and parameterized surfaces.

What makes a manifold N-dimensional is that points in it can be labeled by a system of N real coordinates $x^1, x^2, \ldots, x^N$, in such a way that the correspondence between the points and the labels is one-to-one. We do not require that the whole of the manifold M should be covered by one system of coordinates, nor do we regard any one system as in some way preferred. The general situation is that we have a collection of coordinate systems, each covering some part of M, and all these are on an equal footing. Where two coordinate systems overlap, there are sets of equations giving each coordinate of one system as a function of the coordinates of the other. So if the coordinates x^a cover the region U and the coordinates $x^{a'}$ cover the region U', where these are overlapping regions, then the coordinates of points in the overlap are related by equations of the form

$$x^{a'} = x^{a'}(x^1, x^2, \ldots, x^N) \qquad (a = 1, \ldots, N), \tag{1.66}$$

giving each $x^{a'}$ as a function of the coordinates x^b, and these have inverses of the form

$$x^a = x^a(x^{1'}, x^{2'}, \ldots, x^{N'}) \qquad (a = 1, \ldots, N), \tag{1.67}$$

giving each x^a as a function of the coordinates $x^{b'}$. We shall assume that the functions involved are differentiable so that the partial derivatives

$$X^{a'}_b \equiv \frac{\partial x^{a'}}{\partial x^b} \quad \text{and} \quad X^a_{b'} \equiv \frac{\partial x^a}{\partial x^{b'}}$$

exist. This means that the manifold M is a *differentiable manifold.*[6] The $N \times N$ matrix $[X^{a'}_b]$ is the *Jacobian matrix* associated with equations (1.66) that give the change of coordinates from x^a to $x^{a'}$. The fact that these equations have inverses (1.67) means that this matrix is nonsingular (i.e., has an inverse), so the Jacobian $\det[X^{a'}_b]$ is nonzero at each point of the overlap region. Similar remarks apply to the Jacobian matrix $[X^a_{b'}]$ and the Jacobian $\det[X^a_{b'}]$. In fact the two matrices are inverses, for the chain rule gives

$$\boxed{X^a_{b'} X^{b'}_c = \delta^a_c,} \tag{1.68}$$

in exactly the same way as it yields equation (1.50) for changes of coordinates in Euclidean space. The formula

$$\boxed{X^{a'}_b X^b_{c'} = \delta^a_c} \tag{1.69}$$

follows on interchanging the roles of primed and unprimed coordinates.

We noted above that at each point of a region where two coordinate systems overlap the Jacobian $\det[X^{a'}_b]$ is nonzero. This result has a kind of converse which gives the condition for a set of equations like (1.66) to define a new system of coordinates.

Suppose we wish to introduce a new system of coordinates by giving $x^{a'}$ as differentiable functions of the old coordinates x^a and that there is some point P where the Jacobian $\det[X^{a'}_b]$ is nonzero. The inverse-function theorem[7] then implies that P has a neighborhood U' in which the mapping between the old x^a and the new $x^{a'}$ is one-to-one. Since the correspondence between the coordinates x^a and the points of the manifold is also one-to-one, it follows that in U' the correspondence between the $x^{a'}$ and the points of the manifold is one-to-one, so that they act as coordinates in U'.

For vectors and vector fields on a manifold, our approach is to define them as objects having N components that, under a change of coordinates, transform in a way that generalizes either equation (1.45) or (1.47) for vector components in Euclidean space. Thus we define a *contravariant vector* at a point P as an object having N components λ^a which, under a change of coordinates about P, transform according to

$$\boxed{\lambda^{a'} = X^{a'}_b \lambda^b,} \tag{1.70}$$

where the partial derivatives are evaluated at P. A *covariant vector* is defined in a similar way by requiring its components μ_a to transform according to

[6] Mathematicians allow manifolds in which the coordinate-transformation functions are merely continuous and call them *topological manifolds.*

[7] See, for example, Munem and Foulis, Chap. 7, §1.

$$\boxed{\mu_{a'} = X^b_{a'}\mu_b.} \tag{1.71}$$

For vector fields, this kind of transformation law for components holds at each point of M where the field is defined.

The basic example of a contravariant vector is the tangent vector to a curve γ, which can be given parametrically by setting $x^a = x^a(t)$, where $x^a(t)$ are differentiable functions of t for t in some interval I. At each point of γ the derivatives $\dot{x}^a(t)$ are the components of a vector, as we now show. Using equation (1.66), we see that, in a primed coordinate system, γ is given by

$$x^{a'}(t) = x^{a'}\left(x^1(t), x^2(t), \ldots, x^N(t)\right) \tag{1.72}$$

and the chain rule for differentiation gives

$$\dot{x}^{a'} \equiv \frac{dx^{a'}}{dt} = \frac{\partial x^{a'}}{\partial x^b}\frac{dx^b}{dt} = X^{a'}_b \dot{x}^b,$$

showing that the quantities $\dot{x}^a$ transform according to equation (1.70), as asserted. The vector with components $\dot{x}^a$ is the tangent vector to γ that arises naturally from the parameterization.

The basic example of a covariant vector field is the gradient of a scalar field ϕ. In any coordinate system, such a field can be regarded as a function $\phi(x^a)$ of the coordinates and we can form the N partial derivatives $\partial_a\phi \equiv \partial\phi/\partial x^a$. In a primed coordinate system, we would regard ϕ as a function $\phi(x^{a'})$ of the primed coordinates $x^{a'}$ and form the partial derivatives $\partial_{a'}\phi \equiv \partial\phi/\partial x^{a'}$. The chain rule gives

$$\partial_{a'}\phi \equiv \frac{\partial\phi}{\partial x^{a'}} = \frac{\partial\phi}{\partial x^b}\frac{\partial x^b}{\partial x^{a'}} = X^b_{a'}\partial_b\phi,$$

showing that the quantities $\partial_a\phi$ transform according to equation (1.71). They are therefore the components of a covariant vector field, which is the *gradient* of ϕ.

These examples based on tangents to curves and gradients of scalar fields extend some of the ideas of Section 1.3 to manifolds. In that section we also considered metric properties, like the length of a curve given by equation (1.38) and the line element (1.39), which involve the metric tensor components g_{ij}. These ideas can also be extended to the more general setting of a manifold, but we postpone doing this until Section 1.9, after we have discussed the essentially algebraic properties of tensor fields on manifolds.

Example 1.7.1

The configuration space of a mechanical system with N degrees of freedom is an N-dimensional manifold. Points in the configuration space are labeled by N "generalized" coordinates, which are usually denoted by $q^1, q^2, \ldots, q^N$. The evolution of the system in time from some set of initial conditions is given by a curve $q^a = q^a(t)$ in the configuration space parameterized by time t.

1.8 Tensor fields on manifolds

Suppose that with each coordinate system about a point P of a manifold M there are associated N^{r+s} quantities $\tau^{a_1 \ldots a_r}_{b_1 \ldots b_s}$ which, under a change of coordinates, transform according to

$$\boxed{\tau^{a'_1 \ldots a'_r}_{b'_1 \ldots b'_s} = X^{a'_1}_{c_1} \cdots X^{a'_r}_{c_r} X^{d_1}_{b'_1} \cdots X^{d_s}_{b'_s} \tau^{c_1 \ldots c_r}_{d_1 \ldots d_s},} \tag{1.73}$$

where the Jacobian matrices $[X^{a'}_c]$, $[X^d_{b'}]$ are evaluated at P. Then the quantities $\tau^{a_1 \ldots a_r}_{b_1 \ldots b_s}$ are the *components of a type* (r, s) *tensor* at P. This terminology includes the special cases in which $r = 0$ or $s = 0$, so that the kernel letter τ carries only subscripts or only superscripts. So, for example, the components of a type $(2, 0)$ tensor transform according to

$$\boxed{\tau^{a'b'} = X^{a'}_c X^{b'}_d \tau^{cd},} \tag{1.74}$$

while the components of a type $(1, 2)$ tensor transform according to

$$\boxed{\tau^{a'}_{b'c'} = X^{a'}_d X^e_{b'} X^f_{c'} \tau^d_{ef}.} \tag{1.75}$$

The sum $(r + s)$ is sometimes referred to as the *rank* or *order* of the tensor. A type $(r, 0)$ tensor might be referred to as a *contravariant tensor* of rank r and a type $(0, s)$ tensor as a *covariant tensor* of rank s. If both $r \neq 0$ and $s \neq 0$, the tensor is described as *mixed.* We now recognize a contravariant vector at P as a tensor of type $(1, 0)$ and a covariant vector as a tensor of type $(0, 1)$. Scalars may be included in the general scheme of things by regarding them as type $(0, 0)$ tensors.

If at each point of an N-dimensional region V in M we have a type (r, s) tensor defined, then the result is a *tensor field* on V. The region V might be the whole of M, or just a part of it. The components of the field can be regarded as functions of the coordinates used to label the points of V. If these functions are differentiable, then the tensor field is said to be *differentiable.* Sometimes we have tensors defined at each point of a curve γ in M, but not throughout an N-dimensional region (such as the tangent vector to a curve). These constitute a *tensor field along* γ and their components can be regarded as functions of the parameter t used to label the points of γ. Similar remarks apply to tensor fields defined over a surface Σ in M; here the components can be regarded as functions of the parameters u, v used to label points of Σ.

The basic requirement for a set of quantities to qualify as the components of a tensor is that they should transform in the right sort of way under a change of coordinates (i.e., according to equation (1.73)). The *quotient theorem* provides a means of establishing this requirement without having to demonstrate the transformation law explicitly. Rather than give a general statement of the theorem and its proof, which tend to be obscured by a mass of suffixes, we shall give an example which illustrates the gist of the theorem.

Example 1.8.1
Suppose that with each system of coordinates about a point P there are associated N^3 numbers τ^a_{bc} and that it is known that, for *arbitrary* contravariant vector components λ^a, the N^2 numbers $\tau^a_{bc}\lambda^c$ transform as the components of a type $(1,1)$ tensor at P under a change of coordinates about P. That is,

$$\tau^{a'}_{b'c'}\lambda^{c'} = X^{a'}_d X^e_{b'} \tau^d_{ef}\lambda^f, \tag{1.76}$$

where $\tau^{a'}_{b'c'}$ are the N^3 numbers associated with the primed coordinate system. Then we may deduce that the τ^a_{bc} are the components of a type $(1,2)$ tensor. Because $\lambda^{c'} = X^{c'}_f \lambda^f$, equation (1.76) yields

$$(\tau^{a'}_{b'c'}X^{c'}_f - X^{a'}_d X^e_{b'} \tau^d_{ef})\lambda^f = 0, \tag{1.77}$$

and this holds for arbitrary vector components λ^f. Now let λ^f be the vector having one as its gth component and the others zero, so that $\lambda^f = \delta^f_g$. Equation (1.77) then gives

$$\tau^{a'}_{b'c'}X^{c'}_g = X^{a'}_d X^e_{b'} \tau^d_{eg},$$

valid for all subscripts g. Multiplying by $X^g_{h'}$ and using $X^{c'}_g X^g_{h'} = \delta^c_h$ gives

$$\tau^{a'}_{b'h'} = X^{a'}_d X^e_{b'} X^g_{h'} \tau^d_{eg},$$

which establishes that the τ^a_{bc} are indeed the components of a type $(1,2)$ tensor.

This example illustrates the gist of the quotient theorem, which is that if numbers which are candidates for tensor components display tensor character when some of their suffixes are "killed off" by summation with the components of *arbitrary* vectors (or tensors), then this is sufficient to establish that the original numbers are the components of a tensor. We shall have occasion to use this theorem from time to time.

We now consider some operations with tensors. The first of these is *addition*: adding corresponding components of two tensors of the same type results in quantities that are the components of a tensor of that type. The second is *multiplication by scalars*: multiplying each component of a tensor by a scalar quantity results in quantities that are the components of a tensor of the same type. The validity of these two operations is clear from the transformation formula for components.

The third operation is *tensor multiplication* which gives the *tensor product* of two tensors. The components of the tensor product are obtained simply by multiplying together the components of the two tensors involved, so as to form all possible products. For example, if we put

$$\sigma^a_{bc} \equiv \lambda_b \tau^a_c,$$

where λ_b are the components of a type $(0,1)$ tensor transforming according to $\lambda_{b'} = X^e_{b'}\lambda_e$ and τ^a_c are those of a type $(1,1)$ tensor transforming according to $\tau^{a'}_{c'} = X^{a'}_d X^f_{c'}\tau^d_f$, then for $\sigma^{a'}_{b'c'} \equiv \lambda_{b'}\tau^{a'}_{c'}$ we have

$$\begin{aligned}\sigma^{a'}_{b'c'} &= (X^e_{b'}\lambda_e)(X^{a'}_d X^f_{c'}\tau^d_f) \\ &= X^{a'}_d X^e_{b'} X^f_{c'}\sigma^d_{ef},\end{aligned}$$

showing that σ^a_{bc} are the components of a type $(1,2)$ tensor.

The fourth operation is that of *contraction*. It is an operation which may be applied to any object characterized by sets of numbers specified by letters carrying superscripts and subscripts, but it takes on a special significance when the numbers are tensor components. The operation amounts to setting a subscript equal to a superscript and summing, as the summation convention requires. If there are r superscripts and s subscripts then there are rs ways that this may be done, each leading to a *contraction* of the original set of numbers. The special significance that this operation has for tensors is that if the original numbers are the components of a type (r,s) tensor, then their contractions are components of a type $(r-1, s-1)$ tensor. The proof in the general case is somewhat cumbersome, but an example gives the gist of it.

Example 1.8.2

Suppose τ^{ab}_c are the components of a type $(2,1)$ tensor, and we form $\sigma^a = \tau^{ab}_b$ by contraction. Using primed coordinates we would analogously form $\sigma^{a'} = \tau^{a'b'}_{b'}$. Then

$$\sigma^{a'} = \tau^{a'b'}_{b'} = \tau^{cd}_e X^{a'}_c X^{b'}_d X^e_{b'} = \tau^{cd}_e X^{a'}_c \delta^e_d = \tau^{cd}_d X^{a'}_c = \sigma^c X^{a'}_c,$$

showing that the numbers σ^a obtained by contraction are the components of a type $(1,0)$ tensor (a contravariant vector).

Contraction may be combined with tensor multiplication. For example, if ρ^a_b are the components of a type $(1,1)$ tensor and σ^a those of a contravariant vector, then the contravariant vector with components $\tau^a = \rho^a_b\sigma^b$ is said to be obtained by *contracting* one of the tensors involved with the other. The tensor with components τ^a is a contraction of the type $(2,1)$ tensor with components $\rho^a_b\sigma^c$.

Certain tensors are special in some way. One of these is the *Kronecker tensor*, which is a type $(1,1)$ tensor with the property that whatever coordinate system is used, its components κ^a_b are given by the Kronecker delta δ^a_b. To see this, suppose that when using an unprimed coordinate system we have $\kappa^a_b = \delta^a_b$. Then, when we transform to a primed coordinate system, the components become

$$\kappa^{a'}_{b'} = X^{a'}_c X^d_{b'} \delta^c_d = X^{a'}_c X^c_{b'} = \delta^a_b.$$

Because of this property, it is usual to denote the components by δ^a_b rather than κ^a_b. (What we have shown here is that the property $\kappa^a_b = \delta^a_b$ enjoyed by the components of the Kronecker tensor is *coordinate-independent.* There is no analog of this result for type $(0,2)$ and type $(2,0)$ tensors. See Exercise 1.8.1.)

Another special property possessed by some tensors is that of *symmetry.* A type $(0,2)$ tensor is *symmetric* if its components satisfy $\tau_{ab} = \tau_{ba}$. It is easily checked that if this holds in one coordinate system, then it holds in all coordinate systems (see Exercise 1.8.2). A symmetric type $(2,0)$ is similarly defined. We describe a type $(0,2)$ tensor as *skew-symmetric* (or *anti-symmetric*) if its components satisfy $\tau_{ab} = -\tau_{ba}$. Again this is a coordinate-independent property, and the concept also extends to type $(2,0)$ tensors. In fact, the idea of symmetry or skew symmetry can be extended to refer to any pair of superscripts or subscripts of a type (r,s) tensor.

We finish this section with an explanation of how an association can be made between tensors of different types by contraction with the covariant metric tensor or the related contravariant metric tensor. This association extends the algebraic formalism developed in Section 1.2 for vectors in Euclidean space to the more general setting of tensors on a manifold. How the metric tensor is used to deal with metric properties is the subject of the next section.

As remarked in Section 1.7, we assume that the manifold has a metric tensor field with components g_{ab}. This tensor is symmetric, so that $g_{ab} = g_{ba}$, and is nonsingular in the sense that the matrix $[g_{ab}]$ has an inverse $[g^{ab}]$ whose elements satisfy

$$g^{ab} g_{bc} = \delta^a_c. \tag{1.78}$$

Since $[g_{ab}]$ is a symmetric matrix, so is the inverse $[g^{ab}]$, and $g^{ab} = g^{ba}$. We now use the quotient theorem to show that g^{ab} are components of a type $(2,0)$ tensor. To this end, let α^a, β^a be the components of arbitrary contravariant vectors, and define λ_a, μ_a by putting

$$\lambda_a \equiv g_{ab}\alpha^b \quad \text{and} \quad \mu_a \equiv g_{ab}\beta^b.$$

Then, because of the nonsingularity of $[g_{ab}]$, λ_a and μ_a are the components of *arbitrary* covariant vectors. Thus for arbitrary covariant vector components λ_a and μ_a,

$$\begin{aligned} g^{ab}\lambda_a\mu_b &= g^{ab} g_{ac}\alpha^c g_{bd}\beta^d \\ &= \delta^a_d g_{ac}\alpha^c\beta^d \quad \text{(on using (1.78))} \\ &= g_{dc}\alpha^c\beta^d, \end{aligned}$$

which is a type $(0,0)$ tensor. So g^{ab} with both superscripts killed off by contraction with arbitrary covariant vectors displays tensor character, and the quotient theorem implies that the quantities g^{ab} are the components of a type $(2,0)$ tensor. We shall refer to this tensor as the *contravariant metric tensor*. This rather back-handed way of introducing this tensor is forced on us because our approach to tensors on a manifold is via the transformation law for their components, and we cannot follow the kind of route that was used to introduce the quantities g^{ij} in Section 1.2.

We now have the metric tensor with components g_{ab} that can be used to lower suffixes and the contravariant metric tensor with components g^{ab} that can be used to raise suffixes. For example, if $\tau^{ab}{}_c$ are the components of a type $(2,1)$ tensor, then using the metric tensor to lower the first superscript gives a type $(1,2)$ tensor whose components $\tau_a{}^b{}_c$ are defined by $\tau_a{}^b{}_c = g_{ad}\tau^{db}{}_c$. If the contravariant metric tensor is now used to raise the lowered suffix, then the original tensor is recovered. For, continuing this example, we have

$$g^{ad}\tau_d{}^b{}_c = g^{ad}g_{de}\tau^{eb}{}_c = \delta^a_e\tau^{eb}{}_c = \tau^{ab}{}_c.$$

Tensors which may be obtained from each other by raising or lowering suffixes are said to be *associated*, and it is conventional to use the same kernel letter for components, as in the example above. However, this usage is ambiguous if we have more than one metric tensor field defined on the manifold; but since this is rarely the case, opportunity for ambiguity seldom arises. (See, however, Sec. 5.1.) Another source of ambiguity is the fact that more than one tensor of the same type may be associated with a given tensor. For example, lowering the first superscript of the components τ^{ab} of a type $(2,0)$ tensor yields a type $(1,1)$ tensor which is in general different from that obtained by lowering the second superscript. The distinction between the two may be made clear by careful spacing of the suffixes:

$$\tau_a{}^b = g_{ac}\tau^{cb}, \qquad \tau^a{}_b = g_{bc}\tau^{ac}.$$

In the case of symmetric tensors this distinction is not necessary (see, e.g., the Ricci tensor in Chap. 3).

Since

$$\delta^a_b = g^{ac}g_{cb} \quad \text{and} \quad g^{ab} = g^{ad}\delta^b_d = g^{ad}g^{bc}g_{cd},$$

the metric tensor, the contravariant metric tensor, and the Kronecker tensor are associated. However, the convention of using the same kernel letter for components of associated tensors is relaxed in the case of the Kronecker tensor because of the special form its components take, and we use δ^a_b rather than g^a_b for its components.

Strictly speaking, we should regard tensors that are associated, but of different types, as different tensors. However, we shall regard them as different versions of the same tensor, and using the same kernel letter for the components supports this point of view. So, for example, we can pass from the contravariant version of a vector with components λ^a to its covariant version with components λ_a by lowering the superscript. This brings us closer to the terminology used in Euclidean space, where we referred to the contravariant and covariant components of the *same* vector $\boldsymbol{\lambda}$. In fact, the contravariant and covariant versions are not the same, even in Euclidean space, although it is normal practice not to make a distinction in the Euclidean context. As we explain more fully in Section 1.10, a covariant vector $\boldsymbol{\mu}$ should really be defined as a *linear function* that acts on a general contravariant vector $\boldsymbol{\lambda}$ to produce a real number $\boldsymbol{\mu}(\boldsymbol{\lambda}) \equiv \mu_a \lambda^a$. In forming the dot product $\boldsymbol{\mu} \cdot \boldsymbol{\lambda}$ in Euclidean space, we are letting the vector $\boldsymbol{\mu}$ act linearly on $\boldsymbol{\lambda}$ to produce a real number, and in this way it acts like a covariant vector. The equation $\tilde{\boldsymbol{\mu}}(\boldsymbol{\lambda}) \equiv \boldsymbol{\mu} \cdot \boldsymbol{\lambda}$ distinguishes between the contravariant version $\boldsymbol{\mu}$ of a vector and its covariant version $\tilde{\boldsymbol{\mu}}$, and at the same time gives the association between the two versions that allows us to identify them.

In this section we have introduced tensors on a manifold and discussed their basic algebraic properties in terms of their components, and we shall continue to work with components when applying tensor methods to general relativity. From here onwards we shall adopt a much-used convention, which is *to confuse a tensor with its components*. This allows us to refer simply to *the tensor* τ^{ab}, rather than *the tensor with components* τ^{ab}.

Exercises 1.8

1. Suppose that in some coordinate system the components τ_{ab} of a type $(0,2)$ tensor satisfy $\tau_{ab} = \delta_{ab}$. Show that this property is *not* coordinate-independent.
   (Use the transformations between spherical and cylindrical coordinates developed in Example 1.4.1 as the basis for a counter example.)

2. Verify that the relationship $\tau^{ab} = \tau^{ba}$, defining a symmetric tensor, is coordinate-independent.

3. Show that if $\sigma_{ab} = \sigma_{ba}$ and $\tau^{ab} = -\tau^{ba}$ for all a, b, then $\sigma_{ab}\tau^{ab} = 0$.

4. Show that any type $(2,0)$ or type $(0,2)$ tensor can be expressed as the sum of a symmetric and a skew-symmetric tensor.

1.9 Metric properties

The metric tensor field g_{ab} provides us with an inner product $g_{ab}\lambda^a\mu^b$ for vectors λ^a, μ^a at each point P of a manifold M. As in Euclidean space (see equation (1.26)), there are four ways of writing this inner product:

$$\boxed{g_{ab}\lambda^a\mu^b = g^{ab}\lambda_a\mu_b = \lambda_a\mu^a = \lambda^a\mu_a.} \tag{1.79}$$

The usual requirement of an inner product is that it should be *positive definite*, which means that $g_{ab}\lambda^a\lambda^b \geq 0$ for all vectors λ^a, with $g_{ab}\lambda^a\lambda^b = 0$ only if $\lambda^a = 0$. However, to provide a model for the spacetime of general relativity, we must relax this condition and we require only that the metric tensor be nonsingular, in the sense that matrix $[g_{ab}]$ has an inverse. This leads to some rather odd metrical properties, such as nonzero vectors having zero length, and the need to include modulus signs where square roots are involved. A manifold that possesses a positive definite metric tensor field is called *Riemannian*. If the metric tensor field is indefinite, the description *pseudo-Riemannian* (or *semi-Riemannian*) is sometimes used, or the meaning of *Riemannian* is extended to include the indefinite case. With these understandings, we can make definitions that allow us to deal with metric properties in a way that extends and generalizes the corresponding properties introduced in Sections 1.3 and 1.6 for Euclidean space and surfaces in Euclidean space.

The *length* of a vector λ^a is given by

$$\left|g_{ab}\lambda^a\lambda^b\right|^{1/2} = \left|g^{ab}\lambda_a\lambda_b\right|^{1/2} = \left|\lambda_a\lambda^a\right|^{1/2}. \tag{1.80}$$

A *unit* vector is one whose length is one. As remarked, if g_{ab} is indefinite, we can have $\left|\lambda_a\lambda^a\right|^{1/2} = 0$ for $\lambda^a \neq 0$, in which case the vector λ^a is described as *null*.

The *angle* θ between two non-null vectors λ^a, μ^a is given by

$$\cos\theta = \frac{g_{ab}\lambda^a\mu^b}{\left|g_{cd}\lambda^c\lambda^d\right|^{1/2}\left|g_{ef}\mu^e\mu^f\right|^{1/2}}, \tag{1.81}$$

which generalizes the formula $\cos\theta = (\boldsymbol{\lambda}\cdot\boldsymbol{\mu})/\left|\boldsymbol{\lambda}\right|\left|\boldsymbol{\mu}\right|$. If the metric tensor is indefinite, this formula can lead to $|\cos\theta| > 1$, resulting in a nonreal value for θ.

Two vectors are *orthogonal* (or *perpendicular*) if their inner product is zero. This definition makes sense even if one or both of the vectors are null. In fact, a null vector is a nonzero vector that is orthogonal to itself. An example in relativity is the wave 4-vector (see Secs. 5.2 and A.6).

As explained in Section 1.7, a curve γ in a manifold M is given by setting $x^a = x^a(t)$, where the parameter t belongs to some interval I, and we noted that at each point of γ a tangent vector is given by $\dot{x}^a \equiv dx^a/dt$. If I is given by $a \leq t \leq b$, then (generalizing equation (1.38)) we can define the *length* of γ to be

$$L = \int_a^b \left|g_{ab}\dot{x}^a\dot{x}^b\right|^{1/2} dt. \tag{1.82}$$

It is clear that this definition is coordinate-independent, but not so clear that it does not depend on the way that the curve is parameterized (see

Exercise 1.9.2). If the metric tensor is indefinite then $g_{ab}\dot{x}^a\dot{x}^b$ may be negative, hence the need for the modulus signs. A further aspect of indefiniteness in the metric tensor is that we may have a curve whose tangent vector at each point is null, so that $g_{ab}\dot{x}^a\dot{x}^b = 0$ at every point, giving a curve of zero length. Such a curve is called a *null curve*.

Note that we only define lengths of curves, and make no attempt to define the distance between a pair of arbitrary points in M. We can, however, define the distance δs between nearby points whose coordinate differences are small. These can be regarded as points on a curve given by parameter values whose difference δt is small, and since to first order $\delta x^a = \dot{x}^a \delta t$, the definition yields $\delta s^2 = |g_{ab}\delta x^a \delta x^b|$. The infinitesimal version of this is

$$\boxed{ds^2 = |g_{ab}dx^a dx^b|} \tag{1.83}$$

(often written without the modulus signs, even in the indefinite case), and defines the line element of the manifold M.

The kind of manifold that we use to model spacetime is a four-dimensional pseudo-Riemannian manifold whose metric tensor field $g_{\mu\nu}$ $(\mu, \nu = 0, 1, 2, 3)$ has an indefiniteness characterized by $(+ - - -)$. What this means is that if at any point P we adopt a coordinate system that gives $[g_{\mu\nu}]_\mathrm{P}$ as a diagonal matrix, then one of the diagonal elements is positive, while the other three are negative.[8] Any nonzero vector is then described as

$$\left\{\begin{array}{l} \textit{timelike} \\ \textit{null} \\ \textit{spacelike} \end{array}\right. \quad \text{if } g_{\mu\nu}\lambda^\mu\lambda^\nu \left\{\begin{array}{l} > 0 \\ = 0 \\ < 0 \end{array}\right. .$$

These descriptions are also applied to curves.

If the tangent vectors to a curve (or part of a curve) are timelike, we describe the curve (or part of the curve) as *timelike*, and extend the descriptions *null* and *spacelike* in a similar way. We shall see in the next chapter, where our model for spacetime is more fully developed, that a particle with mass follows a timelike path, while a photon follows a null path. Because of the association with photons, the term *lightlike* is often used in place of *null*.

Exercises 1.9

1. Show that if the metric tensor g_{ab} is positive definite, then $\cos\theta$, as defined by equation (1.81), satisfies $|\cos\theta| \leq 1$.

2. Show that the definition of the length of a curve given by equation (1.82) is independent of the parameter used.

---

[8]The difference between the number of positive elements and the number of negative elements is called the *signature* of the metric. So our model for spacetime has signature -2. Some authors use $(+ + + -)$, resulting in a signature of $+2$.

3. For $r > 2m$, the Schwarzschild solution has a metric tensor field given by

$$[g_{\mu\nu}] = \text{diag}\left(c^2(1-2m/r), -(1-2m/r)^{-1}, -r^2, -r^2\sin^2\theta\right),$$

where the coordinates are labeled according to $t \equiv x^0$, $r \equiv x^1$, $\theta \equiv x^2$, $\phi \equiv x^3$ (see Sec. 4.1). Find the lengths of the following vectors and the angles between them:

(a) $\lambda^\mu \equiv \delta_0^\mu$; (b) $\mu^\mu \equiv \delta_1^\mu$; (c) $\nu^\mu \equiv \delta_0^\mu + c(1-2m/r)\delta_1^\mu$.

Are any of these vectors null? Are any pairs orthogonal?

1.10 What and where are the bases?

Our way of introducing vector and tensor fields on manifolds relies on the use of components and the way in which they transform under a change of coordinates. It leaves unanswered certain questions about the objects that we are trying to define. In Euclidean space we can refer a vector $\boldsymbol{\lambda}$ to a basis $\{\mathbf{e}_i\}$, or the dual basis $\{\mathbf{e}^i\}$, as in equation (1.16). It is natural to ask whether these equations have analogs for vectors defined on a manifold. If so, then what and where is the basis $\{\mathbf{e}_a\}$? How should we picture the dual basis $\{\mathbf{e}^a\}$? Given that tensors have components, are there bases to which these components refer?

If you are satisfied with the explanation of vector and tensor fields on manifolds given in earlier sections, then you can safely move on to Chapter 2, as the material in this section is not a prerequisite for later chapters. However, if you think these questions need answering, then this section will provide you with a pointer towards the more formal methods of dealing with vector and tensor fields on manifolds. If you want a fuller account of the concepts outlined here, then this is provided by Appendix C.

Just as in Euclidean space, if we let only one coordinate vary, while keeping all the other coordinates fixed, we obtain a *coordinate curve* in the manifold M. We can give a parametric equation for the bth coordinate curve through a point P with coordinates x_0^a by putting

$$x^a = x_0^a + \delta_b^a t,$$

where t is a parameter, from which we see that its tangent vector has components $\dot{x}^a = \delta_b^a$. That is, all the components of the tangent vector to the bth coordinate curve are zero, except for the bth, which is equal to one. We therefore conclude that at each point P of M the tangent vector to the bth coordinate curve through P is, in fact, the bth basis vector $\mathbf{e}_b$ for contravariant vectors at P. These basis vectors define the *tangent space* T_P of M at P in much the same way that the tangent vectors to the two coordinate curves passing through a point P of a surface define the tangent plane to the surface at P (see Fig. 1.4). Just as each point P of a surface has its own tangent plane making contact with the surface at P, each point P of a manifold has a tangent space T_P attached to it at P. We can then picture a contravariant vector $\boldsymbol{\lambda} = \lambda^a \mathbf{e}_a$ at a point as an arrow emanating from P: it is not something *in* the manifold (like a curve is in the manifold), but something *attached* to it at P (like the tangent vectors to a surface). Although the tangent plane to a surface at a point gives a useful way of viewing the tangent space of a manifold at a point, this view can be misleading. An abstract manifold should be regarded as a thing in itself: there is no higher-dimensional space in which it and its tangent spaces are embedded.

To identify the basis vectors for covariant vectors, we proceed in a similar way, but use gradient vectors rather than tangent vectors. The bth coordinate can be regarded as a scalar field ϕ on M, by putting $\phi \equiv x^b$. The gradient of this scalar field has components $\partial_a \phi = \delta^b_a$. That is, all its components are zero, except for the bth, which is equal to one. We therefore conclude that at each point P of M the gradient of the bth coordinate (regarded as a scalar field) is the bth basis vector $\mathbf{e}^b$ for covariant vectors at P. These basis vectors define the *cotangent space* T^*_P of M at P. What exactly is T^*_P and how is it related to T_P? The short answer to both these questions is that T^*_P is the dual of T_P, but this needs some explanation.

The tangent space T_P is a real N-dimensional vector space: the vectors in it can be added and multiplied by scalars that are real, and any basis contains N independent vectors (see, e.g., Halmos, 1974, for a detailed discussion). The set of real-valued functions that act linearly on a real N-dimensional vector space V forms a related vector space V^*, known as the *dual* of V, which also has dimension N (see Halmos, 1974, Chap. I). It is in this sense that the cotangent space T^*_P is the dual of T_P. The way in which a covariant vector $\boldsymbol{\mu}$ in T^*_P acts linearly on a contravariant vector $\boldsymbol{\lambda}$ in T_P is readily given in terms of components:

$$\boldsymbol{\mu}(\boldsymbol{\lambda}) \equiv \mu_a \lambda^a. \tag{1.84}$$

The right-hand side is a real scalar quantity, and it is easily checked that

$$\boldsymbol{\mu}(\alpha\boldsymbol{\lambda} + \beta\boldsymbol{\sigma}) = \alpha\boldsymbol{\mu}(\boldsymbol{\lambda}) + \beta\boldsymbol{\mu}(\boldsymbol{\sigma}),$$

showing that $\boldsymbol{\mu}$ acts linearly. If in equation (1.84) we use basis vectors, setting $\boldsymbol{\mu} = \mathbf{e}^b$ and $\boldsymbol{\lambda} = \mathbf{e}_c$, we get

$$\mathbf{e}^b(\mathbf{e}_c) = \delta^b_a \delta^a_c = \delta^b_c, \tag{1.85}$$

since (as remarked above) the ath component of $\mathbf{e}^b$ is δ^b_a and the ath component of $\mathbf{e}_c$ is δ^a_c. Equation (1.85) expresses the fact that $\{\mathbf{e}^b\}$, the basis of T^*_P given by the gradients of coordinates, is the dual of $\{\mathbf{e}_c\}$, the basis of T_P given by the tangents to coordinate curves (see Halmos, 1974, Chap. I, §15).

In Sections 1.1–1.5, we used the bases $\{\mathbf{e}_i\}$ and $\{\mathbf{e}^j\}$ as if they were alternate bases for the same space, referring a given vector $\boldsymbol{\lambda}$ to one or the other, and distinguishing its components by referring to them as either *contravariant* or *covariant.* This practice is not really correct, though is common in Euclidean space, and has its origins in confusing (or identifying[9]) T_P^* with T_P. Euclidean space has an inner product, the usual dot product $\boldsymbol{\lambda} \cdot \boldsymbol{\mu}$ of vector algebra. This inner product allows us to associate a covariant vector $\boldsymbol{\lambda}^*$ in T_P^* with a given contravariant vector $\boldsymbol{\lambda}$ in T_P, by saying that for all vectors $\boldsymbol{\rho}$ in T_P

$$\boldsymbol{\lambda}^*(\boldsymbol{\rho}) \equiv \boldsymbol{\lambda} \cdot \boldsymbol{\rho}. \tag{1.86}$$

In terms of components, this amounts to saying that $\boldsymbol{\lambda}^*$ has components λ_i given by $g_{ij}\lambda^j$, where λ^j are the components of $\boldsymbol{\lambda}$. The confusion amounts to identifying each $\boldsymbol{\lambda}$ in T_P with its associated vector $\boldsymbol{\lambda}^*$ in T_P^*, as determined by equation (1.86). The confusion is excused by noting that when Cartesian coordinates are used (as is often the case in Euclidean space) $g_{ij} = \delta_{ij}$, giving $\lambda_i = \lambda^i$, showing that the associated covariant vector $\boldsymbol{\lambda}^*$ has the same components as $\boldsymbol{\lambda}$.

With the above remarks in mind, it is possible to "correct" the confused statements in the Euclidean sections. For example, equation (1.17) should be replaced by

$$\mathbf{e}^i(\mathbf{e}_j) = \delta^i_j$$

and equation (1.19) should give the components λ^j of the contravariant vector $\boldsymbol{\lambda}$ by arguing that

$$\mathbf{e}^j(\boldsymbol{\lambda}) = \mathbf{e}^j(\lambda^i\mathbf{e}_i) = \lambda^i\mathbf{e}^j(\mathbf{e}_i) = \lambda^i\delta^j_i = \lambda^j.$$

In Figure 1.5 we drew the dual basis vectors $\{\mathbf{e}^u\}$ and $\{\mathbf{e}^v\}$ as if they were situated in the tangent plane to the surface; that is, as if they were contravariant vectors. We now appreciate that the vectors shown are actually the contravariant vectors whose associated covariant vectors are the dual basis vectors $\{\mathbf{e}^u\}$ and $\{\mathbf{e}^v\}$.

The final question posed in the opening paragraph concerned tensors: if a tensor has components, then is there a basis to which these components refer? In particular, can we write a type $(2, 0)$ tensor in the form

$$\boldsymbol{\tau} = \tau^{ab}\mathbf{e}_{ab},$$

where $\{\mathbf{e}_{ab}\}$ serves as a basis for the space of type $(2, 0)$ tensors? The answer is in the affirmative and the basis tensors are the N^2 *tensor products* (also known as *dyad products*) $\mathbf{e}_a \otimes \mathbf{e}_b$ of pairs of basis vectors $\mathbf{e}_a$ and $\mathbf{e}_b$ from T_P. The idea of a tensor product is often introduced in an informal way, especially in texts on classical mechanics.[10] The more formal mathematical approach is to define the tensor product $V \otimes W$ of two real vector spaces V and W as the space of all real-valued bilinear functions acting on the Cartesian product $V^* \times W^*$

[9] It is *confusion* if done unwittingly, but *identification* if intentional.

[10] See, for example, Goldstein, Poole, and Safko, 2002, and Symon, 1971.

of their dual spaces V^* and W^*. If we were to pursue this approach, then we would recognize a type $(2,0)$ tensor at a point P as belonging to $T_P \otimes T_P$ and the basis tensor $\mathbf{e}_{ab}$ as being the bilinear function with the property that

$$\mathbf{e}_{ab}(\mathbf{e}^c, \mathbf{e}^d) = \delta^c_a \delta^d_b,$$

where $\{\mathbf{e}^a\}$ is the basis of T^*_P dual to the basis $\{\mathbf{e}_a\}$ of T_P. See Section C.3 for more details.

In a similar way, we would recognize a type $(0,2)$ tensor at a point P as belonging to $T^*_P \otimes T^*_P$, and a type $(1,1)$ as belonging to $T_P \otimes T^*_P$ (or $T^*_P \otimes T_P$). General tensors of type (r,s) at P belong to

$$\underbrace{T_P \otimes T_P \otimes \cdots \otimes T_P}_{r \text{ times}} \otimes \underbrace{T^*_P \otimes T^*_P \otimes \cdots \otimes T^*_P}_{s \text{ times}},$$

formed by taking repeated tensor products. These ideas are more fully developed in and Section C.4 of Appendix C.

Exercise 1.10

1. Go through Sections 1.1–1.5 to find every occurrence of a dot product that should be more correctly written as a covariant vector acting on a contravariant vector.

Problems 1

1. In Euclidean space, ellipsoidal coordinates (u,v,w) are defined by

$$x = au \sin v \cos w, \quad y = bu \sin v \sin w, \quad z = cu \cos v,$$

where a, b, c are positive constants and $0 \le u < \infty$, $0 \le v \le \pi$ and $0 \le w < 2\pi$.
(a) Describe the three families of coordinate surfaces.
(b) Obtain expressions for the natural basis vectors $\mathbf{e}_u$, $\mathbf{e}_v$, $\mathbf{e}_w$ and the dual basis vectors $\mathbf{e}^u$, $\mathbf{e}^v$, $\mathbf{e}^w$ in terms of $\mathbf{i}$, $\mathbf{j}$, $\mathbf{k}$.
(c) Verify that (in suffix notation) $\mathbf{e}^i \cdot \mathbf{e}_j = \delta^i_j$.
(As a check on your answers, note that when $a = b = c = 1$ we have spherical coordinates.)

2. Figure 1.6 shows the *torus* generated by revolving the circle $(x-a)^2 + z^2 = b^2$ (where $a > b > 0$) in the plane $y = 0$ about the z axis. Show that when the angles θ and ϕ are used as parameters (see figure) points on the torus are given by

$$x = (a + b\cos\theta)\cos\phi, \quad y = (a + b\cos\theta)\sin\phi, \quad z = b\sin\theta,$$

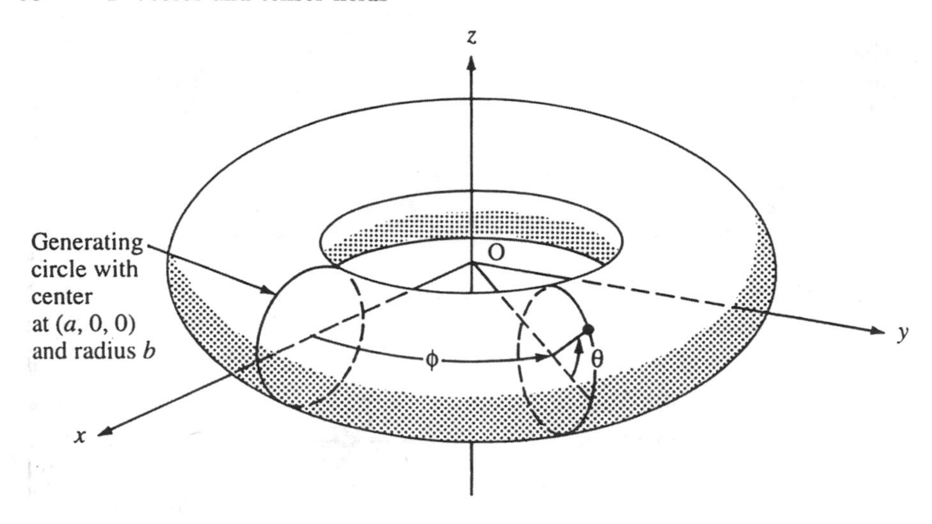

Fig. 1.6. The torus generated by revolving a circle about the z axis.

with $-\pi < \theta \leq \pi$ and $-\pi < \phi \leq \pi$.
Obtain expressions for the natural basis vectors $\mathbf{e}_\theta$ and $\mathbf{e}_\phi$, and hence obtain the metric tensor components g_{AB}.

3. Show that if at a point P of a manifold the contravariant vector λ^a is nonzero, then it is possible to change to a new (primed) coordinate system in which $\lambda^{a'} = \delta^a_1$ at P.
(Use matrix theory to show that if $\lambda^a \neq 0$, then there exists a non-singular matrix $[\mu^a_b]$ such that $\lambda^b \mu^a_b = \delta^a_1$. Then define a new coordinate system about P using $x^{a'} = \mu^a_c x^c$.)

4. If τ^{ab} is a symmetric tensor and λ^a a contravariant vector with the property that
$$\tau^{bc}\lambda^a + \tau^{ca}\lambda^b + \tau^{ab}\lambda^c = 0$$
for all a, b, c, deduce that either $\tau^{ab} = 0$ or $\lambda^a = 0$.
(Hint: If at the point in question $\lambda^a \neq 0$, then we can introduce the special coordinate system of Problem 3.)

5. Suppose that with each coordinate system about a point P of an N-dimensional manifold there are associated N^2 numbers τ_{ab} satisfying $\tau_{ab} = \tau_{ba}$ and it is known that if λ^a are the components of an arbitrary contravariant vector at P, then the expression $\tau_{ab}\lambda^a\lambda^b$ is invariant under a change of coordinates. Show that the numbers τ_{ab} are the components of a type $(0,2)$ tensor at P.

6. A type $(0,4)$ tensor τ_{abcd} satisfies $\tau_{abcd}\lambda^a\mu^b\lambda^c\mu^d = 0$ for all contravariant vectors λ^a and μ^a. Show that its components satisfy

$$\tau_{abcd} + \tau_{cbad} + \tau_{adcb} + \tau_{cdab} = 0.$$

7. Let x^i be a system of Cartesian coordinates in Euclidean space, and let $x^{i'}$ be a new system whose axes are obtained by rotating those of the unprimed system about its x^3 axis through an angle θ in the positive sense.
   (a) Show that at each point of space the new basis vectors are given in terms of the old basis vectors by

$$\mathbf{e}_{1'} = \cos\theta\, \mathbf{e}_1 + \sin\theta\, \mathbf{e}_2, \quad \mathbf{e}_{2'} = -\sin\theta\, \mathbf{e}_1 + \cos\theta\, \mathbf{e}_2, \quad \mathbf{e}_{3'} = \mathbf{e}_3.$$

   What are the transformation matrices $[X^{i'}_j]$ and $[X^i_{j'}]$?
   (b) Recall that, for a rigid body having one of its points fixed at the origin O, its angular momentum L^i about O can be expressed as $L^i = I^i_j \omega^j$, where I^i_j is the *inertia tensor* of the body about O and ω^i is its angular momentum (all regarded as tensors at O). Find $[L^i]$ when

$$[I^i_j] = \begin{bmatrix} 0 & 0 & 0 \\ 0 & m & 0 \\ 0 & 0 & m \end{bmatrix} \quad \text{and} \quad [\omega^i] = \begin{bmatrix} 0 \\ 15 \\ 0 \end{bmatrix}.$$

   (c) Transform the components to find $I^{i'}_{j'}$, $\omega^{i'}$, and $L^{i'}$ relative to the new coordinate system, and check that $L^{i'} = I^{i'}_{j'}\omega^{j'}$.

8. In special relativity the change of coordinates $x^{\mu'} = \Lambda^{\mu'}_{\nu} x^\nu$, where

$$[\Lambda^{\mu'}_{\nu}] = \begin{bmatrix} \cosh\phi & -\sinh\phi & 0 & 0 \\ -\sinh\phi & \cosh\phi & 0 & 0 \\ 0 & 0 & 1 & 0 \\ 0 & 0 & 0 & 1 \end{bmatrix}$$

   $(\mu, \nu = 0, 1, 2, 3)$ gives a *boost* in the x^1 direction. Show that the corresponding matrix $[\Lambda^{\mu}_{\nu'}]$ for the inverse transformation is given by substituting $-\phi$ for ϕ in the matrix for $[\Lambda^{\mu'}_{\nu}]$. Deduce that if $g_{\mu\nu} = \eta_{\mu\nu}$, then after the boost is applied we have $g_{\mu'\nu'} = \eta_{\mu\nu}$, showing that a boost does not change this special form of the metric-tensor components.
   (Here, as usual, $[\eta_{\mu\nu}] = \mathrm{diag}(1, -1, -1, -1)$. See Secs. A.0 and A.1 of Appendix A.)

9. Show that if the metric tensor g_{ab} is positive definite, then it is possible to transform to a new (primed) coordinate system such that, at a given point P, $g_{a'b'} = \delta_{ab}$.
   (We know from matrix theory that if G is a positive definite matrix, then there exists a nonsingular matrix P such that $P^{\mathrm{T}} G P = I$.)

10. Show that if at a point P of spacetime the nonzero vector λ^μ is orthogonal to:

(a) a timelike vector t^μ, then λ^μ is spacelike;

(b) a null vector n^μ, then λ^μ is either spacelike or proportional to n^μ;

(c) a spacelike vector s^μ, then λ^μ may be either timelike, null, or spacelike.

(You can assume that it is possible to introduce a coordinate system such that $g_{\mu\nu} = \eta_{\mu\nu}$ at P, where $[\eta_{\mu\nu}] = \mathrm{diag}(1, -1, -1, -1)$, and that there is then no loss of generality in taking $t^\mu = \delta^\mu_0$, $n^\mu = \delta^\mu_0 + \delta^\mu_1$, and $s^\mu = \delta^\mu_1$.)

11. In Appendix A, we reviewed special relativity using coordinates and components, but not basis vectors. However, given a system of coordinates (t, x, y, z), we can infer the existence of a basis $\{\mathbf{e}_t, \mathbf{e}_x, \mathbf{e}_y, \mathbf{e}_z\}$ at each point of spacetime, where each basis vector is tangential to the corresponding coordinate curve, as explained in Section 1.10. Using this idea, give expressions for the basis vectors $\mathbf{e}_{t'}, \mathbf{e}_{x'}, \mathbf{e}_{y'}, \mathbf{e}_{z'}$ in terms of the basis vectors $\mathbf{e}_t, \mathbf{e}_x, \mathbf{e}_y, \mathbf{e}_z$, where the primed coordinates and unprimed coordinates are related by the boost (A.12).
If $\mathbf{e}_t$ and $\mathbf{e}_x$ are drawn as vertical and horizontal vectors (as in a regular spacetime diagram), what are the directions of $\mathbf{e}_{t'}$ and $\mathbf{e}_{x'}$? Give a sketch which shows how the primed vectors are "squashed up" for $v > 0$, but are "opened out" for $v < 0$.

2

The spacetime of general relativity and paths of particles

2.0 Introduction

Einstein's general theory of relativity postulates that gravitational effects may be explained by the curvature of spacetime (modeled by a four-dimensional pseudo-Riemannian manifold) and that gravity should not be regarded as a force in the conventional sense. To get a preliminary idea of what is involved, we shall follow the practice of a number of authors[1] and consider ants crawling over a curved surface, namely the skin of an apple.

Suppose then that an ant wishes to follow a straight path on the apple's skin. *The straightest path it could take would be achieved by its making its left-hand paces equal to its right-hand ones.* This would clearly generate a straight-line path if it were crawling on a plane, so it is natural to adopt a path generated on a curved surface in this way as the analog of a straight line. These paths are called *geodesics*. If the ant had inky feet, so that it left footprints, then making cuts along the left-hand and right-hand tracks would yield a thin strip of peel which could be removed. If this thin strip were laid flat on a plane it would be straight, confirming that a geodesic, as we have defined it, is the analog of a straight line.

Suppose now that we have several ants crawling over the apple (without colliding) and each follows a geodesic path, leaving a record of its progress on the apple's skin. (A single track rather than a double one: ink on the tip of its abdomen, rather than inky feet.) If we concentrate on a portion of the apple's skin *which is so small that it may be considered flat*, then the tracks of the ants would appear as straight lines on this "flat" portion (see Fig. 2.1). If, however, we take a larger view of things, then the picture is different. For example, suppose two ants leave from nearby points on a starting line at the same time, and move with the same constant speed, following geodesics which are initially perpendicular to the starting line (see Fig. 2.2). Their paths

[1]Notably Misner, Thorne, and Wheeler, on whose well-known illustration our Fig. 2.1 is based. See Misner, Thorne, and Wheeler, 1973, §1.1.

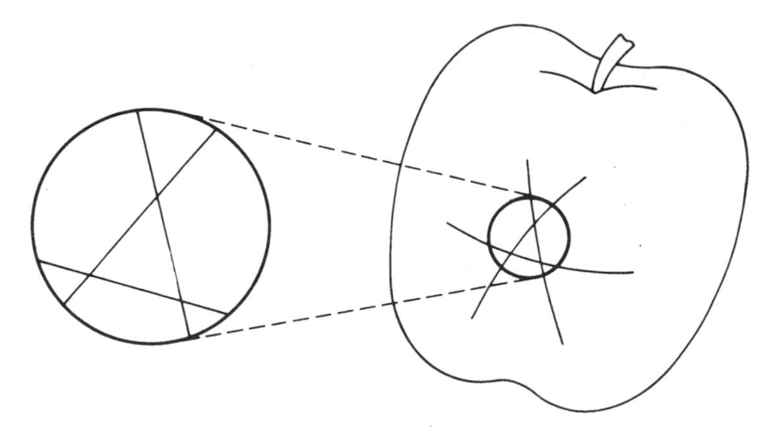

Fig. 2.1. Small portions of an apple's skin may be regarded as flat.

would initially be parallel, but because of the curvature of the apple's skin, they would start to converge. That is, their separation d does not remain constant. More generally, we can see that the relative acceleration of ants which follow neighboring geodesics with constant (but not necessarily equal) speeds is non-zero, if the surface over which they are crawling is curved. In this way, curvature may be detected implicitly by what is called *geodesic deviation.*

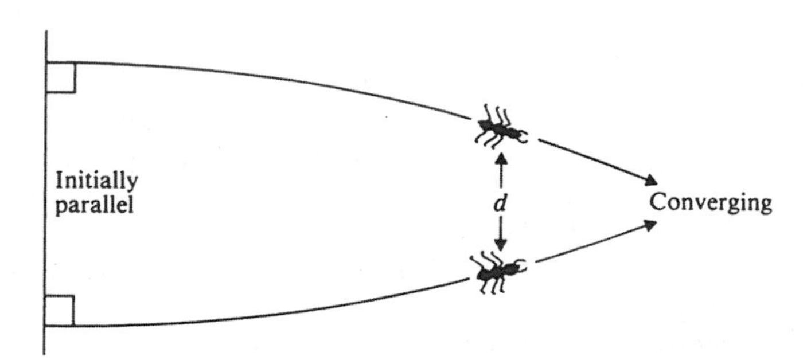

Fig. 2.2. Converging geodesics on an apple's skin.

An apple is not a perfect sphere: there is a dimple caused by the stalk. Should an ant pass near the stalk its geodesic path would suffer a marked deflection, like that of a comet passing near the Sun, and it would look as if the stalk attracted the ant. However, this is not the correct interpretation. The stalk modifies the curvature of the apple's skin in its vicinity, and this produces geodesics which give the effect of an attraction by the stalk.

This allegory may be interpreted in the following way. The curved surface which is the apple's skin represents the curved spacetime of Einstein's general theory, which bears the same relation to the flat spacetime of the special theory as does the apple's skin to a plane. Free particles (i.e., those moving under gravity alone, gravity no longer being a force) follow the straightest paths or geodesics in the curved spacetime, just as the ants follow geodesics on the apple's skin. Locally the spacetime of the general theory is like that of the special theory,[2] but on a larger view it is curved, *and this curvature may be detected implicitly by means of geodesic deviation*, just as the curvature of the apple's skin may be detected by noting the convergence of neighboring geodesics. The way in which the dimple around the stalk gives the impression of attraction corresponds to the fact that massive bodies modify the curvature of spacetime in their vicinity, and this modification affects the geodesics in such a way as to give the impression that free particles are acted on by a force, whereas in actual fact they are following the straightest paths in the curved spacetime.

Given that Einstein's general theory does not involve the idea of gravity as a force, how does the gravitational "force" that is a feature of the Newtonian theory arise? We remarked in the Introduction that in a local inertial frame (a freely falling, nonrotating reference system occupying a small region of spacetime) the laws of physics are those of special relativity, and in particular free particles (those moving under gravity alone) follow straight-line paths with constant speed, so for these frames there is no acceleration and consequently no "force." When discussing gravity in Newtonian terms, it is customary to insist that the frame used is nonrotating (so there are no centrifugal or Coriolis "forces"), but one does not normally use a frame that is freely falling, and it is this use of nonfreely falling frames that gives rise to gravitational forces. Just as the fictitious forces associated with rotation (the centrifugal and the Coriolis forces) can be transformed away (locally) by changing to a nonrotating frame, so can the fictitious force of gravity be transformed away by changing to a freely falling frame.

Newton's theory of gravity is nonrelativistic and uses a model for spacetime that combines three-dimensional Euclidean space with one-dimensional time. Getting the Newtonian theory as an approximation to Einstein's general theory of relativity therefore involves two things: passing from a relativistic to a nonrelativistic way of looking at things and interpreting the effects of the curvature of spacetime in the setting of three-dimensional Euclidean space plus one-dimensional time. The whole process is quite complicated, but is essential for a proper understanding of the relationship between Einstein's theory and the Newtonian theory. We shall perform this approximation later in this chapter and establish various points of contact between the two theories.

Before we can do this, we must explain how our model for spacetime can handle the paths of particles by including the handling of geodesics as part of our mathematical repertoire. The mathematics of geodesics is covered in the next few sections, along with the related concepts of parallelism and

[2]Compare remarks made in the Introduction.

absolute and covariant differentiation, needed for our discussion of curvature in Chapter 3. Note that in the present chapter we are concerned only with the motion of particles in a *given* spacetime: they are test particles responding to a *given* gravitational field. How that field arises is answered in the next chapter, where we relate the curvature of spacetime to the sources of the gravitational field.

Exercise 2.0

1. Ants follow geodesics on a surface which is an infinite cylinder (i.e., the outside of an infinitely long straight pipe).
   Do the geodesics deviate?
   By considering only the paths of itself and its neighbors, can an ant decide whether it is on a cylinder or a plane?

2.1 Geodesics

A geodesic in Euclidean space is simply a straight line, which can be characterized as the shortest curve between two points. Such a characterization could be extended to a geodesic in a manifold, where the metric tensor field gives us the length of a curve via the integral (1.82). However, this approach to geodesics presents some technical difficulties, particularly when the metric tensor field is indefinite (as in the spacetime of general relativity), since in that case we can have curves (or parts of curves) that have zero length. We therefore adopt another characterization of a straight line, namely its *straightness*, and use this as a guide to defining geodesics in a manifold.

What makes a straight line straight is the fact that its tangent vectors all point in the same direction. If we use the arc-length s measured from some base point on the line as a parameter, then the tangent vectors $\boldsymbol{\lambda} \equiv \dot{\mathbf{r}}(s)$ have constant length (as they are unit vectors: see Exercise 1.3.3), so we can express the fact that they have constant direction by stating that

$$d\boldsymbol{\lambda}/ds = \mathbf{0}. \tag{2.1}$$

Let us see what form this equation takes when we use arbitrary coordinates u^i and the related natural basis $\{\mathbf{e}_i\}$.

Putting $\boldsymbol{\lambda} = \lambda^i \mathbf{e}_i$ and using dots to denote differentiation with respect to s give

$$0 = d\boldsymbol{\lambda}/ds = d(\lambda^i \mathbf{e}_i)/ds = \dot{\lambda}^i \mathbf{e}_i + \lambda^i \dot{\mathbf{e}}_i. \tag{2.2}$$

Now

$$\dot{\mathbf{e}}_i = \partial_j \mathbf{e}_i \dot{u}^j$$

and in general the vector fields $\partial_j \mathbf{e}_i$ are nonzero. At each point of space, we can refer $\partial_j \mathbf{e}_i$ to the basis $\{\mathbf{e}_i\}$, so that

$$\partial_j \mathbf{e}_i = \Gamma^k_{ij} \mathbf{e}_k,$$

which gives rise to 27 quantities Γ^k_{ij} defined at each point of space. After some manipulation and relabeling of dummy suffixes we then get

$$(\dot{\lambda}^i + \Gamma^i_{jk} \lambda^j \dot{u}^k) \mathbf{e}_i = \mathbf{0} \tag{2.3}$$

from equation (2.2). Since $\lambda^i = \dot{u}^i = du^i/ds$, we see that the components du^i/ds of the tangent vector to the straight line satisfy

$$\frac{d^2 u^i}{ds^2} + \Gamma^i_{jk} \frac{du^j}{ds} \frac{du^k}{ds} = 0. \tag{2.4}$$

For this last equation to have any meaning, we must obtain an expression for Γ^i_{jk} in terms of known quantities.

We start by noting that[3]

$$\partial_j \mathbf{e}_i = \frac{\partial^2 \mathbf{r}}{\partial u^j \partial u^i} = \frac{\partial^2 \mathbf{r}}{\partial u^i \partial u^j} = \partial_i \mathbf{e}_j,$$

so that $\Gamma^k_{ij} \mathbf{e}_k = \Gamma^k_{ji} \mathbf{e}_k$. Forming the dot product with $\mathbf{e}^l$ then gives the symmetry property

$$\boxed{\Gamma^l_{ij} = \Gamma^l_{ji}.} \tag{2.5}$$

We then use $g_{ij} = \mathbf{e}_i \cdot \mathbf{e}_j$ to get

$$\partial_k g_{ij} = \partial_k \mathbf{e}_i \cdot \mathbf{e}_j + \mathbf{e}_i \cdot \partial_k \mathbf{e}_j = \Gamma^m_{ik} \mathbf{e}_m \cdot \mathbf{e}_j + \mathbf{e}_i \cdot \Gamma^m_{jk} \mathbf{e}_m.$$

So

$$\partial_k g_{ij} = \Gamma^m_{ik} g_{mj} + \Gamma^m_{jk} g_{im}. \tag{2.6}$$

Relabeling suffixes we have

$$\partial_i g_{jk} = \Gamma^m_{ji} g_{mk} + \Gamma^m_{ki} g_{jm} \tag{2.7}$$

and

$$\partial_j g_{ki} = \Gamma^m_{kj} g_{mi} + \Gamma^m_{ij} g_{km}. \tag{2.8}$$

Subtracting equation (2.8) from the sum of equations (2.6) and (2.7), and using the symmetry of both Γ^m_{ij} and g_{ij} give

$$2\Gamma^m_{ki} g_{mj} = \partial_k g_{ij} + \partial_i g_{jk} - \partial_j g_{ki}.$$

Contracting with $\frac{1}{2} g^{lj}$ then gives

[3]Provided that we can change the order of partial differentiation, which is certainly the case if the coordinate functions $x(u^i)$, $y(u^i)$, $z(u^i)$ have continuous second partial derivatives.

$$\boxed{\Gamma^l_{ki} = \tfrac{1}{2}g^{lj}(\partial_k g_{ij} + \partial_i g_{jk} - \partial_j g_{ki}).} \tag{2.9}$$

Equation (2.4), with Γ^i_{jk} given by equation (2.9) is the *geodesic equation* for Euclidean space.

If we use a general parameter t to parameterize the straight line, then the geodesic equation has a more complicated form. However, for parameters related to the arc-length s by an equation of the form

$$t = As + B, \tag{2.10}$$

where A, B are constants ($A \neq 0$), the geodesic equation has basically the same form as when s is used:

$$\frac{d^2u^i}{dt^2} + \Gamma^i_{jk}\frac{du^j}{dt}\frac{du^k}{dt} = 0. \tag{2.11}$$

(See Exercise 2.1.1 for a justification of these claims.) These privileged parameters for which the geodesic equation has the form (2.11) (with Γ^i_{jk} given by equation (2.9)) are known as *affine parameters*. For an affine parameter, ds/dt is constant, so one is taken along the geodesic at a constant sort of rate. (If we think of t as time, then the geodesic is traversed at constant speed.)

Equation (2.11) represents a system of second-order differential equations whose general solution $u^i(t)$ gives the geodesics of Euclidean space (i.e., straight lines) in whatever coordinate system we happen to be using. To obtain a particular solution, six conditions are needed. These might take the form of specifying a starting point and a starting direction, or of specifying a starting point and an ending point for the geodesic.

Using the above ideas as a guide, we can define an *affinely parameterized geodesic* in an N-dimensional Riemannian or pseudo-Riemannian manifold as a curve given by $x^a(u)$ satisfying[4]

$$\boxed{\frac{d^2x^a}{du^2} + \Gamma^a_{bc}\frac{dx^b}{du}\frac{dx^c}{du} = 0,} \tag{2.12}$$

where the N^3 quantities Γ^a_{bc} are given by

$$\boxed{\Gamma^a_{bc} = \tfrac{1}{2}g^{ad}(\partial_b g_{dc} + \partial_c g_{bd} - \partial_d g_{bc}).} \tag{2.13}$$

These quantities are known as *connection coefficients*[5] and, like their three-dimensional Euclidean counterparts, they satisfy the symmetry relation

[4]Note the change of notation from u^i for coordinates and t for parameter in three-dimensional Euclidean space to x^a for coordinates and u for parameter in an N-dimensional manifold.

[5]The reason for this terminology is given in the next section.

$$\boxed{\Gamma^a_{bc} = \Gamma^a_{cb},} \tag{2.14}$$

as is clear from their defining equation. It can be shown that in moving along an affinely parameterized geodesic, the length of the tangent vector $\dot{x}^a \equiv dx^a/du$ remains constant (see Exercise 2.1.2), and it follows that if the geodesic is not null (which could be the case with an indefinite metric tensor field), then the affine parameter is related to the arc-length s by an equation of the form

$$u = As + B, \tag{2.15}$$

where A, B are constants ($A \neq 0$) (see Exercise 2.1.3). If the metric tensor field is indefinite, then we can have affinely parameterized *null* geodesics whose tangent vectors $\dot{x}^a$ satisfy $g_{ab}\dot{x}^a\dot{x}^b = 0$ and for which the arc-length s cannot be used as a parameter.

Example 2.1.1

In this example we show that, of all the circles of latitude on a sphere, only the equator is a geodesic. We take the radius of the sphere to be a, and use $u^1 = \theta$ and $u^2 = \phi$ (borrowed from spherical coordinates) as parameters.

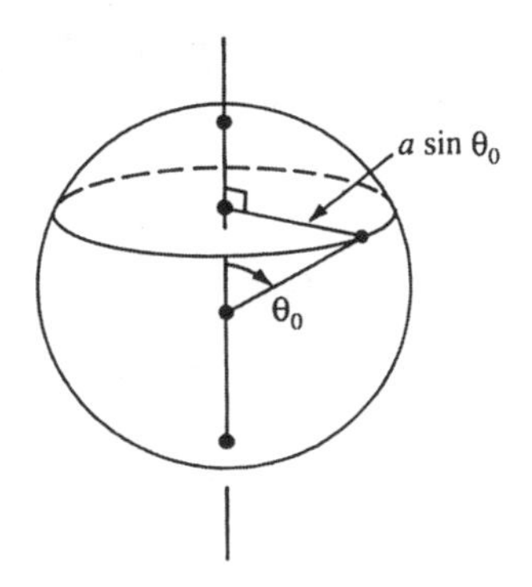

Fig. 2.3. A circle of latitude on a sphere.

The figure shows the circle of latitude given by $\theta = \theta_0$. Since its radius is $a \sin\theta_0$, we can parameterize it by saying that

$$u^1 \equiv \theta = \theta_0, \quad u^2 \equiv \phi = (a\sin\theta_0)^{-1}s,$$

where s is the arc-length measured round from the point where $\phi = 0$. So (for $A = 1, 2$)

$$u^A = \theta_0\delta^A_1 + \frac{s}{a\sin\theta_0}\delta^A_2, \quad \dot{u}^A = \frac{1}{a\sin\theta_0}\delta^A_2, \quad \text{and} \quad \ddot{u}^A = 0,$$

so, for the geodesic equation to be satisfied, we need

$$\ddot{u}^A + \Gamma^A_{BC}\dot{u}^B\dot{u}^C = 0 + \frac{1}{a^2\sin^2\theta_0}\Gamma^A_{22} = 0. \tag{2.16}$$

From Exercise 2.1.5 we have that the only nonzero connection coefficients are

$$\Gamma^1_{22} = -\sin\theta\cos\theta, \quad \Gamma^2_{12} = \Gamma^2_{21} = \cot\theta,$$

so equation (2.16) is satisfied for $A = 2$ (as $\Gamma^2_{22} = 0$), while for $A = 1$ it gives $-\cot\theta_0/a^2 = 0$, which is satisfied only if $\theta_0 = \pi/2$. So, of all the circles of latitude, only the equator is a geodesic.

In order to obtain the parametric equations $x^a = x^a(u)$ of an affinely parameterized geodesic, we must solve the system of differential equations (2.12). These equations are second-order, and require $2N$ conditions to determine a unique solution. Suitable conditions are given by specifying the coordinates x^a_0 of some point on the geodesic, and the components $\dot{x}^a_0$ of the tangent vector at that point. Bearing in mind the equations (2.13) which define the Γ^a_{bc}, it would seem to be a complicated procedure just to set up the geodesic equations, let alone solve them. Fortunately the equations may be generated by a very neat procedure which also produces the Γ^a_{bc} as a spin-off.

Consider the *Lagrangian* $L(\dot{x}^c, x^c) \equiv \frac{1}{2}g_{ab}(x^c)\dot{x}^a\dot{x}^b$, which we regard as a function of $2N$ independent variables x^c and $\dot{x}^c$. The *Euler–Lagrange equations* for a Lagrangian are the equations

$$\boxed{\frac{d}{du}\left(\frac{\partial L}{\partial \dot{x}^c}\right) - \frac{\partial L}{\partial x^c} = 0,} \tag{2.17}$$

and for the given Lagrangian these reduce to the geodesic equations (in covariant rather than contravariant form), as we now show.

Differentiating the Lagrangian we have

$$\frac{\partial L}{\partial \dot{x}^c} = \tfrac{1}{2}g_{ab}\delta^a_c\dot{x}^b + \tfrac{1}{2}g_{ab}\dot{x}^a\delta^b_c = g_{cb}\dot{x}^b$$

and

$$\frac{\partial L}{\partial x^c} = \tfrac{1}{2}\partial_c g_{ab}\dot{x}^a\dot{x}^b,$$

so equations (2.17) are

$$d(g_{cb}\dot{x}^b)/du - \tfrac{1}{2}\partial_c g_{ab}\dot{x}^a\dot{x}^b = 0,$$

or

$$g_{cb}\ddot{x}^b + \partial_a g_{cb}\dot{x}^a\dot{x}^b - \tfrac{1}{2}\partial_c g_{ab}\dot{x}^a\dot{x}^b = 0.$$

But

$$\partial_a g_{cb}\dot{x}^a\dot{x}^b = \tfrac{1}{2}\partial_a g_{cb}\dot{x}^a\dot{x}^b + \tfrac{1}{2}\partial_b g_{ca}\dot{x}^a\dot{x}^b,$$

so we have

$$g_{cb}\ddot{x}^b + \tfrac{1}{2}(\partial_a g_{cb} + \partial_b g_{ca} - \partial_c g_{ab})\dot{x}^a\dot{x}^b = 0.$$

That is, the Euler–Lagrange equations reduce to

$$g_{cb}\ddot{x}^b + \Gamma_{cab}\dot{x}^a\dot{x}^b = 0, \tag{2.18}$$

where $\Gamma_{cab} \equiv \frac{1}{2}(\partial_a g_{cb} + \partial_b g_{ca} - \partial_c g_{ab})$ and raising c gives

$$\ddot{x}^c + \Gamma^c_{ab}\dot{x}^a\dot{x}^b = 0, \tag{2.19}$$

which are the equations of an affinely parameterized geodesic.

Those familiar with the calculus of variations or Lagrangian mechanics will know that the Euler–Lagrange equations give the solution to the problem of finding the curve (with fixed endpoints) which extremizes the integral $\int_{u_1}^{u_2} L(\dot{x}^c, x^c)\, du$. While there is some connection with the characterization of a geodesic as an extremal of length, it should be noted that the integral involved does not give the length of the curve. For reasons stated earlier, we shall not pursue this approach any further, but simply regard the Euler–Lagrange equations as a useful device for generating the geodesic equations and the connection coefficients which may be extracted from them.

Demonstrating the equivalence of the geodesic and the Euler–Lagrange equations allows us to make a useful observation. If g_{ab} does not depend on some particular coordinate x^d, say, then equation (2.17) shows that

$$\frac{d}{du}\left(\frac{\partial L}{\partial \dot{x}^d}\right) = 0,$$

which implies that $\partial L/\partial \dot{x}^d$ is constant along an affinely parameterized geodesic. But $\partial L/\partial \dot{x}^d = g_{db}\dot{x}^b$, so we then have that $p_d \equiv g_{db}\dot{x}^b$ is constant along an affinely parameterized geodesic. The situation is exactly the same as in Lagrangian mechanics where, if the Lagrangian does not contain a particular generalized coordinate, then the corresponding generalized momentum is conserved, and borrowing a term from mechanics we call a coordinate which is absent from g_{ab} *cyclic* or *ignorable*.[6] Being able to say that $p_d =$ constant whenever x^d is cyclic gives us immediate integrals of the geodesic equations, which help with their solution. An example should make some of these ideas clear.

Example 2.1.2

The Robertson–Walker line element is used in cosmology (see Chap. 6). It is defined by

$$g_{\mu\nu}dx^\mu dx^\nu \equiv dt^2 - (R(t))^2\left((1-kr^2)^{-1}dr^2 + r^2 d\theta^2 + r^2\sin^2\theta\, d\phi^2\right),$$

where $\mu, \nu = 0, 1, 2, 3$ (our usual notation for spacetimes), k is a constant, and $x^0 \equiv t$, $x^1 \equiv r$, $x^2 \equiv \theta$, $x^3 \equiv \phi$.

[6]See, for example, Goldstein, Poole, and Safko, 2002, §2–6, or Symon, 1971, §9–6.

So the Lagrangian is

$$L(\dot{x}^\sigma, x^\sigma) \equiv \tfrac{1}{2}\left\{\dot{t}^{\,2} - (R(t))^2\left((1-kr^2)^{-1}\dot{r}^2 + r^2\dot{\theta}^2 + r^2\sin^2\theta\,\dot{\phi}^2\right)\right\},$$

where dots denote differentiation with respect to an affine parameter u. Partial differentiation gives

$$\partial L/\partial\dot{t} = \dot{t},$$

$$\partial L/\partial\dot{r} = -R^2(1-kr^2)^{-1}\dot{r},$$

$$\partial L/\partial\dot{\theta} = -R^2r^2\dot{\theta},$$

$$\partial L/\partial\dot{\phi} = -R^2r^2\sin^2\theta\,\dot{\phi},$$

$$\partial L/\partial t = -RR'[(1-kr^2)^{-1}\dot{r}^2 + r^2\dot{\theta}^2 + r^2\sin^2\theta\,\dot{\phi}^2]$$

(where $R' = dR/dt$),

$$\partial L/\partial r = -R^2(1-kr^2)^{-2}kr\dot{r}^2 - R^2r\dot{\theta}^2 - R^2r\sin^2\theta\,\dot{\phi}^2,$$

$$\partial L/\partial\theta = -R^2r^2\sin\theta\cos\theta\,\dot{\phi}^2,$$

$$\partial L/\partial\phi = 0.$$

Substitution of these derivatives in the Euler–Lagrange equations (2.17) gives

$$\begin{aligned}
&\ddot{t} + RR'[(1-kr^2)^{-1}\dot{r}^2 + r^2\dot{\theta}^2 + r^2\sin^2\theta\,\dot{\phi}^2] = 0,\\
&-R^2(1-kr^2)^{-1}\ddot{r} - 2RR'(1-kr^2)^{-1}\dot{t}\dot{r}\\
&\qquad\qquad - R^2(1-kr^2)^{-2}kr\dot{r}^2 + R^2r\dot{\theta}^2 + R^2r\sin^2\theta\,\dot{\phi}^2 = 0,\\
&-R^2r^2\ddot{\theta} - 2RR'r^2\dot{t}\dot{\theta} - 2R^2r\dot{r}\dot{\theta} + R^2r^2\sin\theta\,\cos\theta\,\dot{\phi}^2 = 0,\\
&-R^2r^2\sin^2\theta\,\ddot{\phi} - 2RR'r^2\sin^2\theta\,\dot{t}\dot{\phi}\\
&\qquad\qquad - 2R^2r\sin^2\theta\,\dot{r}\dot{\phi} - 2R^2r^2\sin\theta\,\cos\theta\,\dot{\theta}\dot{\phi} = 0.
\end{aligned}$$

The above comprise the covariant version of the geodesic equations, as given by equation (2.18). Because $[g_{\mu\nu}]$ is diagonal, it is a simple matter to obtain the contravariant form of the geodesic equations (as given by equation (2.19)). All we have to do is to divide each equation as necessary, so as to make the coefficients of $\ddot{t}$, $\ddot{r}$, $\ddot{\theta}$ and $\ddot{\phi}$ equal to one. We thus arrive at the geodesic equations for the Robertson–Walker spacetime in standard form:

$$\begin{aligned}
&\ddot{t} + RR'[(1-kr^2)^{-1}\dot{r}^2 + r^2\dot{\theta}^2 + r^2\sin^2\theta\,\dot{\phi}^2] = 0,\\
&\ddot{r} + 2R'R^{-1}\dot{t}\dot{r} + kr(1-kr^2)^{-1}\dot{r}^2\\
&\qquad\qquad - r(1-kr^2)\dot{\theta}^2 - r(1-kr^2)\sin^2\theta\,\dot{\phi}^2 = 0,\\
&\ddot{\theta} + 2R'R^{-1}\dot{t}\dot{\theta} + 2r^{-1}\dot{r}\dot{\theta} - \sin\theta\,\cos\theta\,\dot{\phi}^2 = 0,\\
&\ddot{\phi} + 2R'R^{-1}\dot{t}\dot{\phi} + 2r^{-1}\dot{r}\dot{\phi} + 2\cot\theta\,\dot{\theta}\dot{\phi} = 0.
\end{aligned} \tag{2.20}$$

Comparing these with equations (2.19) allows us to pick out the connection coefficients. These are zero except for the following:

$$\Gamma^0_{11} = RR'/(1-kr^2), \qquad \Gamma^0_{22} = RR'r^2, \qquad \Gamma^0_{33} = RR'r^2\sin^2\theta,$$

$$\Gamma^1_{01} = R'/R, \qquad \Gamma^1_{11} = kr/(1-kr^2),\ \Gamma^1_{22} = -r(1-kr^2),$$

$$\Gamma^1_{33} = -r(1-kr^2)\sin^2\theta,$$

$$\Gamma^2_{02} = R'/R, \qquad \Gamma^2_{12} = 1/r, \qquad \Gamma^2_{33} = -\sin\theta\cos\theta,$$

$$\Gamma^3_{03} = R'/R, \qquad \Gamma^3_{13} = 1/r, \qquad \Gamma^3_{23} = \cot\theta.$$

Note, for example, that in the second geodesic equation $2R'R^{-1}\dot{t}\dot{r}$ includes two terms of the sum $\Gamma^1_{\mu\nu}\dot{x}^\mu\dot{x}^\nu$, namely $\Gamma^1_{01}\dot{x}^0\dot{x}^1$ and $\Gamma^1_{10}\dot{x}^1\dot{x}^0$, and one must remember to halve the multipliers of the cross terms $\dot{x}^\mu\dot{x}^\nu$ ($\mu \neq \nu$) when extracting the connection coefficients from the geodesic equations.

Note also that in the example above ϕ is a cyclic coordinate, so one may say immediately that $\partial L/\partial\dot{\phi}$ is constant; that is, $R^2r^2\sin^2\theta\,\dot{\phi} = A$. Differentiation with respect to u results in the last geodesic equation, showing that we do indeed have an integral.

Exercises 2.1

1. Show that if a general parameter $t = f(s)$ is used to parameterize a straight line in Euclidean space, then the geodesic equation takes the form

$$\frac{d^2u^i}{dt^2} + \Gamma^i_{jk}\frac{du^j}{dt}\frac{du^k}{dt} = h(s)\frac{du^i}{dt},$$

   where $h(s) = -\dfrac{d^2t}{ds^2}\left(\dfrac{dt}{ds}\right)^{-2}$.

   Deduce that this reduces to the simple form (2.11) if, and only if, $t = As + B$, where A, B are constants ($A \neq 0$).

2. The aim of this exercise is to show that the length L of the tangent vector $\dot{x}^a$ to an affinely parameterized geodesic is constant.
   (a) Start by arguing that $\pm L^2 = g_{ab}\dot{x}^a\dot{x}^b$.
   (b) Differentiate this equation to obtain an expression for $\pm 2L\dot{L}$ in terms of the quantities g_{ab}, $\dot{g}_{ab}$, $\dot{x}^a$, and $\ddot{x}^a$.
   (c) Put $\dot{g}_{ab} = \partial_c g_{ab}\dot{x}^c$ and use the geodesic equation (2.12) to express the second derivatives $\ddot{x}^a$ in terms of the connection coefficients Γ^a_{bc} and the first derivatives $\dot{x}^a$.
   (d) Then use equation (2.13) to express the Γ^a_{bc} in terms of the metric tensor components and their derivatives.

(e) Simplify to obtain $2L\dot{L} = 0$, from which it follows that $\dot{L} = 0$ and L is constant.

(See Exercise 2.3.4 for a much shorter derivation of this result.)

3. Use the result of Exercise 2 to show that, for a non-null geodesic affinely parameterized by u, $u = As + B$, where A, B are constants ($A \neq 0$).

4. Show that for any geodesic (non-null or null) any two affine parameters u and u' are related by an equation of the form $u' = Au + B$, where A, B are constants with $A \neq 0$.

5. Use the result of Exercise 1.6.2(a) to show that, for a sphere of radius a parameterized in the usual way by $u^1 \equiv \theta$, $u^2 \equiv \phi$ (borrowed from spherical coordinates), the metric tensor components are given by

$$[g_{AB}] = \begin{bmatrix} a^2 & 0 \\ 0 & a^2 \sin^2 \theta \end{bmatrix}.$$

Deduce that the only nonzero connection coefficients are

$$\Gamma^1_{22} = -\sin\theta\cos\theta, \quad \Gamma^2_{12} = \Gamma^2_{21} = \cot\theta.$$

6. Show that all lines of longitude on a sphere (curves given by $\phi =$ constant) are geodesics.

7. In a Robertson–Walker spacetime, a coordinate curve for which r, θ, ϕ are constant and t varies is given by

$$x^\mu(u) = u\delta^\mu_0 + r_0\delta^\mu_1 + \theta_0\delta^\mu_2 + \phi_0\delta^\mu_3,$$

where r_0, θ_0, ϕ_0 are constants and u is a parameter. Verify that all such coordinate curves are geodesics affinely parameterized by u.

(See Example 2.1.2 for the connection coefficients for a Robertson–Walker spacetime.)

2.2 Parallel vectors along a curve

Our way of characterizing a straight line in Euclidean space and (by extension) a geodesic in a manifold is related to the idea of parallelly transporting a vector along a curve.

Let γ be a curve in three-dimensional Euclidean space given parametrically by $u^i(t)$ and let P_0 with parameter t_0 be some initial point on γ where we give a vector $\boldsymbol{\lambda}_0$. We can think of transporting $\boldsymbol{\lambda}_0$ along γ *without any change to its length or direction* so as to obtain a parallel vector $\boldsymbol{\lambda}(t)$ at each point of γ (see Fig. 2.4). The result is a *parallel field of vectors along* γ generated

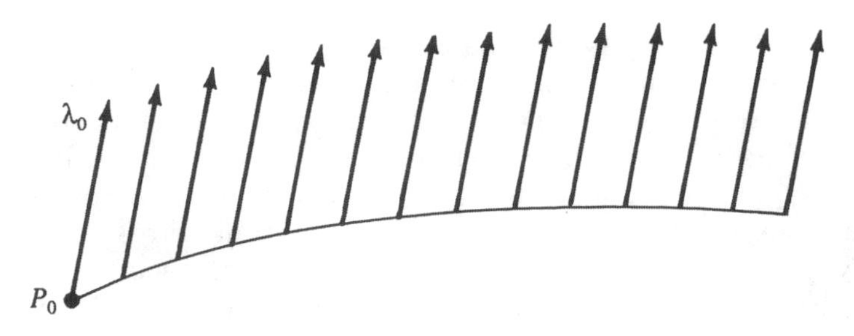

Fig. 2.4. A parallel field of vectors generated by parallel transport.

by the *parallel transport* of $\boldsymbol{\lambda}_0$ along γ. Since there is no change in the length or direction of $\boldsymbol{\lambda}(t)$ along γ, it satisfies the differential equation

$$d\boldsymbol{\lambda}/dt = \mathbf{0}, \tag{2.21}$$

for which $\boldsymbol{\lambda}(t_0) = \boldsymbol{\lambda}_0$ is the initial condition. If we work on equation (2.21) like we did on equation (2.1), then we can deduce from an equation like equation (2.3) that the components λ^i of the transported vector satisfy

$$\dot{\lambda}^i + \Gamma^i_{jk}\lambda^j\dot{u}^k = 0, \tag{2.22}$$

where the connection coefficients are given by equation (2.9).

Equation (2.22) is the component version of the equation for parallelly transporting a vector along a curve in Euclidean space. Its generalization for the parallel transport of a contravariant vector λ^a along a curve γ in an N-dimensional manifold with metric tensor field g_{ab} is clearly

$$\boxed{\dot{\lambda}^a + \Gamma^a_{bc}\lambda^b\dot{x}^c = 0,} \tag{2.23}$$

where the connection coefficients are given by equation (2.13) and $\dot{x}^a$ is the tangent vector arising from the parameterization $x^a(u)$ of γ. We now see that our definition of an affinely parameterized geodesic in the previous section amounts to saying that it is a curve characterized by the fact that its tangent vectors $\dot{x}^a$ form a parallel field of vectors along itself.

Parallel transport along curves in a curved manifold is significantly different from that along curves in flat Euclidean space in that it is *path-dependent*: the vector obtained by transporting a given vector from a point P to a remote point Q depends on the route taken from P to Q. This path dependence also shows up in transporting a vector around a closed loop, where on returning to the starting point the direction of the transported vector is (in general) different from the vector's initial direction. This path dependence can be demonstrated on a curved surface, in both practical and mathematical terms.

In Appendix B we describe the construction of a machine that gives a practical means of transporting a vector parallelly along a curve on a surface. It is a small two-wheeled vehicle carrying a pointer (which represents the vector) equipped with some rather clever gearing that receives input from the two wheels and outputs adjustments to the direction of the pointer. These adjustments ensure that the pointer is parallelly transported along the path taken by the vehicle. If we were to take this parallel transporter for walks on various surfaces, we would confirm that, for a curved surface, parallel transport is (in general) path-dependent, while, for a plane, it is path-independent. We would also observe that on completing a closed loop on a curved surface, the final direction of the pointer is (in general) different from its initial direction. The following example illustrates in mathematical terms this phenomenon for curves on a sphere.

Example 2.2.1
Consider a sphere of radius a, coordinatized in the usual way using $u^1 \equiv \theta$,

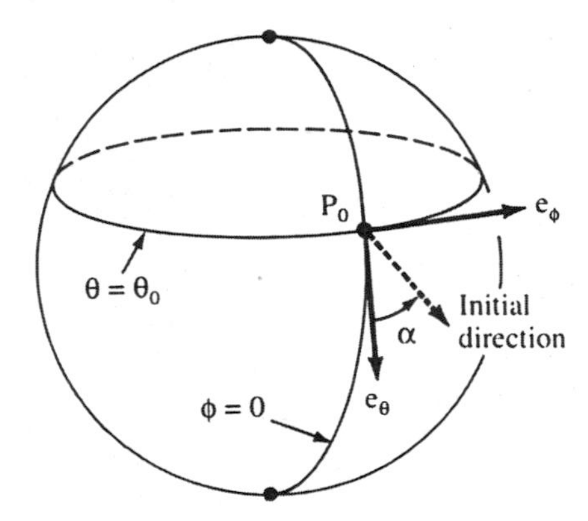

Fig. 2.5. Parallel transport around a circle of latitude.

$u^2 \equiv \phi$, where θ, ϕ are polar angles borrowed from spherical coordinates, with $0 \leq \theta \leq \pi$ and (for convenience) $0 \leq \phi \leq 2\pi$. Then

$$[g_{AB}] = \begin{bmatrix} a^2 & 0 \\ 0 & a^2 \sin^2 \theta \end{bmatrix}$$

and the only nonzero connection coefficients are

$$\Gamma^1_{22} = -\sin\theta\cos\theta, \quad \Gamma^2_{12} = \Gamma^2_{21} = \cot\theta$$

(see Exercise 2.1.5). Let us transport a vector $\boldsymbol{\lambda}$ parallelly around the circle of latitude γ given by $\theta = \theta_0$ ($\theta_0 = \text{const}$), starting and ending at the point P_0 where $\phi = 0$ or 2π (see Fig. 2.5). The circle is given parametrically by

$$u^A(t) = \theta_0\delta_1^A + t\delta_2^A, \quad 0 \le t \le 2\pi,$$

so $\dot{u}^A = \delta_2^A$ and the equation for parallel transport becomes $\dot{\lambda}^A + \Gamma^A_{B2}\lambda^B = 0$, which is equivalent to the pair

$$\begin{cases} \dot{\lambda}^1 - \sin\theta_0 \cos\theta_0\, \lambda^2 = 0 \\ \dot{\lambda}^2 + \cot\theta_0\, \lambda^1 = 0 \end{cases}. \tag{2.24}$$

Suppose initially that $\boldsymbol{\lambda}$ is a unit vector whose direction makes an angle α east of south. Then

$$\lambda^1(0) = a^{-1}\cos\alpha, \quad \lambda^2(0) = (a\sin\theta_0)^{-1}\sin\alpha, \tag{2.25}$$

as may be checked by noting that these must satisfy

$$g_{AB}\lambda^A(0)\lambda^B(0) = 1 \quad \text{and} \quad g_{AB}\lambda^A(0)S^B = \cos\alpha,$$

where $S^A \equiv a^{-1}\delta_1^A$ is the south-pointing unit vector at P_0.

We have an initial-value problem comprising the pair of equations (2.24) with initial conditions (2.25). Its solution is

$$\begin{cases} \lambda^1 = a^{-1}\cos(\alpha - \omega t) \\ \lambda^2 = (a\sin\theta_0)^{-1}\sin(\alpha - \omega t) \end{cases}, \tag{2.26}$$

where $\omega = \cos\theta_0$, as may be checked (see Exercise 2.2.1). On completing the circuit of γ, the vector obtained by parallel transport has components

$$\begin{cases} \lambda^1(2\pi) = a^{-1}\cos(\alpha - 2\pi\omega) \\ \lambda^2(2\pi) = (a\sin\theta_0)^{-1}\sin(\alpha - 2\pi\omega) \end{cases}.$$

We see that $g_{AB}\lambda^A(2\pi)\lambda^B(2\pi) = 1$, so $\lambda^A(2\pi)$ is a unit vector (as it should be), but its direction is not that of the initial vector $\lambda^A(0)$ (unless ω happens to be zero, as on the equator). Noting that

$$\begin{aligned} g_{AB}\lambda^A(0)\lambda^B(2\pi) &= \cos\alpha\cos(\alpha - 2\pi\omega) + \sin\alpha\sin(\alpha - 2\pi\omega) \\ &= \cos(\alpha - (\alpha - 2\pi\omega)) \\ &= \cos 2\pi\omega, \end{aligned}$$

we see that the final vector makes an angle of $2\pi\omega$ with the initial vector, where $\omega \equiv \cos\theta_0$. For example, for $\theta_0 = 85°$ the vector has twisted through 31.4°, whereas for $\theta_0 = 5°$ (near the North Pole) the angle between the final and initial vectors is 1.4°.

The above example can be used to illustrate two further points concerning parallel transport. The first of these is that if the curve along which the vector is transported is a geodesic, then the angle between the transported vector and the tangent to the geodesic remains constant. This is clearly the case when the geodesic is a straight line in Euclidean space and we shall obtain it as a general result for a manifold in the next section. The verification of this result for the sphere is left as an exercise for the reader (see Exercise 2.2.3).

The second point concerns transporting a vector parallelly around a closed curve that is "small." If in the example above θ_0 is small, then γ is a small circle about the North Pole, $\omega \equiv \cos\theta_0 \approx 1$ and the angle between the initial direction and the final direction is approximately 2π, which amounts to a negligible discrepancy between the initial and final vectors. This illustrates the fact that, by sticking to a small portion of a curved surface, we tend not to pick up its curvature by parallel transport around a closed curve.[7] Locally the surface behaves much as if it were flat, and experimentation over an extended region gives us a better chance of detecting curvature. The same is true for manifolds in general.

The connection coefficients Γ^a_{bc} are said to define a *connection* on the manifold. The reason for this kind of terminology is because it provides us with a connection between tangent spaces at different points of a manifold, enabling us to associate a vector in the tangent space at one point with the vector parallel to it at another point. For widely separated points, this association depends on the path used to connect the points, but for neighboring points (separated by small coordinate differences) the association is unique (up to first order in the small coordinate differences), as we now go on to show.

Suppose that P with coordinates x^a and Q with coordinates $x^a + \delta x^a$ are nearby points. Let γ be any parameterized curve through P and Q, with P having parameter u and Q having parameter $u + \delta u$, and let $\bar{\lambda}^a \equiv \lambda^a + \delta\lambda^a$ be the vector at Q parallel to a given vector λ^a at P. Since the vector at Q is obtained by the parallel transport of λ^a at P along the short piece of curve from P to Q, we have that

$$\delta\lambda^a \approx \frac{d\lambda^a}{du}\delta u,$$

where (from equation (2.23)) $d\lambda^a/du = -\Gamma^a_{bc}\lambda^b dx^c/du$, which gives

$$\bar{\lambda}^a \approx \lambda^a - \Gamma^a_{bc}\lambda^b \frac{dx^c}{du}\delta u \approx \lambda^a - \Gamma^a_{bc}\lambda^b \delta x^c. \tag{2.27}$$

So to first order in δx^a, we have a linear mapping from the the tangent space T_P to the tangent space T_Q in which the vector at P with components λ^a is mapped into the parallel vector at Q with components $\bar{\lambda}^a = A^a_b\lambda^b$, where

$$A^a_b \equiv \delta^a_b - \Gamma^a_{bc}\delta x^c. \tag{2.28}$$

We shall make use of this mapping in the next section when defining absolute and covariant differentiation.

In adopting equation (2.23) as the equation defining the parallel transport of a contravariant vector along a curve in a manifold, we completely ignored the

[7] This is because it is a second-order effect. See Sec. 3.3.

question of coordinate independence. If we were to use a primed coordinate system and perform parallel transport by having $\lambda^{a'}$ satisfy

$$\dot{\lambda}^{a'} + \Gamma^{a'}_{b'c'}\lambda^{b'}\dot{x}^{c'} = 0, \tag{2.29}$$

where

$$\Gamma^{a'}_{b'c'} \equiv \tfrac{1}{2}g^{a'd'}(\partial_{b'}g_{d'c'} + \partial_{c'}g_{b'd'} - \partial_{d'}g_{b'c'}), \tag{2.30}$$

would we get the same parallel field of vectors along the curve? We can answer this question in an indirect sort of way by showing that (to first order in small coordinate differences) the mapping from T_P to T_Q given by equation (2.27) does not depend on the coordinate system used. Thus we need to show that if

$$\bar{\lambda}^{a'} \equiv \lambda^{a'} - \Gamma^{a'}_{b'c'}\lambda^{b'}\delta x^{c'},$$

then

$$\bar{\lambda}^{a'}(X^e_{a'})_Q = \bar{\lambda}^e,$$

where $(X^e_{a'})_Q$ denotes $\partial x^e/\partial x^{a'}$ evaluated at Q. In terms of values at P, we can say that (to first order)

$$(X^e_{a'})_Q = X^e_{a'} + X^e_{d'a'}\delta x^{d'},$$

where $X^e_{d'a'} = \partial^2 x^e/\partial x^{d'}\partial x^{a'}$, so we need to show that

$$(\lambda^{a'} - \Gamma^{a'}_{b'c'}\lambda^{b'}\delta x^{c'})(X^e_{a'} + X^e_{d'a'}\delta x^{d'}) = \lambda^e - \Gamma^e_{fg}\lambda^f\delta x^g.$$

But $\lambda^{a'}X^e_{a'} = \lambda^e$, so (to first order) the above condition reduces to

$$X^e_{d'a'}\lambda^{a'}\delta x^{d'} - \Gamma^{a'}_{b'c'}X^e_{a'}\lambda^{b'}\delta x^{c'} = -\Gamma^e_{fg}\lambda^f\delta x^g,$$

which is equivalent to

$$(\Gamma^{a'}_{b'c'} - \Gamma^d_{fg}X^{a'}_d X^f_{b'}X^g_{c'} - X^d_{c'b'}X^{a'}_d)X^e_{a'}\lambda^{b'}\delta x^{c'} = 0, \tag{2.31}$$

since $X^f_{b'}\lambda^{b'} = \lambda^f$ and (to first order) $X^g_{c'}\delta x^{c'} = \delta x^g$. Using the defining equations (2.13) and (2.30), we can show that the connection coefficients transform according to

$$\boxed{\Gamma^{a'}_{b'c'} = \Gamma^d_{fg}X^{a'}_d X^f_{b'}X^g_{c'} + X^d_{c'b'}X^{a'}_d} \tag{2.32}$$

(see Exercise 2.2.4), so condition (2.31) is satisfied and the coordinate independence of parallel transport is established. (See Exercise 2.2.6 for a more direct way of establishing this result.) Since we can express the definition of a geodesic in terms of parallel transport, it follows that this definition is also coordinate-independent.

We finish this section by establishing a few formulae involving the connection coefficients Γ^a_{bc} and the related quantities Γ_{abc} defined by

$$\boxed{\Gamma_{abc} \equiv \tfrac{1}{2}(\partial_b g_{ac} + \partial_c g_{ba} - \partial_a g_{bc}).} \tag{2.33}$$

The traditional names for Γ_{abc} and Γ^a_{bc} are *Christoffel symbols of the first and second kinds*, respectively, and the notation $\Gamma_{abc} \equiv [bc, a]$, $\Gamma^a_{bc} \equiv \{{a \atop bc}\}$ is often used, especially in older texts.

From equation (2.13) we see that

$$\boxed{\Gamma^a_{bc} = g^{ad}\Gamma_{dbc}} \tag{2.34}$$

and a short calculation shows that

$$\boxed{\Gamma_{abc} = g_{ad}\Gamma^d_{bc}.} \tag{2.35}$$

Adding Γ_{bac} to Γ_{abc} gives

$$\boxed{\partial_c g_{ab} = \Gamma_{abc} + \Gamma_{bac},} \tag{2.36}$$

allowing us to express the partial derivatives of the metric tensor components in terms of the connection coefficients. If we denote the value of the determinant $|g_{ab}|$ by g, then the cofactor of g_{ab} in this determinant is gg^{ab}. (Note that g is not a scalar: changing coordinates changes the value of g at any point.) It follows that $\partial_c g = (\partial_c g_{ab})gg^{ab}$, so from equations (2.36) and (2.34) we have

$$\partial_c g = gg^{ab}(\Gamma_{abc} + \Gamma_{bac}) = g(\Gamma^b_{bc} + \Gamma^a_{ac}) = 2g\Gamma^a_{ac}.$$

So the contraction Γ^a_{ab} of the connection coefficients is given by

$$\Gamma^a_{ab} = \tfrac{1}{2}g^{-1}\partial_b g = \tfrac{1}{2}\partial_b \ln|g|, \tag{2.37}$$

the modulus signs being needed as g is not necessarily positive in the indefinite case. Alternative expressions are

$$\Gamma^a_{ab} = \partial_b \ln|g|^{1/2} \quad \text{and} \quad \Gamma^a_{ab} = |g|^{-1/2}\,\partial_b\,|g|^{1/2}. \tag{2.38}$$

Exercises 2.2

1. Verify that the initial-value problem comprising the pair of equations (2.24) with initial conditions (2.25) has a solution given by equations (2.26).

2. For what circle(s) of latitude is the final direction of the transported vector in Example 2.2.1 exactly opposite to that of the initial direction?

3. Noting the result of Example 2.1.1, verify that for parallel transport along a geodesic the angle between the transported vector of Example 2.2.1 and the tangent to the geodesic is constant.

4. Verify that the connection coefficients transform according to equation (2.32).

5. Show that an alternative form for the transformation formula (2.32) is

$$\Gamma^{a'}_{b'c'} = \Gamma^{d}_{ef} X^{a'}_{d} X^{e}_{b'} X^{f}_{c'} - X^{e}_{b'} X^{f}_{c'} X^{a'}_{ef}. \tag{2.39}$$

6. By transforming the left-hand side of equation (2.23) to a primed coordinate system, show that this defining equation for the parallel transport of a contravariant vector along a curve is coordinate-independent.

2.3 Absolute and covariant differentiation

In this section, we turn our attention to the effect of differentiation on tensor fields on a manifold M. Initially we shall consider fields defined along a curve, rather than throughout a region U or throughout the whole manifold M. Here we can regard the components of the field as functions of the parameter u used to label points on the curve, and we can consider their derivatives with respect to u. As we shall see, these derivatives are *not* the components of a tensor, which may come as a surprise to those used to differentiating the velocity components of a particle with respect to time t (which acts as a parameter along the particle's path) to obtain its acceleration. To make differentiation respect the tensor character of fields it needs to be modified, which, for differentiation along curves, leads to the idea of the *absolute derivative*. Having made this modification for fields along curves, we shall then go on to consider tensor fields defined throughout a region covered by a coordinate system, where the components can be regarded as functions of the coordinates. For these there is a corresponding modification of partial differentiation, called *covariant differentiation*, which is defined so that it respects tensor character. Both absolute and covariant differentiation depend on the notion of parallelism introduced in the previous section. These ideas play a crucial role in the formulation of the general theory of relativity and, because of this, this section and the following one are particularly important.

Suppose that we have a vector field $\lambda^a(u)$ defined along a curve γ given parametrically by $x^a(u)$. As we remarked above, the N quantities $d\lambda^a/du$ are *not* the components of a vector. To see this we use another (primed) coordinate system and look at the corresponding primed quantities $d\lambda^{a'}/du$ to see how they are related to the unprimed quantities $d\lambda^a/du$. These primed quantities are given by

$$d\lambda^{a'}/du = d(X^{a'}_{b}\lambda^{b})/du = X^{a'}_{b}(d\lambda^{b}/du) + X^{a'}_{bc}(dx^{c}/du)\lambda^{b}, \tag{2.40}$$

and the term involving $X^{a'}_{bc} \equiv \partial^2 x^{a'}/\partial x^b \partial x^c$ would be absent if the $d\lambda^a/du$ were the components of a vector. The reason for the presence of this term is that in the defining equation

$$\frac{d\lambda^a}{du} \equiv \lim_{\delta u \to 0} \frac{\lambda^a(u+\delta u) - \lambda^a(u)}{\delta u}, \tag{2.41}$$

we take differences of components at *different* points of γ, and here is the origin of our problem. Because in general the transformation coefficients depend on position, we have $(X^{a'}_b)_u \neq (X^{a'}_b)_{u+\delta u}$, which means that these differences in components are not the components of a vector (at either of the points in question). In the limit the difference between $(X^{a'}_b)_u$ and $(X^{a'}_b)_{u+\delta u}$ shows up as $X^{a'}_{bc}$. For differentiation to yield a vector, we must take component differences at the *same* point of γ, and we can do this by exploiting the notion of parallelism introduced in the previous section.

Let P be the point on γ with parameter value u and Q be a neighboring point with parameter value $u+\delta u$. Then $\lambda^a(u+\delta u)$ is a vector at Q, as is the vector $\bar{\lambda}^a$ obtained by the parallel transport of $\lambda^a(u)$ at P to Q. The difference $\lambda^a(u+\delta u) - \bar{\lambda}^a$ is then a vector at Q, and so is the quotient $(\lambda^a(u+\delta u) - \bar{\lambda}^a)/\delta u$. It is the limit of this quotient (as $\delta u \to 0$) that gives the *absolute derivative* $D\lambda^a/du$ of $\lambda^a(u)$ along γ. Now

$$\lambda^a(u+\delta u) \approx \lambda^a(u) + \frac{d\lambda^a}{du}\delta u$$

and, from equation (2.27),

$$\bar{\lambda}^a \approx \lambda^a(u) - \Gamma^a_{bc}\lambda^b(u)\delta x^c,$$

so

$$\frac{\lambda^a(u+\delta u) - \bar{\lambda}^a}{\delta u} \approx \frac{d\lambda^a}{du} + \Gamma^a_{bc}\lambda^b(u)\frac{\delta x^c}{\delta u}.$$

As $\delta u \to 0$, the point Q tends to P, and the limit of the quotient is

$$\boxed{\frac{D\lambda^a}{du} \equiv \frac{d\lambda^a}{du} + \Gamma^a_{bc}\lambda^b \frac{dx^c}{du},} \tag{2.42}$$

where all quantities are evaluated at the same point P of γ. Thus the absolute derivative of a vector field λ^a along a curve γ (which is a vector field along γ) involves not only the total derivative $d\lambda^a/du$ (which is not a vector field along γ), but also the connection coefficients Γ^a_{bc}.

The claim that the absolute derivative is a vector field along γ is justified by the way in which it is defined. It can also be justified by checking that

$$\left(\frac{d\lambda^{a'}}{du} + \Gamma^{a'}_{b'c'}\lambda^{b'}\frac{dx^{c'}}{du}\right) = X^{a'}_d\left(\frac{d\lambda^d}{du} + \Gamma^d_{ef}\lambda^e\frac{dx^f}{du}\right), \tag{2.43}$$

using the transformation equations (2.40) and (2.39) for $d\lambda^a/du$ and Γ^a_{bc}. Both of these involve second derivatives of the form $X^{a'}_{bc}$, but in such a way that they cancel when used to transform the quantities $D\lambda^a/du$. (See Exercise 2.3.1.)

We now see that equation (2.23) for parallelly transporting a contravariant vector along a curve can be written as $D\lambda^a/du = 0$ and that $\lambda^a(u)$ form a parallel field of vectors along γ if, and only if, $D\lambda^a/du = 0$. By extending the definition of absolute differentiation to general tensor fields $\tau^{a_1 \ldots a_r}_{b_1 \ldots b_s}(u)$ defined along a curve γ, we can also extend the notion of parallel transport along γ by requiring that $D\tau^{a_1 \ldots a_r}_{b_1 \ldots b_s}/du = 0$.

There are two approaches which may be taken to defining the absolute derivative of general tensor fields along a curve. One is to extend the notion of parallelism between neighboring tangent spaces T_P and T_Q to one between the space of type (r, s) tensors at P and the space of type (r, s) at Q, while the other is to demand that the operation of absolute differentiation satisfies certain reasonable conditions which allow us to extend the concept to general tensor fields along curves. We shall take the latter course, and impose the following conditions on the differential operator D/du applied to tensor fields defined along a curve parameterized by u:

(a) When applied to a tensor field, D/du yields a tensor field of the same type.
(b) D/du is a linear operation.
(c) D/du obeys Leibniz' rule with respect to tensor products.
(d) For any scalar field ϕ, $D\phi/du = d\phi/du$.

Condition (b) is a normal requirement of a differential operator and simply means that we are allowed to say things like $D(\sigma^{ab}_c + \tau^{ab}_c)/du = D\sigma^{ab}_c/du + D\tau^{ab}_c/du$, and $D(k\tau^a_{bc})/du = k(D\tau^a_{bc}/du)$ for constant k, while condition (c) allows us to say things like $D(\sigma^a_b \tau^{cd})/du = (D\sigma^a_b/du)\tau^{cd} + \sigma^a_b(D\tau^{cd}/du)$.

We now show how, by using conditions (a)–(d) and the expression already obtained for a contravariant vector field, we can obtain expressions for the absolute derivatives of tensor fields of any type. We shall do this in detail for some simple fields of specific type, from which we shall be able to infer the general pattern for a field of any type.

The absolute derivative of a scalar field
From condition (d) above, we have

$$\boxed{D\phi/du \equiv d\phi/du.} \tag{2.44}$$

The absolute derivative of a contravariant vector field
With the dot-notation for derivatives, equation (2.42) takes the form

$$\boxed{D\lambda^a/du \equiv \dot{\lambda}^a + \Gamma^a_{bc}\lambda^b\dot{x}^c.} \tag{2.45}$$

The absolute derivative of a covariant vector field
If μ_a is a covariant vector field along a curve γ, then for any contravariant vector field λ^a along γ, $\lambda^a\mu_a$ is a scalar field, so using equation (2.44) we have

$$d(\lambda^a\mu_a)/du = D(\lambda^a\mu_a)/du. \tag{2.46}$$

Then using Leibniz' rule (condition (c)) on the contracted tensor product we get

$$\begin{aligned}\frac{d\lambda^a}{du}\mu_a + \lambda^a\frac{d\mu_a}{du} &= \frac{D\lambda^a}{du}\mu_a + \lambda^a\frac{D\mu_a}{du}\\ &= \mu_a\left(\frac{d\lambda^a}{du} + \Gamma^a_{bc}\lambda^b\frac{dx^c}{du}\right) + \lambda^a\frac{D\mu_a}{du},\end{aligned}$$

which implies that

$$\lambda^a\frac{D\mu_a}{du} = \lambda^a\frac{d\mu_a}{du} - \Gamma^a_{bc}\lambda^b\frac{dx^c}{du}\mu_a = \lambda^a\left(\dot{\mu}_a - \Gamma^d_{ac}\mu_d\dot{x}^c\right).$$

Since this holds for arbitrary vector fields λ^a, we deduce that

$$\boxed{D\mu_a/du \equiv \dot{\mu}_a - \Gamma^d_{ac}\mu_d\dot{x}^c,} \tag{2.47}$$

and in this way our conditions yield the absolute derivative of a covariant vector field. (As we note below, when forming the absolute derivative of a tensor field, a Γ term with a minus sign is included for each subscript. As a reminder, we can extend our mnemonic to "co-below and minus.")

The absolute derivative of a type $(2, 0)$ tensor field
As a guide to obtaining an expression for the absolute derivative of a type $(2, 0)$ tensor field, we consider the special case in which $\tau^{ab} = \lambda^a\mu^b$, where λ^a, μ^a are contravariant vector fields along the curve. Then using condition (c) we have

$$D\tau^{ab}/du = D(\lambda^a\mu^b)/du = (D\lambda^a/du)\mu^b + \lambda^a(D\mu^b/du).$$

Inserting appropriate expressions for $D\lambda^a/du$ and $D\mu^b/du$, and recombining $\lambda^a\mu^b$ as τ^{ab} results in

$$\boxed{D\tau^{ab}/du \equiv \dot{\tau}^{ab} + \Gamma^a_{cd}\tau^{cb}\dot{x}^d + \Gamma^b_{cd}\tau^{ac}\dot{x}^d,} \tag{2.48}$$

which we take to be the formula for the absolute derivative of a type $(2, 0)$ tensor field.

The absolute derivative of a type $(0,2)$ tensor field
Similarly, by considering $\tau_{ab} = \lambda_a \mu_b$, we can arrive at

$$D\tau_{ab}/du \equiv \dot{\tau}_{ab} - \Gamma^c_{ad}\tau_{cb}\dot{x}^d - \Gamma^c_{bd}\tau_{ac}\dot{x}^d \tag{2.49}$$

as the formula for the absolute derivative of a type $(0,2)$ tensor field. (See Exercise 2.3.2.)

The absolute derivative of a type $(1,1)$ tensor field
Likewise, by considering $\tau^a_b = \lambda^a \mu_b$, we get

$$D\tau^a_b/du \equiv \dot{\tau}^a_b + \Gamma^a_{cd}\tau^c_b\dot{x}^d - \Gamma^c_{bd}\tau^a_c\dot{x}^d \tag{2.50}$$

for the absolute derivative of a type $(1,1)$ tensor field. (Again, see Exercise 2.3.2.)

The pattern should now be clear. The absolute derivative of a type (r,s) tensor field $\tau^{a_1\ldots a_r}_{b_1\ldots b_s}$ along a curve γ is given by the sum of the total derivative $\dot{\tau}^{a_1\ldots a_r}_{b_1\ldots b_s}$ of its components, r terms of the form $\Gamma^{a_k}_{cd}\tau^{\ldots c\ldots}_{\ldots}\dot{x}^d$ and s terms of the form $-\Gamma^c_{b_k d}\tau^{\ldots}_{\ldots c\ldots}\dot{x}^d$. For example,

$$D\tau^{ab}_c/du \equiv \dot{\tau}^{ab}_c + \Gamma^a_{de}\tau^{db}_c\dot{x}^e + \Gamma^b_{de}\tau^{ad}_c\dot{x}^e - \Gamma^d_{ce}\tau^{ab}_d\dot{x}^e. \tag{2.51}$$

As we remarked above, we can extend the notion of parallel transport to a tensor of any type, simply by requiring that its absolute derivative along the curve be zero. Again we emphasize that the *parallel transport of tensors is in general path-dependent.* Scalar fields are, of course, excepted, since $D\phi/du = 0$ implies that $d\phi/du = 0$, which in turn implies that ϕ is constant along the curve.

We are now in a position to introduce the *covariant derivative* of a tensor field, which is closely related to the absolute derivative. For absolute differentiation, the tensor fields involved need only be defined along the curve in question. The covariant derivative arises where we have a tensor field defined throughout M (or throughout a region of M).

Suppose, for example, we have a contravariant vector field λ^a defined throughout a region U. If γ is a curve in U, we can restrict λ^a to γ, and define its absolute derivative:

$$D\lambda^a/du \equiv \dot{\lambda}^a + \Gamma^a_{bc}\lambda^b\dot{x}^c. \tag{2.52}$$

But $\dot{\lambda}^a = \dfrac{\partial\lambda^a}{\partial x^c}\dot{x}^c$, so this may be written

$$\frac{D\lambda^a}{du} = \left(\frac{\partial\lambda^a}{\partial x^c} + \Gamma^a_{bc}\lambda^b\right)\dot{x}^c.$$

The bracketed expression on the right of this last equation does not depend on γ but only on the components λ^a and their derivatives at the point in question, and the equation is true for arbitrary tangent vectors $\dot{x}^a$ at the point in question. The usual argument involving the quotient theorem entitles us to deduce that $\dfrac{\partial \lambda^a}{\partial x^c} + \Gamma^a_{bc}\lambda^b$ are the components of a type $(1,1)$ tensor field. This tensor field is the *covariant derivative* of the vector field λ^a, and we denote it by $\lambda^a{}_{;c}$.

It is convenient at this point to introduce some more notation. We have already used ∂_a as an abbreviation for $\partial/\partial x^a$ and we shall continue to use it when dealing with covariant derivatives. We shall also use a comma followed by a subscript a written after the object on which it is acting to mean the same thing. So the covariant derivative of λ^a may be written as

$$\boxed{\lambda^a{}_{;c} = \partial_c \lambda^a + \Gamma^a_{bc}\lambda^b = \lambda^a{}_{,c} + \Gamma^a_{bc}\lambda^b.} \tag{2.53}$$

This notation extends naturally to repeated derivatives. For example, we write $\partial^2\lambda^a/\partial x^b \partial x^c$ as $\partial_b\partial_c\lambda^a$ or as $\lambda^a{}_{,cb}$. In a similar way we shall use $\lambda^a{}_{;cb}$ to denote the repeated covariant derivative $(\lambda^a{}_{;c})_{;b}$.

Returning now to covariant differentiation, we see that the argument above may be applied to a type (r,s) tensor field so as to define its covariant derivative, and it is clear that the resulting tensor field is of type $(r, s+1)$. It follows that covariant differentiation satisfies conditions analogous to (a)–(d) stipulated for absolute differentiation. Expressions for the covariant derivatives of general lower-rank tensor fields are noted below.

Covariant derivatives of lower-rank tensor fields

For a scalar field ϕ, covariant differentiation is simply partial differentiation:

$$\boxed{\phi_{;a} \equiv \partial_a \phi.} \tag{2.54}$$

For a contravariant vector field λ^a, we have

$$\boxed{\lambda^a{}_{;b} \equiv \partial_b \lambda^a + \Gamma^a_{cb}\lambda^c.} \tag{2.55}$$

For a covariant vector field μ_a, we have

$$\boxed{\mu_{a;c} \equiv \partial_c \mu_a - \Gamma^b_{ac}\mu_b.} \tag{2.56}$$

For a type $(2,0)$ tensor field τ^{ab}, we have

$$\boxed{\tau^{ab}{}_{;c} \equiv \partial_c \tau^{ab} + \Gamma^a_{dc}\tau^{db} + \Gamma^b_{dc}\tau^{ad}.} \tag{2.57}$$

For a type $(0,2)$ tensor field τ_{ab}, we have

$$\boxed{\tau_{ab;c} \equiv \partial_c \tau_{ab} - \Gamma^d_{ac}\tau_{db} - \Gamma^d_{bc}\tau_{ad}.} \tag{2.58}$$

For a type $(1,1)$ tensor field τ^a_b, we have

$$\boxed{\tau^a_{b;c} \equiv \partial_c \tau^a_b + \Gamma^a_{dc}\tau^d_b - \Gamma^d_{bc}\tau^a_d.} \tag{2.59}$$

Again, the mnemonic "co-below and minus" is a useful reminder for the sign of a Γ term.

The essential property common to both covariant and absolute differentiation is that when the operation is applied to a tensor field it produces a tensor field, while the operations of partial and total differentiation do not (i.e., the partial derivatives and total derivatives of tensor components do not obey transformation laws of the kind (1.73)). Another way in which the covariant derivative differs from the partial derivative is that in repeated differentiation the order matters. Thus for a vector field λ^a, we must acknowledge that even if ${\lambda^a}_{,bc} = {\lambda^a}_{,cb}$ holds, in general, ${\lambda^a}_{;bc} \neq {\lambda^a}_{;cb}$. We shall have more to say on this matter in the next chapter. We finish this section by considering the derivatives of the metric tensor field and its associated fields, noting in particular a special property that they possess.

Using equation (2.35) and the fact that $\Gamma^a_{bc} = \Gamma^a_{cb}$, we can rewrite equation (2.36) as

$$\partial_c g_{ab} - \Gamma^d_{ca} g_{db} - \Gamma^d_{cb} g_{ad} = 0,$$

which shows that $g_{ab;c} = 0$. That is, *the covariant derivative of the metric tensor field is identically zero.* The Kronecker tensor field with components δ^a_b and the contravariant metric tensor field with components g^{ab} also have covariant derivatives that are zero, as we now show. For the Kronecker tensor field, we simply note that

$$\delta^a_{b;c} = \partial_c \delta^a_b + \Gamma^a_{dc}\delta^d_b - \Gamma^d_{bc}\delta^a_d = 0 + \Gamma^a_{bc} - \Gamma^a_{bc} = 0,$$

while for the contravariant metric tensor field, we use the result just established to argue that

$$\begin{aligned} 0 = \delta^a_{b;c} &= (g^{ad} g_{db})_{;c} \\ &= {g^{ad}}_{;c} g_{db} + g^{ad} g_{db;c} \quad &&\text{(by Leibniz' rule)} \\ &= {g^{ad}}_{;c} g_{db} &&(\text{as } g_{db;c} = 0). \end{aligned}$$

Then contraction with g^{be} gives

$$0 = {g^{ad}}_{;c}\delta^e_d = {g^{ae}}_{;c},$$

as claimed. Along any curve γ, where we can regard the components g_{ab} as functions of the parameter u, we have that $Dg_{ab}/du = g_{ab;c}\dot{x}^c = 0$, establishing that the absolute derivative of the metric tensor field along γ is zero. We can argue similarly that the absolute derivatives of the Kronecker tensor field and the contravariant metric tensor field are also zero along any curve.

To sum up, we have shown that *the metric tensor field g_{ab}, the Kronecker tensor field with components δ^a_b and the contravariant metric tensor field g^{ab} have covariant derivatives that are zero:*

$$\boxed{g_{ab;c} = 0, \quad \delta^a_{b;c} = 0, \quad g^{ab}{}_{;c} = 0;} \tag{2.60}$$

and that along any curve γ their absolute derivatives are also zero:

$$\boxed{Dg_{ab}/du = 0, \quad D\delta^a_b/du = 0, \quad Dg^{ab}/du = 0.} \tag{2.61}$$

These special properties of the metric tensor field and its associated fields allow us to establish the important result that *inner products are preserved under parallel transport.* What we mean by this is that if two vector fields λ^a, μ^a are parallelly transported along a curve γ, then the inner product $g_{ab}\lambda^a\mu^b$ is constant along γ. We prove this by noting that

$$\begin{aligned} d(g_{ab}\lambda^a\mu^b)/du &= D(g_{ab}\lambda^a\mu^b)/du \\ &= (Dg_{ab}/du)\lambda^a\mu^b + g_{ab}(D\lambda^a/du)\mu^b + g_{ab}\lambda^a(D\mu^b/du) \\ &= 0, \end{aligned}$$

since $D\lambda^a/du = D\mu^b/du = 0$ (because the vectors are parallelly transported) and $Dg_{ab}/du = 0$ (established above). It follows that the length of a parallelly transported vector is constant, and also that the angle between two parallelly transported vectors is constant. Since the tangent vector to an affinely parameterized geodesic is parallelly transported along the geodesic, we can deduce that *if a vector is parallelly transported along a geodesic, then the angle between the transported vector and the tangent to the geodesic remains constant.* (See the remarks after Example 2.2.1 and Exercise 2.2.3.)

Having defined covariant differentiation, we can extend the familiar notion of the divergence of a vector field in Euclidean space to vector and tensor fields on a manifold. For a contravariant vector field λ^a we define its *divergence* to be the scalar field $\lambda^a{}_{;a}$. This definition is reasonable, for in a Cartesian coordinate system in Euclidean space $g_{ij} = \delta_{ij}$, so $\partial_k g_{ij} = 0$ giving $\Gamma^i_{jk} = 0$, and $\lambda^i{}_{;i}$ reduces to $\lambda^i{}_{,i}$. The *divergence* of a covariant vector field μ_a is defined to be that of the associated contravariant vector field $\mu^a \equiv g^{ab}\mu_b$. For a type (r, s) tensor field we may define $(r + s)$ divergences,

$$\tau^{a_1 \ldots c \ldots a_r}_{b_1 \ldots b_s}{}_{;c}, \quad (\tau^{a_1 \ldots a_r}_{b_1 \ldots c \ldots b_s} g^{cd})_{;d},$$

although these will not be distinct if the tensor field possesses symmetries. We can use this approach to calculate the divergence of a vector field in Euclidean space using curvilinear coordinate systems, as in the following example.

Example 2.3.1
In spherical coordinates the position vector field is $\mathbf{r} = r\mathbf{e}_r = r\mathbf{e}_1$ (on labeling the coordinates according to $u^1 \equiv r$, $u^2 \equiv \theta$, $u^3 \equiv \phi$), so its components are $r^i = r\delta^i_1$. Its divergence can then be calculated by saying

$$\begin{aligned}
\nabla \cdot \mathbf{r} = r^i{}_{;i} &= \partial_i r^i + \Gamma^i_{ji} r^j = \partial_i(r\delta^i_1) + \Gamma^i_{ji}(r\delta^j_1) = \partial r/\partial r + r\Gamma^i_{1i} \\
&= 1 + \tfrac{1}{2} r g^{-1} \partial_1 g \qquad \text{(using equation (2.37))} \\
&= 1 + \tfrac{1}{2} r (r^4 \sin^2\theta)^{-1} \partial(r^4 \sin^2\theta)\,/\partial r \qquad (\text{as } g = \det[g_{ij}] = r^4 \sin^2\theta) \\
&= 1 + 2 = 3.
\end{aligned}$$

Exercises 2.3

1. Check formula (2.43).
(Most of the work was done in Exercise 2.2.6.)

2. Obtain formulae (2.49) and (2.50), using methods similar to that used in deriving the result (2.48).

3. Show that equation (2.12) for an affinely parameterized geodesic can be written as $D\dot{x}^a/du = 0$.

4. Prove that the length of the tangent vector $\dot{x}^a$ to an affinely parameterized geodesic is constant.

2.4 Geodesic coordinates

It can be seen from equation (2.13) that if we could introduce a coordinate system throughout which the metric tensor components were constant, then the connection coefficients would be zero, and the mathematics of parallel transport, absolute differentiation, and covariant differentiation would be much simpler. It is possible to introduce such a coordinate system in Euclidean space, for example, by using Cartesian coordinates in which $g_{ij} = \delta_{ij}$, but in a general curved manifold it is not. However, it is possible to introduce a system of coordinates in which $\Gamma^a_{bc} = 0$ at a given point O, and such systems have their uses in simplifying some calculations involving the connection coefficients (see, e.g., Sec. 3.2 where the Bianchi identity is established). Such coordinates are generally referred to as *geodesic coordinates* with origin O, but this is not always appropriate, as they need not be based on geodesics.

Suppose we start with some system of coordinates in which O has coordinates x^a_O. Let us define a new system of (primed) coordinates by means of the equation

$$x^{a'} \equiv x^a - x^a_O + \tfrac{1}{2}(\Gamma^a_{bc})_O(x^b - x^b_O)(x^c - x^c_O), \tag{2.62}$$

where $(\Gamma^a_{bc})_O$ are the connection coefficients at O, as given in the original (unprimed) coordinate system. Differentiation with respect to x^d gives

$$\begin{aligned} X^{a'}_d &= \delta^a_d + \tfrac{1}{2}(\Gamma^a_{bc})_O\delta^b_d(x^c - x^c_O) + \tfrac{1}{2}(\Gamma^a_{bc})_O(x^b - x^b_O)\delta^c_d \\ &= \delta^a_d + (\Gamma^a_{dc})_O(x^c - x^c_O), \end{aligned}$$

so $(X^{a'}_d)_O = \delta^a_d$ and $\det[X^{a'}_d]_O \neq 0$. This means that equation (2.62) defines a new system of coordinates in some neighborhood U' of O, as claimed (see Sec. 1.7). A second differentiation gives

$$X^{a'}_{ed} = (\Gamma^a_{dc})_O\delta^c_e = (\Gamma^a_{de})_O,$$

showing that $(X^{a'}_{ed})_O = (\Gamma^a_{de})_O$. If we now use the transformation equation of Exercise 2.2.5 (noting that $(X^d_{a'})_O = \delta^d_a$ as a consequence of $(X^{a'}_d)_O = \delta^a_d$), we get

$$(\Gamma^{a'}_{b'c'})_O = (\Gamma^d_{ef})_O\delta^a_d\delta^e_b\delta^f_c - \delta^e_b\delta^f_c(\Gamma^a_{fe})_O = (\Gamma^a_{bc})_O - (\Gamma^a_{bc})_O = 0.$$

So in the new (primed) coordinate system the connection coefficients at O are zero, and we have a system of geodesic coordinates with origin O.

Geodesic coordinates can be used to construct a system of *local Cartesian coordinates* about a point O. These are an approximation to Cartesian coordinates, valid near O in a region of limited extent where the curvature of the manifold can be neglected. To get at such a system, we make use of a second coordinate transformation that brings the metric tensor at O to a simple diagonal form, while keeping the connection coefficients at O zero. To this end, we introduce a third (double-primed) system of coordinates about O defined by

$$x^{a''} = p^a_b x^{b'}, \tag{2.63}$$

where p^a_b are constants such that the matrix $P \equiv [p^a_b]$ is nonsingular. Differentiation of equation (2.63) shows that

$$X^{a''}_{c'} = p^a_b\delta^b_c = p^a_c,$$

so that the matrix version of

$$(g_{a''b''})_O = (g_{c'd'})_O(X^{c'}_{a''})_O(X^{d'}_{b''})_O$$

is

$$G''_O = P^{\mathrm{T}} G'_O P.$$

This means that, in matrix terms, G''_O is obtained from G'_O by a similarity transformation using the matrix P (see, e.g., Birkhoff and Mac Lane, 1977, §2–6). Matrix theory tells us that there exists a matrix P that brings G''_O to diagonal form in which each diagonal entry is either $+1$ or -1. If the metric tensor is positive definite, then all these entries are $+1$, but if it is indefinite, some will be $+1$ and others will be -1. In the latter case it is usual to use a diagonalizing matrix P that gives a diagonal form for G''_O with all the positive entries preceding the negative ones, so that

$$[g_{a''b''}]_O = G''_O = \mathrm{diag}(1, \ldots, 1, -1, \ldots, -1).$$

This second transformation has $X^{a''}_{d'c'} = 0$, so it follows from the equation of Exercise 2.2.5 (adapted for primed and double-primed coordinates) that if $(\Gamma^{a'}_{b'c'})_O = 0$, then $(\Gamma^{a''}_{b''c''})_O = 0$, which is what we required of it. Note that $x^{a'}_O = 0$ (from equation (2.62)), so $x^{a''}_O = 0$ (from equation (2.63)), showing that the point O is the "origin" of the double-primed coordinate system.

Dropping the double primes, we see that about O we have introduced a system of coordinates in which

$$x^a_O = 0, \qquad (\Gamma^a_{bc})_O = 0,$$

and

$$[g_{ab}]_O = \mathrm{diag}(1, \ldots, 1, -1, \ldots, -1)$$

(where the negative entries are absent in the positive definite case), so that

$$\Gamma^a_{bc} \approx 0, \qquad [g_{ab}] \approx \mathrm{diag}(1, \ldots, 1, -1, \ldots, -1), \tag{2.64}$$

in some neighborhood of O. These are *local Cartesian coordinates*, and the extent of the region in which the approximation (2.64) is valid depends (in a way to be made precise later) on the curvature of the manifold in the vicinity of O.

The implication of this for general relativity is that about each point of spacetime we can introduce a coordinate system in which

$$\Gamma^\mu_{\nu\sigma} \approx 0, \qquad g_{\mu\nu} \approx \eta_{\mu\nu}, \tag{2.65}$$

where $[\eta_{\mu\nu}] = \mathrm{diag}(1, -1, -1, -1)$, showing that locally the spacetime of general relativity looks like that of special relativity. This observation is a key factor in our discussion of the spacetime of general relativity in the next section.

Exercise 2.4

1. Show that, as a result of the coordinate transformation leading to geodesic coordinates (equation (2.62)), $(g_{a'b'})_O = (g_{ab})_O$.

2.5 The spacetime of general relativity

The spacetime of special relativity is discussed in Appendix A. In the language of Section 1.9, it is a four-dimensional pseudo-Riemannian manifold with the property that there exist global coordinate systems in which the metric tensor takes the form

$$[\eta_{\mu\nu}] \equiv \begin{bmatrix} 1 & 0 & 0 & 0 \\ 0 & -1 & 0 & 0 \\ 0 & 0 & -1 & 0 \\ 0 & 0 & 0 & -1 \end{bmatrix}$$

and we call such coordinate systems *Cartesian*. As explained in Section 1.2, we use x^μ to label points in spacetime, where the Greek suffixes, μ, ν, etc., have the range 0, 1, 2, 3, there being a certain convenience in counting from zero rather than one. As is customary in relativity we shall frequently refer to a point in spacetime as an *event*. Cartesian coordinates are related to the more familiar coordinates, t, x, y, z of special relativity by $x^0 \equiv ct$, $x^1 \equiv x$, $x^2 \equiv y$, $x^3 \equiv z$, c being the speed of light. We may, of course, use non-Cartesian coordinates, where the metric tensor $g_{\mu\nu} \neq \eta_{\mu\nu}$ but the essential feature of the spacetime of special relativity is that we may always introduce a Cartesian coordinate system about any point, so that $g_{\mu\nu} = \eta_{\mu\nu}$, and this coordinate system is global in the sense that it covers the whole of spacetime.

One of our guiding requirements for the spacetime of general relativity is that locally it should be like the spacetime of special relativity. We therefore assume that it is a four-dimensional pseudo-Riemannian manifold with the property that about any point there exists a system of local Cartesian coordinates in which the metric tensor field $g_{\mu\nu}$ is approximately $\eta_{\mu\nu}$. Note that we do *not* assert the existence of coordinate systems in which $g_{\mu\nu} = \eta_{\mu\nu}$ exactly, and this is the essential difference between the spacetimes of general and special relativity.

As explained in the previous section, we can construct a coordinate system about any point P of general-relativistic spacetime in which $(\Gamma^\mu_{\nu\sigma})_P = 0$, and $(x^\mu)_P = (0, 0, 0, 0)$. This means that $(\partial_\sigma g_{\mu\nu})_P = 0$, and so for points near to P, where the coordinates x^μ are small, Taylor's theorem gives

$$g_{\mu\nu} \approx \eta_{\mu\nu} + \tfrac{1}{2}(\partial_\alpha \partial_\beta g_{\mu\nu})_P x^\alpha x^\beta, \tag{2.66}$$

and this approximation is valid for small x^μ.

If we are sufficiently close to P for the second term on the right of equation (2.66) to be neglected, we have a coordinate system in which $g_{\mu\nu} = \eta_{\mu\nu}$ approximately, and the extent of the region in which this approximation is valid will depend on the sizes of the second derivatives $(\partial_\alpha \partial_\beta g_{\mu\nu})_P$, and also on the accuracy of our measuring procedures. It should be stressed that in special relativity we have *global* Cartesian coordinate systems, where $g_{\mu\nu} = \eta_{\mu\nu}$ *exactly*, whereas in general relativity we have only *local* Cartesian coordinate systems of limited extent, where $g_{\mu\nu} = \eta_{\mu\nu}$ *approximately*. We distinguish the two by saying that the spacetime of special relativity is *flat*, while that of general relativity is *curved*. The above discussion shows that the departure from flatness is connected with the nonvanishing of the second derivatives $\partial_\alpha \partial_\beta g_{\mu\nu}$, and we shall see the significance of this in Chapter 3, when we give a more formal definition of flatness in terms of the curvature tensor.

The purpose of the above discussion was to show, by introducing local Cartesian coordinates, the sense in which the spacetime of general relativity is locally like that of special relativity. However, it is not sensible to work in terms of local Cartesian coordinates as these involve approximations which amount to neglecting gravity, nor is it often convenient, since more suitable coordinates may be defined in a natural way. *We therefore use general coordinates, and formulate things in ways which are valid in any coordinate system.*

Another feature of the above discussion is that it gives us a means of generalizing to general relativity results which are valid in special relativity. For example, it is shown in Appendix A that in a Cartesian coordinate system of special-relativistic spacetime, Maxwell's equations may be written in the form

$$\begin{aligned} F^{\mu\nu}{}_{,\nu} &= \mu_0 j^\mu, \\ F_{\mu\nu,\sigma} + F_{\nu\sigma,\mu} + F_{\sigma\mu,\nu} &= 0. \end{aligned} \tag{2.67}$$

where a comma denotes partial differentiation. We may adopt

$$\begin{aligned} F^{\mu\nu}{}_{;\nu} &= \mu_0 j^\mu, \\ F_{\mu\nu;\sigma} + F_{\nu\sigma;\mu} + F_{\sigma\mu;\nu} &= 0. \end{aligned} \tag{2.68}$$

where a semicolon now denotes covariant differentiation, as the general-relativistic version of these, for in a local Cartesian coordinate system (where $g_{\mu\nu} = \eta_{\mu\nu}$ approximately, and we can neglect $\Gamma^\mu_{\nu\sigma}$) equations (2.68) reduce to equations (2.67). There are really two points to note here. The first is that if any physical quantity can be defined as a Cartesian tensor in special relativity, then we can give its definition in general relativity by defining it in exactly the same way in a local Cartesian coordinate system; its components in any other coordinate system are then given by the usual transformation formulae (1.73). Given this first point, the second is that any Cartesian tensor equation valid in special relativity may be converted to an equation valid in general relativity in any coordinate system, simply by replacing partial differentiation with respect to coordinates by covariant differentiation, total derivatives along curves by absolute derivatives, and $\eta_{\mu\nu}$ by $g_{\mu\nu}$. (Compare remarks made in the Introduction.)

As an example of this, consider the path of a particle (with mass) in special relativity. Its world velocity is $u^\mu \equiv dx^\mu/d\tau$ (see Sec. A.5), where the proper time τ for the particle is defined by (see Sec. A.0)

$$c^2 d\tau^2 \equiv \eta_{\mu\nu} dx^\mu dx^\nu.$$

Its equation of motion is then (equation (A.29))

$$dp^\mu/d\tau = f^\mu,$$

where $p^\mu \equiv mu^\mu$, m being the proper mass of the particle and f^μ the 4-force acting on it. The generalization of these ideas to general relativity gives $u^\mu \equiv dx^\mu/d\tau$ as the definition of the world velocity of the particle, where now the proper time τ is defined by

$$\boxed{c^2 d\tau^2 \equiv g_{\mu\nu} dx^\mu dx^\nu} \tag{2.69}$$

and

$$Dp^\mu/d\tau = f^\mu, \tag{2.70}$$

as the equation of motion, where $p^\mu \equiv mu^\mu$, and the definitions of m and f^μ are taken over from special relativity as explained above. Moreover, these equations are valid in any coordinate system.

As in special relativity we assume that a clock measures its own proper time. In particular, if the particle is a pulsating atom, the proper time interval between events on the atom's path where successive pulses occur is constant.

In the case of a free particle for which $f^\mu = 0$, equation (2.70) reduces to $D(dx^\mu/d\tau)/d\tau = 0$, or

$$\boxed{\frac{d^2x^\mu}{d\tau^2} + \Gamma^\mu_{\nu\sigma}\frac{dx^\sigma}{d\tau}\frac{dx^\nu}{d\tau} = 0.} \tag{2.71}$$

This reinforces our assertion that the path of a free particle is a geodesic in spacetime, and establishes that the proper time experienced by the particle is an affine parameter along it. This result is often stated as an explicit postulate of general relativity (the *geodesic postulate*), but it emerges here as a natural consequence of the way in which we generalize special-relativistic concepts. It is a perfectly natural generalization, for the path of a free particle in the flat spacetime of special relativity is a straight line and this generalizes to a geodesic in curved spacetime.

The path of a photon (or any other zero-rest-mass particle) in the spacetime of special relativity is also a straight line, and this also generalizes to a geodesic in curved spacetime. However, there is no change in proper time along the path of a photon, so τ cannot be used as a parameter. But we can still use an affine parameter u so that the analog of equation (2.71) for a photon is

$$\boxed{\frac{d^2x^\mu}{du^2} + \Gamma^\mu_{\nu\sigma}\frac{dx^\sigma}{du}\frac{dx^\nu}{du} = 0.} \tag{2.72}$$

The fact that the photon's speed is c finds expression as

$$\boxed{g_{\mu\nu}\frac{dx^\mu}{du}\frac{dx^\nu}{du} = 0,} \tag{2.73}$$

which generalizes the relation $\eta_{\mu\nu}\dfrac{dx^\mu}{du}\dfrac{dx^\nu}{du} = 0$ (equivalent to $c^2dt^2 - dx^2 - dy^2 - dz^2 = 0$) of special relativity.

We have already remarked on the characterization of a vector λ^μ as

$$\begin{Bmatrix} \text{timelike} \\ \text{null} \\ \text{spacelike} \end{Bmatrix} \quad \text{if } g_{\mu\nu}\lambda^\mu\lambda^\nu \begin{cases} > 0 \\ = 0 \\ < 0 \end{cases}$$

(see Sec. 1.9). One should note that at any point of spacetime the null cone of vectors given by $g_{\mu\nu}\lambda^\mu\lambda^\nu = 0$ lies in the tangent space at that point and not in the manifold. This fact is not readily appreciated in the flat spacetime of special relativity, because its basic linear structure allows one to regard the tangent space at each point as being embedded in the spacetime.

At any point on the path of a particle (with mass) its world velocity is a tangent vector to the path, and equation (2.69) tells us that this tangent vector is timelike. So a particle with mass follows a timelike path through spacetime, and in particular a free particle follows a timelike geodesic. A photon, however, follows a null geodesic, as equation (2.73) tells us that the tangent vectors to its path are null. Spacelike paths and spacelike geodesics may also be defined, but these have no physical significance.[8]

In moving from the flat spacetime of special relativity to the curved spacetime of general relativity we hope somehow to incorporate the effects of gravity, and the point of view we are adopting is that gravity is not a force, and that gravitational effects may be explained in terms of the curvature of spacetime. It should therefore be understood that by free particles we mean particles moving under gravity alone. Comparing equation (2.71) with its special-relativistic analog $d^2x^\mu/d\tau^2 = 0$ indicates that the connection coefficients play an important role in explaining gravitational effects. Since these are given by derivatives of the metric tensor field, we see that it is this tensor field which, in a sense, carries the gravitational content of spacetime. For the moment we shall take the metric tensor field as given, and postpone until Chapter 3 the question of how it is determined by the distribution of matter and energy in spacetime. In the rest of this chapter we take a closer look at equations (2.70) and (2.71), and relate them to some familiar Newtonian ideas.

Exercises 2.5

1. Is the world velocity of a stationary chair (in the lab) timelike or spacelike? Is its world line a geodesic?

2. Deduce the geodesic equation (2.71) from equation (2.70).

---

[8] Unless one believes in tachyons.

2.6 Newton's laws of motion

Newton's first law that "every body perseveres in its state of rest, or of uniform motion in a right [straight] line, unless it is compelled to change that state by forces impressed thereon" clearly has its counterpart in the statement that "every particle follows a geodesic in spacetime."[9] Indeed, in a local inertial coordinate system where we may neglect the $\Gamma^{\mu}_{\nu\sigma}$, the geodesic equation reduces to $d^2x^{\mu}/d\tau^2 = 0$. For nonrelativistic speeds $d\tau/dt$ is approximately one, so the geodesic equation yields $d^2x^i/dt^2 = 0$ $(i = 1, 2, 3)$, the familiar Newtonian equation of motion of a free particle. Newton's second law that

Newton	Einstein
Free particles move in straight lines through space.	Free particles follow geodesics through spacetime.
$\mathbf{F} = m\dfrac{d^2\mathbf{x}}{dt^2}$	$f^{\mu} = m\left(\dfrac{d^2x^{\mu}}{d\tau^2} + \Gamma^{\mu}_{\nu\sigma}\dfrac{dx^{\nu}}{d\tau}\dfrac{dx^{\sigma}}{d\tau}\right)$
To every action there is always opposed an equal reaction.	The third law is true for non-gravitational forces, just as in Newtonian physics (but see text for gravitational interaction).

Table 2.1. Newton's laws and their relativistic counterparts.

"the alteration of motion is ever proportional to the motive force impressed; and is made in the right line in which that force is impressed" is usually rendered as the 3-vector equation

$$d\mathbf{p}/dt = \mathbf{F},$$

where $\mathbf{p}$ is the momentum and $\mathbf{F}$ the applied force. This clearly has its counterpart in equation (2.70).

Newton's third law that "to every action there is always opposed an equal reaction: or the mutual actions of two bodies upon each other are always equal, and directed to contrary parts" is true in general relativity also. However, we must be careful, because Newton's gravitational force is now replaced by Einstein's idea that a massive body causes curvature of the spacetime around it, and a free particle responds by moving along a geodesic in that spacetime. It should be noted that this viewpoint ignores any curvature produced by the particle following the geodesic. That is, the particle is a *test particle*, and there

[9] The versions of Newton's laws quoted here are from Andrew Motte's translation (London, 1729) of Newton's *Principia*.

is no question of its having any effect on the body producing the gravitational field.

The gravitational interaction of two large bodies is not directly addressed by Einstein's theory, although it is of importance in astronomy, as for example in the famous pair of orbiting neutron stars PSR 1913+16. Approximation methods for such cases were studied in the 1980s,[10] but are beyond the scope of our book.

2.7 Gravitational potential and the geodesic

Suppose we have a coordinate system in which the metric tensor field is given by

$$g_{\mu\nu} \equiv \eta_{\mu\nu} + h_{\mu\nu}, \tag{2.74}$$

where the $h_{\mu\nu}$ are small, but not so small that they may be neglected. Our aim in this section is to obtain a Newtonian approximation to the geodesic equation given by the metric tensor field (2.74) valid for a particle whose velocity components dx^i/dt ($i = 1, 2, 3$) are small compared with c. We shall assume that the gravitational field, as expressed by $h_{\mu\nu}$, is quasi-static in the sense that $\partial_0 h_{\mu\nu} \equiv c^{-1}\partial h_{\mu\nu}/\partial t$ is negligible when compared with $\partial_i h_{\mu\nu}$.

If instead of the proper time τ we use the coordinate time t (defined by $x_0 \equiv ct$) as a parameter, then the geodesic equation giving the path of a free particle has the form

$$\frac{d^2x^\mu}{dt^2} + \Gamma^\mu_{\nu\sigma}\frac{dx^\nu}{dt}\frac{dx^\sigma}{dt} = h(t)\frac{dx^\mu}{dt}, \tag{2.75}$$

where

$$h(t) \equiv -\frac{d^2t}{d\tau^2}\left(\frac{dt}{d\tau}\right)^{-2} = \frac{d^2\tau}{dt^2}\left(\frac{d\tau}{dt}\right)^{-1}. \tag{2.76}$$

This can be deduced by an argument like that used in Exercise 2.1.1 and by noting that $\dfrac{d}{d\tau}\left(\dfrac{dt}{d\tau}\right) = \dfrac{dt}{d\tau}\dfrac{d}{dt}\left(\dfrac{d\tau}{dt}\right)^{-1}$. On dividing by c^2, the spatial part of equation (2.75) may be written

$$\frac{1}{c^2}\frac{d^2x^i}{dt^2} + \Gamma^i_{00} + 2\Gamma^i_{0j}\left(\frac{1}{c}\frac{dx^j}{dt}\right) + \Gamma^i_{jk}\left(\frac{1}{c}\frac{dx^j}{dt}\right)\left(\frac{1}{c}\frac{dx^k}{dt}\right) = \frac{1}{c}h(t)\left(\frac{1}{c}\frac{dx^i}{dt}\right), \tag{2.77}$$

and the last term on the left is clearly negligible.

If we put $h^{\mu\nu} \equiv \eta^{\mu\sigma}\eta^{\nu\rho}h_{\sigma\rho}$, then a short calculation shows that, to first order in the small quantities $h_{\mu\nu}$ and $h^{\mu\nu}$,

$$g^{\mu\nu} = \eta^{\mu\nu} - h^{\mu\nu} \quad \text{and} \quad \Gamma^\mu_{\nu\sigma} = \tfrac{1}{2}\eta^{\mu\rho}(\partial_\nu h_{\sigma\rho} + \partial_\sigma h_{\nu\rho} - \partial_\rho h_{\nu\sigma}). \tag{2.78}$$

---

[10] See Damour and Deruelle, 1986.

So to first order,

$$\begin{aligned}\Gamma^i_{00} &= \tfrac{1}{2}\eta^{i\rho}(\partial_0 h_{0\rho} + \partial_0 h_{0\rho} - \partial_\rho h_{00}) \\ &= -\tfrac{1}{2}\eta^{ij}\partial_j h_{00} = \tfrac{1}{2}\delta^{ij}\partial_j h_{00},\end{aligned}$$

on neglecting $\partial_0 h_{\mu\nu}$ in comparison with $\partial_i h_{\mu\nu}$. Also to first order,

$$\begin{aligned}\Gamma^i_{0j} &= \tfrac{1}{2}\eta^{i\rho}(\partial_0 h_{j\rho} + \partial_j h_{0\rho} - \partial_\rho h_{0j}) \\ &= -\tfrac{1}{2}\delta^{ik}(\partial_j h_{0k} - \partial_k h_{0j}),\end{aligned}$$

again on neglecting $\partial_0 h_{\mu\nu}$.

We have now approximated all the terms on the left-hand side of equation (2.77), and there remains the right-hand side to deal with. Working to the same level of approximation as above, and neglecting squares and products of $c^{-1}dx^i/dt$, we find from

$$\left(\frac{d\tau}{dt}\right)^2 = \frac{1}{c^2}g_{\mu\nu}\frac{dx^\mu}{dt}\frac{dx^\nu}{dt}$$

that

$$d\tau/dt = (1 + h_{00})^{1/2} = 1 + \tfrac{1}{2}h_{00}, \tag{2.79}$$

so

$$d^2\tau/dt^2 = \tfrac{1}{2}ch_{00,0}$$

and

$$\frac{1}{c}h(t) = \tfrac{1}{2}h_{00,0}(1 - \tfrac{1}{2}h_{00}) = \tfrac{1}{2}h_{00,0}$$

from equation (2.76).

It follows that the right-hand side of equation (2.77) is negligible, and our approximation gives

$$\frac{1}{c^2}\frac{d^2x^i}{dt^2} + \tfrac{1}{2}\delta^{ij}\partial_j h_{00} - \delta^{ik}(\partial_j h_{0k} - \partial_k h_{0j})\frac{1}{c}\frac{dx^j}{dt} = 0.$$

Introducing the mass m of the particle and rearranging gives

$$m\frac{d^2x^i}{dt^2} = -m\delta^{ij}\partial_j(\tfrac{1}{2}c^2h_{00}) + mc\delta^{ik}(\partial_j h_{0k} - \partial_k h_{0j})\frac{dx^j}{dt}. \tag{2.80}$$

Let us now interpret this in Newtonian terms. The left-hand side is mass × acceleration, so the right-hand side is the "gravitational force" on the particle. The first term on the right is the force $-m\nabla V$ arising from a potential V given by $V \equiv \frac{1}{2}c^2h_{00}$, while the second term on the right is velocity-dependent and clearly smacks of rotation.[11] This is not surprising, for the principle of equivalence asserts that the forces of acceleration, such as the velocity-dependent

[11] Most authors assume that the derivatives $\partial_i h_{\mu\nu}$ are small along with the $h_{\mu\nu}$, and therefore do not obtain these velocity-dependent rotational terms. However, the fact that the $h_{\mu\nu}$ are small does not mean that their derivatives are also small (see Sec. 2.9).

Coriolis force[12] which would arise from using a rotating reference system, are on the same footing as gravitational forces. If we agree to call a nearly inertial coordinate system in which $\partial_j h_{0k} - \partial_k h_{0j}$ is zero *nonrotating*, then we have for a slowly moving particle in a nearly inertial, nonrotating, coordinate system, in which the quasi-static condition holds, the approximation

$$d^2x^i/dt^2 = -\delta^{ij}\partial_j V, \tag{2.81}$$

where

$$V \equiv \tfrac{1}{2}c^2 h_{00} + \text{const.} \tag{2.82}$$

This is the Newtonian equation of motion for a particle moving in a gravitational field of potential V, provided we make the identification (2.82). This gives

$$g_{00} = 2V/c^2 + \text{const},$$

and if we choose the constant to be 1, then g_{00} reduces to its flat spacetime value when $V = 0$. This gives

$$\boxed{g_{00} = 1 + 2V/c^2} \tag{2.83}$$

as the relation between g_{00} and the Newtonian potential V in this approximation.

Exercises 2.7

1. Check approximations (2.78) and (2.79).

2. Show that equation (2.81) is equivalent to $m\mathbf{a} = \mathbf{F} = -m\nabla V$, where $\mathbf{a}$ is the acceleration and $\mathbf{F}$ is the force on the particle.

2.8 Newton's law of universal gravitation

Newton's law of universal gravitation does not survive intact in general relativity, which is after all a new theory replacing the Newtonian theory. However, we should be able to recover it as an approximation.

The Schwarzschild solution is an exact solution of the field equations of general relativity, and it may be identified as representing the field produced by a massive body. This solution is derived in the next chapter (see Sec. 3.7), and its line element is

$$c^2d\tau^2 = (1 - 2GM/rc^2)c^2dt^2 - (1 - 2GM/rc^2)^{-1}dr^2 - r^2d\theta^2 - r^2\sin^2\theta\, d\phi^2,$$

---

[12]See, for example, Goldstein, Poole, and Safko, 2002, §4–10.

where M is the mass of the body and G the gravitational constant. For small values of GM/rc^2 this is close to the line element of flat spacetime in spherical coordinates, and r then behaves like radial distance. If we were to put

$$x^0 \equiv ct, \quad x^1 \equiv r\sin\theta\cos\phi, \quad x^2 \equiv r\sin\theta\sin\phi, \quad x^3 \equiv r\cos\theta,$$

we would obtain a line element whose metric tensor had the form $g_{\mu\nu} = \eta_{\mu\nu} + h_{\mu\nu}$, where, for large values of rc^2/GM, the $h_{\mu\nu}$ are small and $g_{00} = 1 - 2GM/rc^2$. This gives $h_{00} = -2GM/rc^2$, and according to the results of the last section, a Newtonian potential $V = -GM/r$. The 3-vector form of equation (2.81) gives

$$m\, d^2\mathbf{r}/dt^2 = -m\nabla V = -GMmr^{-2}\hat{\mathbf{r}},$$

where $\mathbf{r} \equiv (x^1, x^2, x^3)$, m is the mass of the test particle, and $\hat{\mathbf{r}}$ is a unit vector in the direction of $\mathbf{r}$. The "force" on the test particle is in agreement with that given by Newton's law, and in this way the law is recovered as an approximation valid for large values of rc^2/GM and slowly moving particles.

2.9 A rotating reference system

The principle of equivalence (see the Introduction) implies that the "fictitious" forces of accelerating coordinate systems are essentially in the same category as the "real" forces of gravity. Put another way, if the geodesic equation contains gravity in the $\Gamma^\mu_{\nu\sigma}$ it must also contain any accelerations which may have been built in by choice of coordinate system. In a curved spacetime it is not always easy, and often impossible, to sort these forces out, but in flat spacetime we have only the fictitious forces of acceleration and these should be included in the $\Gamma^\mu_{\nu\sigma}$. As an example of this, let us consider a rotating reference system in flat spacetime.

Starting with a nonrotating system K with coordinates (T, X, Y, Z) and line element

$$c^2 d\tau^2 = c^2 dT^2 - dX^2 - dY^2 - dZ^2, \tag{2.84}$$

let us define new coordinates (t, x, y, z) by (see Fig. 2.6)

$$\begin{aligned} T &\equiv t, \\ X &\equiv x\cos\omega t - y\sin\omega t, \\ Y &\equiv x\sin\omega t + y\cos\omega t, \\ Z &\equiv z. \end{aligned} \tag{2.85}$$

Note that at this point we are only defining a change of coordinates and we are not too concerned (yet) about their physical meanings. Points given by x, y, z constant rotate with angular speed ω about the Z axis of K, and this

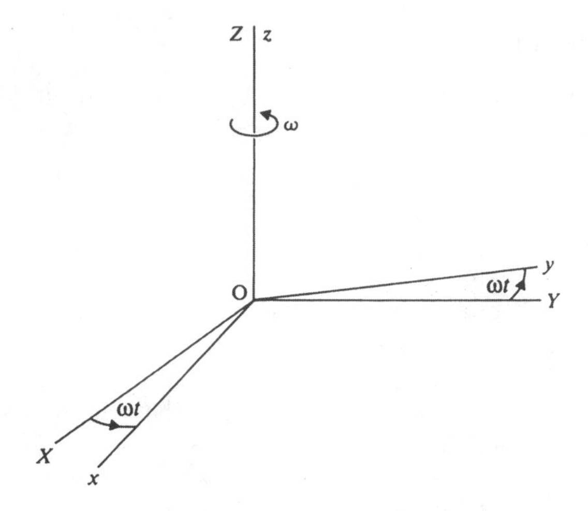

Fig. 2.6. Coordinate system $K'(t, x, y, z)$ rotating relative to the coordinate system $K(T, X, Y, Z)$.

defines the rotating system K' (see Fig. 2.6). In terms of the new coordinates the line element is

$$c^2 d\tau^2 = [c^2 - \omega^2(x^2 + y^2)]dt^2 + 2\omega y\, dx\, dt - 2\omega x\, dy\, dt - dx^2 - dy^2 - dz^2, \quad (2.86)$$

and the geodesic equations are

$$\begin{aligned} \ddot{t} &= 0, \\ \ddot{x} - \omega^2 x\dot{t}^2 - 2\omega\dot{y}\dot{t} &= 0, \\ \ddot{y} - \omega^2 y\dot{t}^2 + 2\omega\dot{x}\dot{t} &= 0, \\ \ddot{z} &= 0. \end{aligned} \quad (2.87)$$

where dots denote differentiation with respect to proper time (see Exercises 2.9.1 and 2.9.2). These constitute the equation of motion of a free particle (with mass).

The first of equations (2.87) implies that $dt/d\tau$ is constant, so the remaining equations may be written as

$$\begin{aligned} d^2x/dt^2 - \omega^2 x - 2\omega\, dy/dt &= 0, \\ d^2y/dt^2 - \omega^2 y + 2\omega\, dx/dt &= 0, \\ d^2z/dt^2 &= 0. \end{aligned}$$

Introducing the mass m of the particle and rearranging gives

$$\begin{aligned} m\, d^2x/dt^2 &= m\omega^2 x + 2m\omega\, dy/dt, \\ m\, d^2y/dt^2 &= m\omega^2 y - 2m\omega\, dx/dt, \\ m\, d^2z/dt^2 &= 0. \end{aligned} \quad (2.88)$$

or, in 3-vector notation,

$$m\,d^2\mathbf{r}/dt^2 = -m\boldsymbol{\omega} \times (\boldsymbol{\omega} \times \mathbf{r}) - 2m\boldsymbol{\omega} \times (d\mathbf{r}/dt), \tag{2.89}$$

where $\mathbf{r} \equiv (x, y, z)$ and $\boldsymbol{\omega} \equiv (0, 0, \omega)$.

An observer using t for time would interpret the left-hand side of equation (2.89) as mass $\times$ acceleration, and would therefore assert the existence of a "gravitational force" as given by the right-hand side. This "force" is, of course, the sum of the centrifugal force $-m\boldsymbol{\omega} \times (\boldsymbol{\omega} \times \mathbf{r})$ and the Coriolis force $-2m\boldsymbol{\omega} \times (d\mathbf{r}/dt)$, and this would seem to bear out our assertion that the geodesic equation does indeed include the forces of acceleration in the $\Gamma^{\mu}_{\nu\sigma}$. However, such an observer would be using the time associated with the nonrotating system K, because $t \equiv T$ and T is the time measured by clocks at rest in K. It is possible to define a time for K' based on a system of clocks at rest in K', but we shall not follow that course, as it would involve replacing equations (2.88) and (2.89) by more complicated ones that tend to conceal the Coriolis and centrifugal forces. Note that t is *exactly* the proper time for an observer situated at the common origin O of the two systems, so observers close to O who are at rest in the rotating system would accept equations (2.88) and (2.89) as approximately valid and recognize the terms on the right as forces of acceleration.

We can relate the situation described above to the approximation methods of Section 2.7 by putting $x^0 \equiv ct$, $x^1 \equiv x$, $x^2 \equiv y$, $x^3 \equiv z$, and noting that the line element (2.86) then gives $g_{\mu\nu} = \eta_{\mu\nu} + h_{\mu\nu}$, where

$$[h_{\mu\nu}] \equiv \begin{bmatrix} -\omega^2(x^2+y^2)/c^2 & \omega y/c & -\omega x/c & 0 \\ \omega y/c & 0 & 0 & 0 \\ -\omega x/c & 0 & 0 & 0 \\ 0 & 0 & 0 & 0 \end{bmatrix}.$$

The $h_{\mu\nu}$ are small, provided we restrict ourselves to the region near the z axis where $\omega^2(x^2+y^2)/c^2$ is small. Moreover, $\partial_0 h_{\mu\nu} = 0$, so the quasi-static condition is fulfilled. However, our system is rotating, so we must use the approximation (2.80) rather than (2.81). We see that

$$\tfrac{1}{2}c^2 h_{00} = -\tfrac{1}{2}\omega^2(x^2+y^2),$$

and a straightforward calculation (see Exercise 2.9.4) gives

$$[A^i_j] \equiv \begin{bmatrix} 0 & 2\omega & 0 \\ -2\omega & 0 & 0 \\ 0 & 0 & 0 \end{bmatrix}, \tag{2.90}$$

where $A^i_j \equiv c\delta^{ik}(\partial_j h_{0k} - \partial_k h_{0j})$. Hence the approximation (2.80) gives

$$\begin{aligned} m\,d^2x/dt^2 &= m\omega^2 x + 2m\omega\,dy/dt, \\ m\,d^2y/dt^2 &= m\omega^2 y - 2m\omega\,dx/dt, \\ m\,d^2z/dt^2 &= 0. \end{aligned}$$

These equations are identical with equations (2.88), and may be rearranged to exhibit the centrifugal and Coriolis forces, as before.

Exercises 2.9

1. Check the form of the line element (2.86) and verify that

$$[g^{\mu\nu}] = c^{-2}\begin{bmatrix} c^2 & \omega y c & -\omega x c & 0 \\ \omega y c & \omega^2 y^2 - c^2 & -\omega^2 xy & 0 \\ -\omega x c & -\omega^2 xy & \omega^2 x^2 - c^2 & 0 \\ 0 & 0 & 0 & -c^2 \end{bmatrix}.$$

2. Obtain the geodesic equations (2.87) in three different ways:
   (a) By using the Euler–Lagrange equations (and $[g^{\mu\nu}]$ from Exercise 1).
   (b) By extracting $[g_{\mu\nu}]$ from the line element (2.86), and then calculating the $\Gamma^{\mu}_{\nu\sigma}$ (again using $[g^{\mu\nu}]$ from Exercise 1).
   (c) By substituting for T, X, Y, Z in $\ddot{T} = \ddot{X} = \ddot{Y} = \ddot{Z} = 0$, using equations (2.85).

3. Cylindrical coordinates (ρ, ϕ, z) may be introduced into the rotating system K' by putting $x \equiv \rho\cos\phi$, $y \equiv \rho\sin\phi$. Show that in terms of these the geodesic equations are

$$\begin{aligned} \ddot{t} &= 0, \\ \ddot{\rho} - \rho\omega^2\dot{t}^2 - \rho\dot{\phi}^2 - 2\omega\rho\dot{\phi}\dot{t} &= 0, \\ \ddot{\phi} + 2\rho^{-1}\dot{\rho}\dot{\phi} + 2\omega\rho^{-1}\dot{\rho}\dot{t} &= 0, \\ \ddot{z} &= 0, \end{aligned}$$

so that corresponding to equations (2.88) one has

$$\begin{aligned} m\left[\frac{d^2\rho}{dt^2} - \rho\left(\frac{d\phi}{dt}\right)^2\right] &= m\rho\omega^2 + 2m\omega\rho\frac{d\phi}{dt}, \\ m\left(\rho\frac{d^2\phi}{dt^2} + 2\frac{d\rho}{dt}\frac{d\phi}{dt}\right) &= -2m\omega\frac{d\rho}{dt}, \\ m\frac{d^2 z}{dt^2} &= 0. \end{aligned}$$

Interpret these in terms of the radial, transverse, and axial components of acceleration, centrifugal, and Coriolis forces.

4. Check that the matrix $[A^i_j]$ is as given by equation (2.90), and that the approximation (2.80) does give equations (2.88).

Problems 2

1. Obtain the geodesic equations (using arc-length s as a parameter) for the hyperbolic paraboloid of Example 1.6.1.

   Deduce that all parametric curves are geodesics.

2. Using polar coordinates $\rho \equiv u^1$, $\phi \equiv u^2$, obtain the geodesic equations for the plane and verify that the ray $\phi = \phi_0$ ($\phi_0 =$ constant) is a geodesic. Use the equations of parallel transport to show that if λ^A is parallelly transported along this ray from its initial value λ_0^A at (ρ_0, ϕ_0), then

$$\lambda^1 = \lambda_0^1, \quad \lambda^2 = (\rho_0/\rho)\lambda_0^2.$$

   Verify that its length is constant and that it makes a constant angle with the ray.

3. If in spherical coordinates we set $\theta = \theta_0$, where θ_0 is a constant between 0 and $\pi/2$, we get a cone, and the remaining coordinates (r, ϕ) act as parameters on the cone. Show that the line element of the cone is

$$ds^2 = dr^2 + \omega^2 r^2 d\phi^2,$$

   where $\omega \equiv \sin\theta_0$, and that the Euler–Lagrange equations for geodesics on the cone yield

$$\ddot{r} - \omega^2 r\dot{\phi}^2 = 0, \quad \dot{\phi} = k/r^2,$$

   where k is constant. By eliminating the parameter, show that the geodesics satisfy

$$\frac{d}{d\phi}\left(\frac{1}{r^2}\frac{dr}{d\phi}\right) = \frac{\omega^2}{r}.$$

   Use the substitution $u = 1/r$ to solve this equation and hence show that the geodesics are given by $1 = Ar\cos(\omega\phi) + Br\sin(\omega\phi)$, where A and B are constants of integration.

   Use this result to show that (as intuition suggests) if the cone is cut along a generator and flattened to lie in a plane, then the geodesics are straight lines on the resulting flat surface.

4. The *curl* of a covariant field λ_a is the skew-symmetric tensor field A_{ab} defined by

$$A_{ab} \equiv \lambda_{a;b} - \lambda_{b;a}.$$

   Show that $A_{ab} = \lambda_{a,b} - \lambda_{b,a}$.

5. If A_{ab} is a skew-symmetric type $(0, 2)$ tensor field, prove that

$$B_{abc} \equiv A_{ab,c} + A_{bc,a} + A_{ca,b}$$

   are the components of a type $(0, 3)$ tensor field.

   (Hint: Put $A_{ab,c} = A_{ab;c} + \Gamma^d_{ac}A_{db} + \Gamma^d_{bc}A_{ad}$.)

6. Show that when spherical coordinates are used the line element of flat spacetime is

$$c^2 d\tau^2 = c^2 dt^2 - dr^2 - r^2 d\theta^2 - r^2 \sin^2\theta \, d\phi^2.$$

7. The line element of a static spherically symmetric spacetime is

$$c^2 d\tau^2 = A(r)dt^2 - B(r)dr^2 - r^2 d\theta^2 - r^2 \sin^2\theta \, d\phi^2.$$

Use the Euler–Lagrange equations to obtain the geodesic equations, and hence show that the only nonvanishing connection coefficients are:

$$\begin{array}{lll} \Gamma^0_{01} = A'/2A, & \Gamma^1_{00} = A'/2B, & \Gamma^1_{11} = B'/2B, \\ \Gamma^1_{22} = -r/B, & \Gamma^1_{33} = -(r\sin^2\theta)/B, & \Gamma^2_{12} = 1/r, \\ \Gamma^2_{33} = -\sin\theta\cos\theta, & \Gamma^3_{13} = 1/r, & \Gamma^3_{23} = \cot\theta, \end{array}$$

where primes denote derivatives with respect to r, and

$$x^0 \equiv t, \quad x^1 \equiv r, \quad x^2 \equiv \theta, \quad x^3 \equiv \phi.$$

8. One can conceive of an observer in a swivel chair located above the Sun, looking down on the plane of the Earth's orbit. If the chair rotates at the rate of one revolution a year, then to the observer the Earth appears stationary. If for some reason all heavenly bodies other than the Earth and the Sun are invisible, how does the observer explain why the Earth does not collapse in towards the Sun, there being no detectable orbit?

3

Field equations and curvature

3.0 Introduction

The main purpose of this chapter is to establish the field equations of general relativity, which couple the gravitational field (contained in the curvature of spacetime) with its sources. We start by discussing a tensor which effectively and concisely describes the sources, and follow that with a discussion of curvature, then bring these together in the field equations.

The field equation of Newton's theory is Poisson's equation, which is a scalar equation, valid in the Euclidean space of Newtonian gravitation theory. Einstein sought a replacement and in 1915 obtained a tensor equation, valid in the curved spacetime of general relativity. In our discussion, we use Poisson's equation as a guide in constructing the field equations of general relativity, which should, and do, yield Poisson's equation as an approximation.

The chapter finishes with an exact solution of the field equations representing the gravitational field of spherically symmetric massive body. This solution forms the basis of our discussions in Chapter 4.

3.1 The stress tensor and fluid motion

Except at the very end, where we take the step to curved spacetime, our discussion in this introductory section takes place in flat spacetime, where we use inertial coordinate systems. We shall be dealing with 4-vectors and 3-vectors, and we shall use bold-faced type to denote the latter. With some abuses of notation, we shall write

$$\lambda^{\mu} \equiv (\lambda^0, \lambda^1, \lambda^2, \lambda^3) \equiv (\lambda^0, \boldsymbol{\lambda}).$$

We start by considering a particle, and some quantities we make use of are:

$m \equiv$ rest or proper mass of a particle,[1]

[1] We use m rather than the more usual notation m_0 for rest mass. Similarly we use ρ rather than ρ_0 or ρ_{00} for the proper rest-mass density.

$t \equiv$ coordinate time,
$\tau \equiv$ proper time,
$\gamma \equiv dt/d\tau = (1 - v^2/c^2)^{-1/2}$, where v is the particle's speed,
$E \equiv \gamma mc^2 \equiv$ energy of particle,
$u^\mu \equiv dx^\mu/d\tau \equiv$ world velocity,
$v^\mu \equiv dx^\mu/dt \equiv u^\mu/\gamma \equiv$ coordinate velocity,
$p^\mu \equiv mu^\mu \equiv$ 4-momentum of particle.

So in our notation, $v^\mu \equiv (c, \mathbf{v})$, where $\mathbf{v}$ is the particle's 3-velocity, so that v occurring in the formula for γ is $|\mathbf{v}|$ (see Appendix A for details). Of the quantities listed above, only m and τ are scalars, and only u^μ and p^μ are vectors.[2]

A stationary particle situated at the point with position vector $\mathbf{x}_0$ has

$$u^\mu \equiv dx^\mu/d\tau = d(c\tau, \mathbf{x}_0)/d\tau = (c, \mathbf{0})$$

and

$$p^\mu = m(c, \mathbf{0}).$$

The zeroth component of p^μ is in this case the *rest energy* of the particle (up to a factor c). For a moving particle we have

$$p^\mu \equiv mu^\mu = \gamma m v^\mu = (\gamma mc, \gamma m\mathbf{v}) = (E/c, \mathbf{p}). \tag{3.1}$$

Equation (3.1) emphasizes the fact that in relativity, energy and momentum are the temporal and spatial parts of a single 4-vector p^μ. They always maintain this distinction, even after (Lorentz) transformations, just as the 4-vector $x^\mu \equiv (ct, \mathbf{x})$ is split distinctly into two parts, time and position, with time always being the zeroth component.

Let us now pass to a continuous distribution of matter, and for simplicity we shall take it to be a perfect fluid, which is characterized by two scalar fields, namely its density ρ and its pressure p, and a vector field, namely its world velocity u^μ. In order that ρ be a scalar field, one must define it to be the *proper density*, that is, the rest mass per unit rest volume. In place of the particle 4-momentum $p^\mu \equiv mu^\mu$, we now have the 4-momentum density ρu^μ.

What we wish to do is to exhibit some tensor which in some way represents the energy content of the fluid, and which, when taken over to curved spacetime, can act as the source of the gravitational field. Since in relativity we lose the distinction between mass and energy, all forms of energy should produce a gravitational field. Moreover, energy is not a scalar, but only the zeroth component of the 4-momentum, so we expect our source to contain the 4-momentum density of the fluid. Rather than try to construct a suitable

[2]The differential dt and the energy E are not scalar quantities (as in Newtonian physics): cdt and E are components of dx^μ and the momentum 4-vector p^μ, respectively. See Sec. A.6 for details.

tensor, let us simply write one down, and then discuss its physical significance. The tensor in question is the *energy-momentum-stress tensor* (or *stress tensor* for short), and for a perfect fluid is defined to be

$$T^{\mu\nu} \equiv (\rho + p/c^2)u^\mu u^\nu - p\eta^{\mu\nu}. \tag{3.2}$$

The first thing to note is that $T^{\mu\nu}$ is symmetric, and is made up from ρ, p, and u^μ, the scalar and vector fields which characterize the fluid. The pressure of the fluid makes some contribution to its energy content, and so should find a place in the tensor. The next thing to note is that

$$T^{\mu\nu}u_\nu = c^2(\rho + p/c^2)u^\mu - pu^\mu = c^2\rho u^\mu,$$

so $T^{\mu\nu}u_\nu$ is (up to a factor c^2) the 4-momentum density of the fluid. Finally we assert that setting its divergence $T^{\mu\nu}{}_{,\mu}$ equal to zero yields two important equations, namely the continuity equation and the equation of motion. (Since $T^{\mu\nu}$ is symmetric, it has only one divergence.) To prove our assertion would involve us in a lengthy digression into relativistic fluid mechanics, so we will simply derive the two equations, and give supporting arguments for their validity.

Setting the divergence $T^{\mu\nu}{}_{,\mu}$ equal to zero gives

$$(\rho u^\mu)_{,\mu}u^\nu + \rho u^\mu u^\nu{}_{,\mu} + (p/c^2)u^\mu{}_{,\mu}u^\nu + (p/c^2)u^\mu u^\nu{}_{,\mu} + c^{-2}p_{,\mu}u^\mu u^\nu - p_{,\mu}\eta^{\mu\nu} = 0. \tag{3.3}$$

Now the world velocity u^ν satisfies $u^\nu u_\nu = c^2$, and differentiation gives

$$u^\nu{}_{,\mu}u_\nu + u^\nu u_{\nu,\mu} = 0, \tag{3.4}$$

which implies that $u^\nu{}_{,\mu}u_\nu = 0$ (see Exercise 3.1.4). So contracting equation (3.3) with u_ν and dividing by c^2 gives

$$(\rho u^\mu)_{,\mu} + (p/c^2)u^\mu{}_{,\mu} = 0. \tag{3.5}$$

Equation (3.3) therefore simplifies to

$$(\rho + p/c^2)u^\nu{}_{,\mu}u^\mu = (\eta^{\mu\nu} - c^{-2}u^\mu u^\nu)p_{,\mu}. \tag{3.6}$$

Note that in obtaining equation (3.5) we contracted the equation $T^{\mu\nu}{}_{,\mu} = 0$ with u_ν, which is equivalent to taking its zeroth component in an instantaneous rest system of the fluid at the point in question.

We now justify our assertion concerning $T^{\mu\nu}{}_{,\mu} = 0$, by arguing that equation (3.5) is the equation of continuity of the fluid, while equation (3.6) is its equation of motion. We do this by showing that, for slowly moving fluids and small pressures, they reduce to the classical equations. To this end let us put $u^\mu = \gamma v^\mu = \gamma(c, \mathbf{v})$. Then by a slowly moving fluid we mean one for which we may neglect v/c, and so take $\gamma = 1$, and by small pressures we mean that p/c^2 is negligible compared to ρ. Equation (3.5) then reduces to

$$(\rho v^\mu)_{,\mu} = 0,$$

which gives

$$(\rho c)_{,0} + (\rho v^i)_{,i} = 0.$$

In 3-vector notation this is

$$\boxed{\partial\rho/\partial t + \nabla \cdot (\rho \mathbf{v}) = 0,} \tag{3.7}$$

which is the classical continuity equation,[3] the difference between proper density and density disappearing in the classical limit.

As for equation (3.6), this reduces to

$$\rho v^\nu{}_{,\mu} v^\mu = (\eta^{\mu\nu} - c^{-2} v^\mu v^\nu) p_{,\mu}, \tag{3.8}$$

and in our approximation

$$[\eta^{\mu\nu} - c^{-2} v^\mu v^\nu] = \begin{bmatrix} 0 & 0 & 0 & 0 \\ 0 & -1 & 0 & 0 \\ 0 & 0 & -1 & 0 \\ 0 & 0 & 0 & -1 \end{bmatrix},$$

so the zeroth components of the left-hand and right-hand sides are both zero. Its nonzeroth components reduce to

$$\rho v^i{}_{,\mu} v^\mu = -\delta^{ji} p_{,j},$$

which gives

$$\rho[\partial v^i/\partial t + v^i{}_{,j} v^j] = -\delta^{ji} p_{,j}.$$

In 3-vector notation this is

$$\boxed{\rho(\partial/\partial t + \mathbf{v} \cdot \nabla)\mathbf{v} = -\nabla p,} \tag{3.9}$$

which is Euler's classical equation of motion (1755) for a perfect fluid.[4]

Returning to the relativistic continuity equation (3.5), we see that it contains the pressure as well as the density, but this is not surprising, for we know that in relativity it is energy rather than mass which is conserved, and for a fluid under pressure, the pressure makes a contribution to the energy content. The relativistic equation (3.6) may be written in the form

[3]See, for example, Landau and Lifshitz, 1987, §1.

[4]See, for example, Landau and Lifshitz, 1987, §2.

$$(\rho + p/c^2)d^2x^\nu/d\tau^2 = (\eta^{\mu\nu} - c^{-2}u^\mu u^\nu)p_{,\mu},$$

because

$$u^\nu{}_{,\mu}u^\mu = \left(\frac{\partial}{\partial x^\mu}\frac{dx^\nu}{d\tau}\right)\frac{dx^\mu}{d\tau} = \frac{d^2x^\nu}{d\tau^2}.$$

In this form it looks more like an equation of motion, for it shows that the fluid particles are pushed off geodesics ($d^2x^\nu/d\tau^2 = 0$) by the pressure gradient $p_{,\mu}$.

If we were to accept equation (3.5) as the relativistic continuity equation and equation (3.6) as the equation of motion of a perfect fluid, then we could reverse our argument, and claim that $T^{\mu\nu}{}_{,\mu} = 0$ by virtue of the continuity equation and the equation of motion. It is, in fact, possible to give more complicated expressions representing the stress tensors of imperfect fluids and charged fluids, and even an electromagnetic field. These tensors are all symmetric, and all have zero divergence by virtue of equations such as continuity equations, equations of motion, or Maxwell's equations (see Problem 3.4).

Let us now take the step to the curved spacetime of general relativity. Our discussion in Section 2.5 gave us a prescription for taking over definitions and tensor equations from flat spacetime. In particular, we replace $\eta_{\mu\nu}$ by $g_{\mu\nu}$, and partial by covariant derivatives, so our defining equation for the stress tensor of a perfect fluid becomes

$$\boxed{T^{\mu\nu} \equiv (\rho + p/c^2)u^\mu u^\nu - pg^{\mu\nu},} \tag{3.10}$$

and the vanishing of its divergence is expressed as

$$\boxed{T^{\mu\nu}{}_{;\mu} = 0.} \tag{3.11}$$

With suitable definitions of $T^{\mu\nu}$ equation (3.11) is valid for all fluids and fields, not just perfect fluids. It is the stress tensor which we take as the source of the gravitational field, and the result (3.11) plays an important role in formulating the field equations; but before doing that we must take a closer look at curvature.

Exercises 3.1

1. Show that in a Cartesian coordinate system which brings the velocity of the fluid at a point P to rest (i.e., in an instantaneous rest system for the fluid at P) the components of the stress tensor (as defined by equation (3.2)) are given by

$$[T^{\mu\nu}] = \begin{bmatrix} \rho c^2 & 0 & 0 & 0 \\ 0 & p & 0 & 0 \\ 0 & 0 & p & 0 \\ 0 & 0 & 0 & p \end{bmatrix}.$$

2. A particle of 4-momentum p^μ is travelling through space and just misses an observer with world velocity U^μ. Show that he assigns an energy $p_\mu U^\mu$ (evaluated at the event of near-collision) to the particle.

3. Check that all the terms on the right-hand side of equation (3.2) have the same dimension.

4. Verify that $u^\nu u_\nu = c^2$ implies that $u^\nu{}_{;\mu} u_\nu = 0$.

3.2 The curvature tensor and related tensors

The material of this section is applicable to any N-dimensional manifold, so we use suffices a, b, etc., which have the range 1 to N, rather than μ, ν, etc, with the range 0 to 3. Of course, the requirements of general relativity will govern the scope of our results.

Covariant differentiation is clearly a generalization of partial differentiation. However, there is one important respect in which it differs from partial differentiation: the order in which covariant differentiations are done matters, and changing the order (in general) changes the result. We start by taking a closer look at this question.

The covariant derivative of a covariant vector field λ_a is

$$\lambda_{a;b} \equiv \partial_b \lambda_a - \Gamma^d_{ab} \lambda_d,$$

and a further covariant differentiation gives

$$\begin{aligned} \lambda_{a;bc} &= \partial_c(\lambda_{a;b}) - \Gamma^e_{ac}\lambda_{e;b} - \Gamma^e_{bc}\lambda_{a;e} \\ &= \partial_c\partial_b\lambda_a - (\partial_c\Gamma^d_{ab})\lambda_d - \Gamma^d_{ab}\partial_c\lambda_d - \Gamma^e_{ac}(\partial_b\lambda_e - \Gamma^d_{eb}\lambda_d) - \Gamma^e_{bc}(\partial_e\lambda_a - \Gamma^d_{ae}\lambda_d). \end{aligned}$$

Interchanging b and c, and then subtracting gives

$$\lambda_{a;bc} - \lambda_{a;cb} = R^d{}_{abc}\lambda_d, \tag{3.12}$$

where

$$\boxed{R^d{}_{abc} \equiv \partial_b\Gamma^d_{ac} - \partial_c\Gamma^d_{ab} + \Gamma^e_{ac}\Gamma^d_{eb} - \Gamma^e_{ab}\Gamma^d_{ec}.} \tag{3.13}$$

The left-hand side of equation (3.12) is a tensor for arbitrary vectors λ_a, so the contraction of $R^d{}_{abc}$ with λ_d is a tensor, and since $R^d{}_{abc}$ does not depend on λ_a, the quotient theorem entitles us to conclude that $R^d{}_{abc}$ is a type $(1,3)$ tensor. It is called the *curvature tensor* (or *Riemann–Christoffel curvature tensor* or *Riemann tensor*), and equation (3.13) indicates that it is defined in terms of the metric tensor and its derivatives.

So the necessary and sufficient condition that the order of covariant differentiations of any type $(0,1)$ tensor field can be interchanged is that $R^a{}_{bcd} = 0$.

This is in fact the necessary and sufficient condition for interchanging the order of covariant differentiations of fields of all types (see Exercise 3.2.1).

In the flat spacetime of special relativity we know that coordinate systems exist in which $g_{\mu\nu} = \eta_{\mu\nu}$, and in these coordinate systems $\Gamma^{\mu}_{\nu\sigma} = 0$, and hence the curvature tensor is identically zero. However, this does not entitle us to assume that the curvature tensor field of an arbitrary manifold is zero, and in Problem 3.1 we give an example of a two-dimensional manifold with nonvanishing curvature tensor field. We can now give a more formal definition of flatness. A manifold is *flat* if at each point of it $R^a{}_{bcd} = 0$, otherwise it is *curved.* (We may also speak of flat regions of a manifold.) It may be shown that in any region where $R^a{}_{bcd} = 0$ it is possible to introduce a coordinate system in which the components g_{ab} are constants, and hence a Cartesian coordinate system (one in which $[g_{ab}]$ is a diagonal matrix with ± 1 as its diagonal entries).[5]

On the face of it, $R^a{}_{bcd}$ has N^4 components. However, it possesses a number of symmetries and its components satisfy a certain identity, and it may be shown that these cut the number down to $N^2(N^2 - 1)/12$ independent components. The identity is given by the relation

$$R^a{}_{bcd} + R^a{}_{cdb} + R^a{}_{dbc} = 0, \tag{3.14}$$

and is known as the *cyclic identity.* Its proof is left as an exercise. The symmetries possessed by the curvature tensor are best expressed in terms of the associated type $(0, 4)$ tensor

$$R_{abcd} \equiv g_{ae} R^e{}_{bcd}.$$

Making use of equations (2.33), (2.35), and (2.36) gives, after extensive manipulation,

$$R_{abcd} \equiv \tfrac{1}{2}(\partial_d \partial_a g_{bc} - \partial_d \partial_b g_{ac} + \partial_c \partial_b g_{ad} - \partial_c \partial_a g_{bd}) - g^{ef}(\Gamma_{eac}\Gamma_{fbd} - \Gamma_{ead}\Gamma_{fbc}). \tag{3.15}$$

From this form for R_{abcd} it is a simple matter to check the following symmetry properties:

$$\text{(a)} \quad R_{abcd} = -R_{bacd}, \tag{3.16}$$

$$\text{(b)} \quad R_{abcd} = -R_{abdc}, \tag{3.17}$$

$$\text{(c)} \quad R_{abcd} = R_{cdab}. \tag{3.18}$$

It follows from (a) that

$$R^a{}_{acd} = 0. \tag{3.19}$$

The covariant derivatives $R^a{}_{bcd;e}$ also satisfy an identity, namely

$$R^a{}_{bcd;e} + R^a{}_{bde;c} + R^a{}_{bec;d} = 0. \tag{3.20}$$

[5]See, for example, Møller, 1972, Appendix 5.

It is known as the *Bianchi identity*, and may easily be proved in the following way. About any point P we can construct a coordinate system with $(\Gamma^a_{bc})_P = 0$, as explained in Section 2.4. Differentiating equation (3.13) and then evaluating at P gives, in this coordinate system,

$$(R^a{}_{bcd;e})_P = (\partial_e \partial_c \Gamma^a_{bd} - \partial_e \partial_d \Gamma^a_{bc})_P.$$

Cyclically permuting c, d, and e and adding gives the result at P. But P is arbitrary, so the result holds everywhere.

Equation (3.19) states that the contraction $R^a{}_{acd}$ is zero. However, in general the contraction $R^a{}_{bca}$ is nonzero, and this leads to a new tensor, the *Ricci tensor*. It is traditional to use the same kernel letter for the Ricci tensor as for the curvature tensor, so we denote its components by[6]

$$\boxed{R_{ab} \equiv R^c{}_{abc}.} \tag{3.21}$$

The Ricci tensor is in fact symmetric, as may be shown by contracting the cyclic identity (see Exercise 3.2.4). Since R_{ab} is symmetric, $R^a_{\ b} = R_b{}^a$ and we can denote both by R^a_b. A further contraction gives the *curvature scalar*

$$\boxed{R \equiv g^{ab} R_{ab} = R^a_a,} \tag{3.22}$$

and again the same kernel letter is used.

One final tensor, which is of some importance for later work, is the *Einstein tensor* G_{ab} defined by

$$\boxed{G_{ab} \equiv R_{ab} - \tfrac{1}{2} R g_{ab}.} \tag{3.23}$$

It is clearly symmetric, and this means that it possesses only one divergence $G^{ab}{}_{;a}$. The reason for the importance of the Einstein tensor is that this divergence is zero. Contracting a with d in the Bianchi identity (3.20) gives

$$R_{bc;e} + R^a{}_{bae;c} + R^a{}_{bec;a} = 0,$$

or, on using equation (3.17),

$$R_{bc;e} - R_{be;c} + R^a{}_{bec;a} = 0.$$

If we now raise b and contract with e, we get

$$R^b_{c;b} - R_{;c} + R^{ab}{}_{bc;a} = 0.$$

[6] There is wide disagreement over the sign of the curvature tensor, many authors giving it the opposite sign to ours. There is less disagreement over the sign of the Ricci tensor, agreement being effected by defining it according to $R_{ab} = R^c{}_{acb}$ in the case of opposite sign.

But, from equation (3.18),

$$R^{ab}{}_{bc;a} = R^{ba}{}_{cb;a} = R^{a}_{c;a} = R^{b}_{c;b},$$

so the above reduces to

$$2R^{b}_{c;b} - R_{;c} = 0,$$

which gives

$$(R^{b}_{c} - \tfrac{1}{2}R\delta^{b}_{c})_{;b} = 0,$$

on dividing by two and using the second of equations (2.60). We thus have $G^{b}_{c;b} = 0$, which implies that $G^{bc}{}_{;b} = 0$, as asserted.

Exercises 3.2

1. (a) Show that for a contravariant vector field λ^a,

$$\lambda^{a}{}_{;bc} - \lambda^{a}{}_{;cb} = -R^{a}{}_{dbc}\lambda^{d}.$$

(b) Show that for a type $(2,0)$ tensor field τ^{ab},

$$\tau^{ab}{}_{;cd} - \tau^{ab}{}_{;dc} = -R^{a}{}_{ecd}\tau^{eb} - R^{b}{}_{ecd}\tau^{ae}.$$

(Without loss of generality take $\tau^{ab} = \lambda^a\mu^b$.)

(c) Guess the corresponding expression for a type $(2,1)$ tensor field τ^{ab}_{c}.

2. Prove the cyclic identity (3.14).

3. Verify the form of R_{abcd} given by equation (3.15).

4. By contracting the cyclic identity (3.14) prove that the Ricci tensor is symmetric.

3.3 Curvature and parallel transport

We remarked in Section 2.2 that parallel transport in a curved manifold was path-dependent, and showed by example that this was indeed the case for paths on a sphere. Since we have now adopted a more formal approach to the notion of curvature by asserting that a manifold is flat if its curvature tensor vanishes, but is curved otherwise, we are clearly implying that there is some connection between the curvature tensor and parallel transport. Our aim in this section is to make that connection clear, and we shall show explicitly how the change $\Delta\lambda^a$ that results from parallelly transporting a vector λ^a around a small closed loop near a point P depends on the curvature tensor at P.

Suppose then that λ^a is transported parallelly along a curve γ from some initial point O where it is equal to λ^a_0. If γ is parameterized by t, then λ^a satisfies the differential equation

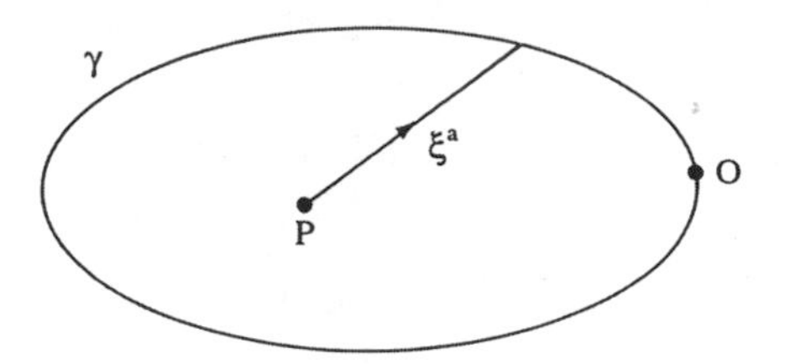

Fig. 3.1. A small loop near a point P.

$$\frac{d\lambda^a}{dt} = -\Gamma^a_{bc}\lambda^b \frac{dx^c}{dt}, \tag{3.24}$$

from which we see that λ satisfies the integral equation

$$\lambda^a = \lambda^a_0 - \int \Gamma^a_{bc}\lambda^b dx^c, \tag{3.25}$$

where the integral is taken along γ from the initial point O. We can use this equation to calculate the change $\Delta\lambda^a$ in λ as it is transported around a small loop close to a point P. If P has coordinates x^a_{P}, then points on the loop will have coordinates x^a given by

$$x^a = x^a_{\text{P}} + \xi^a,$$

where the ξ^a are small. The small coordinate differences ξ^a can be thought of as a vector extending from P to a general point on γ (see Fig. 3.1). Since the x^a_{P} are constant, equation (3.25) can be written

$$\lambda^a = \lambda^a_0 - \int \Gamma^a_{bc}\lambda^b d\xi^c. \tag{3.26}$$

This equation does not give λ^a in terms of λ^a_0 in a straightforward manner, because the transported vector also occurs in the integral on the right. However, it can be used to give successively better approximations that are valid when λ^a does not differ much from its initial value λ^a_0, which will be the case for our small loop γ close to P.

As a first approximation, we can put $\lambda^b = \lambda^b_0$ in the integral on the right-hand side which then yields the better approximation

$$\begin{aligned} \lambda^a &= \lambda^a_0 - \int \Gamma^a_{bc}\lambda^b_0 \, d\xi^c \\ &= \lambda^a_0 - \lambda^b_0 \int \Gamma^a_{bc} \, d\xi^c. \end{aligned}$$

This can then be fed into the right-hand side to yield an even better approximation:

$$\begin{aligned}\lambda^a &= \lambda_0^a - \int \Gamma^a_{bc}\left(\lambda_0^b - \lambda_0^d \int \Gamma^b_{de}\, d\xi^e\right) d\xi^c \\ &= \lambda_0^a - \lambda_0^b \int \Gamma^a_{bc}\, d\xi^c + \lambda_0^d \int \Gamma^a_{bc}\left(\int \Gamma^b_{de}\, d\xi^e\right) d\xi^c.\end{aligned} \tag{3.27}$$

This process can be repeated indefinitely, but the approximation given by equation (3.27) is enough for our purpose, which involves working to second order in ξ^a.

In the integral in the second term on the right, we can use the first-order approximation

$$\Gamma^a_{bc} = (\Gamma^a_{bc})_{\mathrm{P}} + (\partial_d \Gamma^a_{bc})_{\mathrm{P}}\, \xi^d,$$

as this gives second-order accuracy when integrated with respect to ξ^c. For the repeated integral in the third term, we can approximate Γ^a_{bc} by $(\Gamma^a_{bc})_{\mathrm{P}}$, since integrating twice gives second-order accuracy. Using these approximations to integrate from O around γ and back to O again, we get

$$\oint \Gamma^a_{bc} d\xi^c = (\Gamma^a_{bc})_{\mathrm{P}} \oint d\xi^c + (\partial_d \Gamma^a_{bc})_{\mathrm{P}} \oint \xi^d d\xi^c = (\partial_d \Gamma^a_{bc})_{\mathrm{P}} \oint \xi^d d\xi^c$$

(as $\oint d\xi^c = 0$) and

$$\oint \Gamma^a_{bc}\left(\int \Gamma^b_{de}\, d\xi^e\right) d\xi^c = (\Gamma^a_{bc}\Gamma^b_{de})_{\mathrm{P}} \oint \left(\int d\xi^e\right) d\xi^c = (\Gamma^a_{bc}\Gamma^b_{de})_{\mathrm{P}} \oint \xi^e d\xi^c,$$

so, from equation (3.27), the change in λ^a on transporting it around γ is

$$\Delta\lambda^a = -\lambda_0^b\, (\partial_d \Gamma^a_{bc})_{\mathrm{P}} \oint \xi^d d\xi^c + \lambda_0^d\, (\Gamma^a_{bc}\Gamma^b_{de})_{\mathrm{P}} \oint \xi^e d\xi^c,$$

which reduces to

$$\Delta\lambda^a = -\lambda_0^b\, (\partial_c \Gamma^a_{bd} - \Gamma^a_{ed}\Gamma^e_{bc})_{\mathrm{P}} \oint \xi^c d\xi^d, \tag{3.28}$$

on doing some relabeling of suffixes. Since $\oint d(\xi^c\xi^d) = 0$, one may show that

$$f^{cd} \equiv \oint \xi^c d\xi^d = \tfrac{1}{2} \oint (\xi^c d\xi^d - \xi^d d\xi^c) \tag{3.29}$$

(see Exercise 3.3.1), which is antisymmetric in its suffixes. Hence the coefficient of the integral on the right-hand side of equation (3.28) can be antisymmetrized in c, d to obtain

$$\Delta\lambda^a = -\tfrac{1}{2}(\partial_c \Gamma^a_{bd} - \partial_d \Gamma^a_{bc} + \Gamma^a_{ec}\Gamma^e_{bd} - \Gamma^a_{ed}\Gamma^e_{bc})_{\mathrm{P}} \lambda_0^b f^{cd}.$$

That is,

$$\boxed{\Delta\lambda^a = -\tfrac{1}{2}(R^a{}_{bcd})_{\mathrm{P}} \lambda_0^b f^{cd},} \tag{3.30}$$

where $f^{cd} \equiv \frac{1}{2}\oint(\xi^c d\xi^d - \xi^d d\xi^c)$.

Equation (3.30) establishes the basic relationship between the curvature tensor at a point P and parallel transport about a small loop close to P. By suitable choices for γ, this relationship can be exploited to investigate the components of the curvature tensor at P.

Suppose we restrict ourselves to points near P with coordinates

$$x^a = x^a_{\mathrm{P}} + xi^a + yj^a,$$

where $\{i^a, j^a\}$ are an orthogonal pair of unit vectors at P and x, y are small. Such points lie in a surface Σ embedded in the manifold and the pair (x, y) act as locally Cartesian coordinates on Σ, with P as the origin and "axes" given by the unit vectors i^a, j^a. If we then take γ to be a loop lying in Σ and surrounding P, we can express ξ^a in the form $\xi^a = xi^a + yj^a$ and arrive at

$$f^{cd} = \tfrac{1}{2}\oint(x\,dy - y\,dx)(i^c j^d - i^d j^c) \tag{3.31}$$

(see Exercise 3.3.1). Since (x, y) are locally Cartesian coordinates, $\frac{1}{2}\oint(x\,dy - y\,dx)$ is (very nearly) the area A enclosed by γ (provided its sense of description is related to i^a and j^a in the conventional way). We deduce that

$$\Delta\lambda^a/A = -\tfrac{1}{2}(R^a{}_{bcd})_{\mathrm{P}}\lambda^b_0(i^c j^d - i^d j^c) = -(R^a{}_{bcd})_{\mathrm{P}}\lambda^b_0 i^c j^d, \tag{3.32}$$

which in principle allows us to determine the components of the curvature tensor at P by appropriately choosing λ^a_0, i^a, j^a and noting the change $\Delta\lambda^a$ in transporting λ^a_0 around a small loop γ enclosing the area A. Equation (3.32) lets us do this approximately; for an exact result we should take a limit in which $A \to 0$ by letting γ shrink to the point P.

If the manifold is a surface, then (as Problem 3.1 asks you to show) all components of R_{ABCD} are either zero or equal to $\pm R_{1212}$. The component R_{1212} can be obtained as indicated above, the method being particularly straightforward if the coordinates are orthogonal: the embedded surface Σ is the surface itself and for i^A, j^A we can use normalized vectors tangential to the coordinate curves through P. The following example obtains R_{1212} at a general point on a sphere by this method.

Example 3.3.1

Let P be the point with coordinates (θ_0, ϕ_0) on a sphere with radius a (where we use the usual system of coordinates based on spherical coordinates). For the loop γ, take sections of the circles of latitude $\theta = \theta_0 \pm \varepsilon$ and the circles of longitude $\phi = \phi_0 \pm \varepsilon$ (where ε is small), as shown in the figure.

Let us start with a unit vector λ^A_0 that initially points south and transport it parallelly around γ via Q, R, S and back to the initial point O (see the figure). Then, from equation (2.25) of Example 2.2.1 with $\alpha = 0$,

$$\lambda^1_0 = a^{-1}, \quad \lambda^2_0 = 0.$$

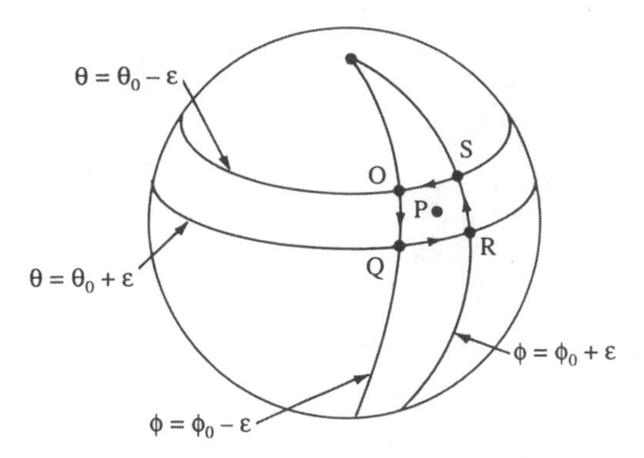

Fig. 3.2. A small loop on a sphere.

Since the section OQ is a geodesic (see Exercise 2.1.6), λ^A arrives at Q still pointing south, so its components are

$$\lambda^1_{\mathrm{Q}} = a^{-1}, \quad \lambda^2_{\mathrm{Q}} = 0.$$

After transporting along QR, it arrives at R with components given by

$$\lambda^1_{\mathrm{R}} = \frac{\cos(-2\varepsilon\cos(\theta_0+\varepsilon))}{a}, \quad \lambda^2_{\mathrm{R}} = \frac{\sin(-2\varepsilon\cos(\theta_0+\varepsilon))}{a\sin(\theta_0+\varepsilon)},$$

as can be deduced from equations (2.26) of Example 2.2.1. The section RS is again a geodesic, so it arrives at S with components given by

$$\lambda^1_{\mathrm{S}} = \frac{\cos(-2\varepsilon\cos(\theta_0+\varepsilon))}{a}, \quad \lambda^2_{\mathrm{S}} = \frac{\sin(-2\varepsilon\cos(\theta_0+\varepsilon))}{a\sin(\theta_0-\varepsilon)}.$$

On finally returning to O, its components are

$$\tilde{\lambda}^1_0 = \frac{\cos(-2\varepsilon\cos(\theta_0+\varepsilon)+2\varepsilon\cos(\theta_0-\varepsilon))}{a} = \frac{\cos(4\varepsilon\sin\theta_0\sin\varepsilon)}{a},$$

$$\tilde{\lambda}^2_0 = \frac{\sin(-2\varepsilon\cos(\theta_0+\varepsilon)+2\varepsilon\cos(\theta_0-\varepsilon))}{a\sin(\theta_0-\varepsilon)} = \frac{\sin(4\varepsilon\sin\theta_0\sin\varepsilon)}{a\sin(\theta_0-\varepsilon)},$$

as can again be deduced from equations (2.26).

To use equation (3.32), we need an expression for

$$\Delta\lambda^A \equiv \tilde{\lambda}^A_0 - \lambda^A_0$$

that is accurate to second order in ε. Using the expressions above (see Exercise 3.3.3), we arrive at

$$\Delta\lambda^A = (4\varepsilon^2/a)\delta^A_2.$$

The area A enclosed by γ is (again to second order in ε)

$$A = 4a^2\varepsilon^2 \sin\theta_0,$$

and for i^A, j^A we can use the orthonormal pair

$$i^A = a^{-1}\delta_1^A, \quad j^A = (a\sin\theta_0)^{-1}\delta_2^A.$$

Inserting everything into equation (3.32) gives

$$\frac{4}{4a^3\sin\theta_0}\delta_2^A = -(R^A{}_{BCD})_{\mathrm{P}}(a^{-1}\delta_1^B)(a^{-1}\delta_1^C)(a\sin\theta_0)^{-1}\delta_2^D,$$

which implies that $\delta_2^A = -(R^A{}_{112})_{\mathrm{P}}$ or $(R_{1A12})_{\mathrm{P}} = g_{A2}$ on lowering A. On setting $A = 2$, we get

$$(R_{1212})_{\mathrm{P}} = a^2\sin^2\theta_0,$$

in agreement with the result of Problem 3.1.

Exercises 3.3

1. Check that f^{cd} can be expressed as in equations (3.29) and (3.31).

2. Verify that λ_{Q}^A, λ_{R}^A, λ_{S}^A, and $\tilde{\lambda}_0^A$ are as claimed, using results from Example 2.2.1.

3. Verify that, to second order in ε, $\Delta\lambda^A = (4\varepsilon^2/a)\delta_2^A$.

3.4 Geodesic deviation

Another important place where curvature makes an appearance is in the equation of geodesic deviation, a concept introduced in Section 2.0. The derivation of this equation is somewhat tedious, and the reader short on time may prefer to skip this section at a first reading and return to it later.

Consider two neighboring geodesics, γ given by $x^a(u)$ and $\tilde{\gamma}$ given by $\tilde{x}^a(u)$, both affinely parameterized, and let $\xi^a(u)$ be the small "vector" connecting points with the same parameter value, that is, $\xi^a(u) \equiv \tilde{x}^a(u) - x^a(u)$ (see Fig. 3.3). To ensure that ξ^a as defined above is small, we must be careful with our parameterization. Having parameterized γ, we can use the affine freedom available ($u \to Au + B$, see Exercise 2.1.4) to parameterize $\tilde{\gamma}$ so that (over some range of parameter values) ξ^a is indeed small. If neither geodesic is null, then arc-length s may be used, and care need only be taken in fixing the zero value of s on, say, $\tilde{\gamma}$.

Since γ and $\tilde{\gamma}$ are geodesics we have

$$\frac{d^2\tilde{x}^a}{du^2} + \tilde{\Gamma}^a_{bc}\frac{d\tilde{x}^b}{du}\frac{d\tilde{x}^c}{du} = 0 \tag{3.33}$$

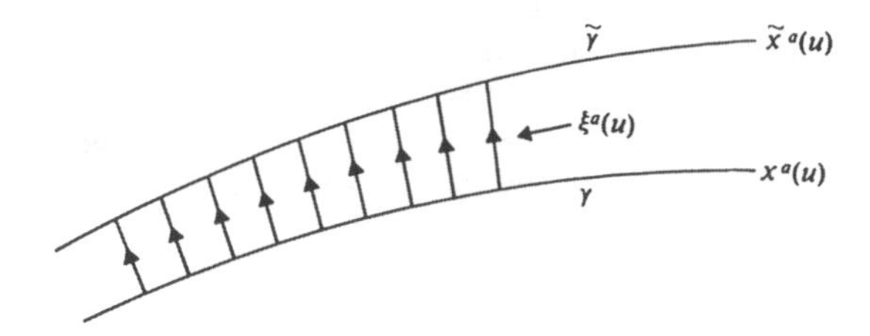

Fig. 3.3. Geodesic deviation.

and

$$\frac{d^2x^a}{du^2} + \Gamma^a_{bc}\frac{dx^b}{du}\frac{dx^c}{du} = 0, \tag{3.34}$$

where the tilde on the connection coefficient in equation (3.33) indicates that it is evaluated at the point with coordinates $\tilde{x}^a(u)$, whereas that in equation (3.34) is evaluated at the point with coordinates $x^a(u)$. But to first order in ξ^a,

$$\tilde{\Gamma}^a_{bc} = \Gamma^a_{bc} + \Gamma^a_{bc,d}\xi^d,$$

and subtracting equation (3.34) from (3.33) gives

$$\ddot{\xi}^a + \Gamma^a_{bc,d}\dot{x}^b\dot{x}^c\xi^d + \Gamma^a_{bc}\dot{x}^b\dot{\xi}^c + \Gamma^a_{bc}\dot{\xi}^b\dot{x}^c = 0,$$

where dots denote derivatives with respect to u, and only first-order terms have been retained. This may be written as

$$d(\dot{\xi}^a + \Gamma^a_{bc}\xi^b\dot{x}^c)/du - \Gamma^a_{bc,d}\xi^b\dot{x}^c\dot{x}^d - \Gamma^a_{bc}\xi^b\ddot{x}^c + \Gamma^a_{bc,d}\dot{x}^b\dot{x}^c\xi^d + \Gamma^a_{bc}\dot{x}^b\dot{\xi}^c = 0.$$

Substitution for $\ddot{x}^c$ from equation (3.34) and some rearrangement gives

$$\begin{aligned} d(\dot{\xi}^a + \Gamma^a_{bc}\xi^b\dot{x}^c)/du + \Gamma^a_{de}(\dot{\xi}^d + \Gamma^d_{bc}\xi^b\dot{x}^c)\dot{x}^e - \Gamma^a_{de}\Gamma^d_{bc}\xi^b\dot{x}^c\dot{x}^e \\ - \Gamma^a_{bc,d}\xi^b\dot{x}^c\dot{x}^d + \Gamma^a_{bc}\xi^b\Gamma^c_{de}\dot{x}^d\dot{x}^e + \Gamma^a_{bc,d}\dot{x}^b\dot{x}^c\xi^d = 0. \end{aligned}$$

On relabeling dummy suffices, this rather complicated expression reduces to

$$D^2\xi^a/du^2 + (\Gamma^a_{cd,b} - \Gamma^a_{bc,d} + \Gamma^a_{be}\Gamma^e_{dc} - \Gamma^a_{ed}\Gamma^e_{bc})\xi^b\dot{x}^c\dot{x}^d = 0,$$

which can be written more compactly as

$$\boxed{D^2\xi^a/du^2 + R^a{}_{cbd}\xi^b\dot{x}^c\dot{x}^d = 0.} \tag{3.35}$$

This is the *equation of geodesic deviation.*

In a flat manifold $R^a_{bcd} = 0$, and in Cartesian coordinates $D/du = d/du$, so equation (3.35) then reduces to $d^2\xi^a/du^2 = 0$, which implies that $\xi^a(u) = A^a u + B^a$, where A^a and B^a are constants. So in a flat manifold the separation vector increases linearly with u (and therefore with s if γ is non-null). However, in a curved manifold $R^a{}_{bcd} \neq 0$, and we do not have this linear relationship. These observations should be compared with the remarks made in Section 2.0.

Exercise 3.4

1. Check the derivation of equation (3.35).

3.5 Einstein's field equations

The field equations of general relativity are variously referred to as *Einstein's equation* or *Einstein's field equations*, and they were obtained by him at the end of 1915, after what he referred to as a period of unremitting labor. During the approximate period 1909–13 Einstein and his friend from undergraduate days, Marcel Grossmann, had realized that the metric tensor $g_{\mu\nu}$ describing the geometry of spacetime seemed to depend on the amount of gravitating matter in the region in question (and so adopted the kernel letter g for *gravity*).[7]

The metric tensor contains two separate pieces of information:

(i) the relatively unimportant information concerning the specific coordinate system used (e.g., spherical coordinates, Cartesian coordinates, etc.);
(ii) the important information regarding the existence of any gravitational potentials.

In Section 2.7 we saw that in a nearly Cartesian coordinate system g_{00} was essentially the Newtonian potential. In a more general coordinate system, this Newtonian potential would be dispersed throughout the $g_{\mu\nu}$, so there is a sense in which all the components $g_{\mu\nu}$ can be regarded as gravitational potentials.

We have seen in Section 3.1 that the matter content of spacetime is concisely summarized in the stress tensor $T^{\mu\nu}$, so if matter causes the geometry, it might be tempting to put

$$g^{\mu\nu} = \kappa T^{\mu\nu}, \tag{3.36}$$

where κ is some coupling constant. This looks plausible, because both $g^{\mu\nu}$ and $T^{\mu\nu}$ are symmetric, and $g^{\mu\nu}{}_{;\mu} = 0$ (see last of equations (2.60)) in agreement with $T^{\mu\nu}{}_{;\mu} = 0$. However, equation (3.36) does not reduce to Poisson's equation $\nabla^2 V = 4\pi G\rho$, in the Newtonian limit. Since the $g_{\mu\nu}$ are the gravitational potentials, it is clear that what is needed in place of $g^{\mu\nu}$ in equation (3.36) is a symmetric tensor involving the second derivatives of $g^{\mu\nu}$.

During the period 1914–15 Einstein made many attempts to find the exact form of the suspected relationship between the metric tensor and matter, and

[7] The story of Einstein's quest for the field equations is told in Hoffmann, 1972, Chap. 8.

in 1915 (by which time he had moved to Berlin, leaving Grossmann in Zurich) he published his belief in the equation

$$R^{\mu\nu} = \kappa T^{\mu\nu}, \tag{3.37}$$

where $R^{\mu\nu}$ is the contravariant Ricci tensor. Again this looks plausible, since $R^{\mu\nu}$ is symmetric and contains second derivatives of $g_{\mu\nu}$. However, $R^{\mu\nu}$ does not satisfy $R^{\mu\nu}{}_{;\mu} = 0$, and later in the same year he modified the equation to

$$\boxed{R^{\mu\nu} - \tfrac{1}{2}Rg^{\mu\nu} = \kappa T^{\mu\nu}.} \tag{3.38}$$

The left-hand side of this equation is the Einstein tensor $G^{\mu\nu}$, and we know from Section 3.2 that $G^{\mu\nu}{}_{;\mu} = 0$, so equation (3.38) looks satisfactory in all respects. We shall see in the next section that it gives Poisson's equations as an approximation, and that this approximation allows us to give the coupling constant κ the value $-8\pi G/c^4$. (Note that we now have ten field equations replacing the single field equation of the Newtonian theory.) An alternative form for the field equations (3.38) is

$$R^{\mu\nu} = \kappa(T^{\mu\nu} - \tfrac{1}{2}Tg^{\mu\nu}), \tag{3.39}$$

where $T \equiv T^{\mu}_{\mu}$ (see Exercise 3.5.1).

Recall that $T^{\mu\nu}$ contains all forms of energy and momentum. For example, if there is electromagnetic radiation present, then this must be included in $T^{\mu\nu}$. A region of spacetime in which $T^{\mu\nu} = 0$ is called *empty*, and such a region is therefore not only devoid of matter, but of radiative energy and momentum also. It can be seen from equation (3.39) that the empty spacetime field equations are

$$R^{\mu\nu} = 0. \tag{3.40}$$

Further support for the correctness of the field equations is given by comparing the equation of geodesic deviation with its Newtonian counterpart. With proper time τ as affine parameter, equation (3.35) of geodesic deviation takes the form

$$D^2\xi^\mu/d\tau^2 + R^{\mu}{}_{\sigma\nu\rho}\xi^\nu\dot{x}^\sigma\dot{x}^\rho = 0, \tag{3.41}$$

where $\xi^\mu(\tau)$ is the small "vector" connecting corresponding points on neighboring geodesics. For comparison with its Newtonian counterpart, let us write this as

$$D^2\xi^\mu/d\tau^2 = -K^\mu_\nu\xi^\nu, \tag{3.42}$$

where

$$K^\mu_\nu \equiv R^{\mu}{}_{\sigma\nu\rho}\dot{x}^\sigma\dot{x}^\rho = -R^{\mu}{}_{\sigma\rho\nu}\dot{x}^\sigma\dot{x}^\rho. \tag{3.43}$$

The corresponding situation in Newtonian gravitation theory is two particles moving under gravity on neighboring paths given by $\tilde{x}^i(t)$ and $x^i(t)$. Their equations of motion are

$$d^2\tilde{x}^i/dt^2 = -\delta^{ik}\tilde{\partial}_k V$$

and

$$d^2x^i/dt^2 = -\delta^{ik}\partial_k V,$$

where $\tilde{\partial}_k$ in the first equation indicates that the gradient of the gravitational potential V is evaluated at $\tilde{x}^i(t)$. If we subtract and put $\xi^i(t) \equiv \tilde{x}^i(t) - x^i(t)$, and make use of the fact that, for small ξ^j,

$$\tilde{\partial}_k V = \partial_k V + (\partial_j \partial_k V)\xi^j,$$

then there results

$$d^2\xi^i/dt^2 = -\delta^{ik}(\partial_j\partial_k V)\xi^j,$$

which gives

$$d^2\xi^i/dt^2 = -K^i_j\xi^j, \tag{3.44}$$

where

$$K^i_j \equiv \delta^{ik}\partial_j\partial_k V. \tag{3.45}$$

Equation (3.44) is the Newtonian counterpart of equation (3.42) and brings out the correspondence:

$$K^\mu_\nu \equiv -R^\mu{}_{\sigma\rho\nu}\dot{x}^\sigma\dot{x}^\rho \quad \leftrightarrow \quad K^i_j \equiv \delta^{ik}\partial_j\partial_k V.$$

Now the empty space field equation of Newtonian gravitation is $\nabla^2 V = 0$, or equivalently $K^i_i = 0$. This suggests that in empty spacetime we should have $K^\mu_\mu = 0$, that is, $R_{\sigma\rho}\dot{x}^\sigma\dot{x}^\rho = 0$. Since this should hold for arbitrary tangent vectors $\dot{x}^\mu$ to geodesics we conclude (because $R_{\mu\nu}$ is symmetric) that $K^\mu_\mu = 0$ is equivalent to $R_{\sigma\rho} = 0$. In this way comparison between the equation of geodesic deviation and its Newtonian counterpart lends support to equation (3.40) as the field equations of empty spacetime. Support for the nonempty spacetime field equations (3.38) or (3.39) is given in the next section.

Exercise 3.5

1. By contracting the mixed form

$$R^\mu_\nu - \tfrac{1}{2}R\delta^\mu_\nu = \kappa T^\mu_\nu$$

of equation (3.38) show that $R = -\kappa T$, where $T \equiv T^\mu_\mu$, and hence verify equation (3.39).

3.6 Einstein's equation compared with Poisson's equation

Poisson's equation may be recovered from Einstein's equation by considering its 00-component in the "weak-field" approximation. We use the covariant version of equation (3.39), and so are interested in

$$R_{00} = \kappa(T_{00} - \tfrac{1}{2}Tg_{00}). \tag{3.46}$$

As in Section 2.7, we use a nearly Cartesian coordinate system in which $g_{\mu\nu} = \eta_{\mu\nu} + h_{\mu\nu}$, where products of the $h_{\mu\nu}$ may be neglected; we also assume that the quasi-static condition of that section holds.

Let us assume that our weak gravitational field arises from a perfect field whose particles have (in our coordinate system) speeds v which are small when compared with c, so we take $\gamma = (1 - v^2/c^2)^{-1/2}$ to be one. For most classical distributions (e.g., water, the Sun, or a gas at high pressure) $p/c^2 \ll \rho$, so we take for the stress tensor

$$T_{\mu\nu} = \rho u_\mu u_\nu.$$

This gives $T = \rho c^2$, and equation (3.46) becomes

$$R_{00} = \kappa\rho(u_0 u_0 - \tfrac{1}{2}c^2 g_{00}).$$

But $u_0 \approx c$ and $g_{00} \approx 1$, so we have

$$R_{00} \approx \tfrac{1}{2}\kappa\rho c^2, \tag{3.47}$$

where, from equation (3.13),

$$R_{00} = \partial_0 \Gamma^\mu_{0\mu} - \partial_\mu \Gamma^\mu_{00} + \Gamma^\nu_{0\mu}\Gamma^\mu_{\nu 0} - \Gamma^\nu_{00}\Gamma^\mu_{\nu\mu}. \tag{3.48}$$

In our nearly Cartesian coordinate system the $\Gamma^\mu_{\nu\sigma}$ are small, so we can neglect the last two terms in equation (3.48),[8] and on using the quasi-static condition we have

$$R_{00} \approx -\partial_i \Gamma^i_{00}.$$

But from Section 2.7, we have that in this approximation

$$\Gamma^i_{00} = \tfrac{1}{2}\delta^{ij}\partial_j h_{00},$$

so equation (3.47) reduces to

$$-\tfrac{1}{2}\delta^{ij}\partial_i\partial_j h_{00} \approx \tfrac{1}{2}\kappa\rho c^2.$$

But $\delta^{ij}\partial_i\partial_j = \nabla^2$, and from equation (2.83) $h_{00} = 2V/c^2$, where V is the gravitational potential, so there results

---

[8] This is equivalent to assuming that the derivatives of $h_{\mu\nu}$ are also small (see footnote 11 of Chap. 2), and it is this assumption that makes the field "weak."

$$\nabla^2 V \approx -\tfrac{1}{2}\kappa\rho c^4, \tag{3.49}$$

which corresponds satisfactorily with Poisson's equation, provided we identify the coupling constant κ in Einstein's equation as $-8\pi G/c^4$. Equation (3.49) then becomes

$$\boxed{\nabla^2 V \approx 4\pi G\rho.} \tag{3.50}$$

3.7 The Schwarzschild solution

Is it possible to solve the field equations and thus discover $g_{\mu\nu}$? If one examines how $g_{\mu\nu}$ enters $R^{\mu\nu}$ and $G^{\mu\nu}$, one readily appreciates the high degree of nonlinearity possessed by the equations, so any solution will not be easy to obtain. The problem becomes easier if one looks for special solutions, for example those representing spacetimes possessing symmetries, and the first exact solution, obtained by K. Schwarzschild in 1916 (astonishingly in the trenches of the First World War), is of this type.

What Schwarzschild sought was the metric tensor field representing the static spherically symmetric gravitational field in the empty spacetime surrounding some massive spherical object like a star. His guiding assumptions were[9]

(a) that the field was static,
(b) that the field was spherically symmetric,
(c) that the spacetime was empty,
(d) that the spacetime was asymptotically flat.

He also assumed that spacetime could be coordinatized by (t, r, θ, ϕ), where t was a timelike coordinate,[10] θ and ϕ were polar angles picking out radial directions in the usual manner, and r was some radial coordinate. He then postulated

$$c^2 d\tau^2 = A(r)dt^2 - B(r)dr^2 - r^2 d\theta^2 - r^2 \sin^2\theta \, d\phi^2 \tag{3.51}$$

as a form for the line element, where $A(r)$ and $B(r)$ were some unknown functions of r to be obtained by solving the field equations.

The fact that none of the $g_{\mu\nu}$ depends on t expresses his assumption (a), and the fact that the surfaces given by r, t constant have line elements given by

$$ds^2 = r^2(d\theta^2 + \sin^2\theta \, d\phi^2) \tag{3.52}$$

[9] Birkhoff's theorem (see, e.g., Misner, Thorne, and Wheeler, 1973, §32.2) states that (b) and (c) imply (a), so condition (a) is, in fact, redundant.

[10] A coordinate is *timelike* if the tangent vector to its coordinate curve is timelike. *Null* and *spacelike coordinates* are correspondingly defined.

(and so have the geometry of spheres, as Exercise 1.6.2 confirms) expresses his assumption (b). Assumption (c) means that $A(r)$ and $B(r)$ are to be found using the empty spacetime field equations $R_{\mu\nu} = 0$, while assumption (d) gives boundary conditions on A and B, namely

$$A(r) \to c^2 \quad \text{and} \quad B(r) \to 1 \quad \text{as} \quad r \to \infty \tag{3.53}$$

(see Example 1.3.1). Note that because $B(r)$ is not necessarily 1, we cannot assume that r is radial distance. In fact the line element (3.52) shows that a surface given by r, t constant has surface area $4\pi r^2$, and at the moment this is the only meaning we can give to r; further discussion on its meaning is given in the next chapter.

Let us now retrace Schwarzschild's solution of the field equations. The idea is to use $g_{\mu\nu}$, obtained from the line element (3.51) as a trial solution for the empty spacetime field equations. As with all trial solutions, the main justification for it is that it works. From equation (3.13) we have

$$R_{\mu\nu} \equiv \partial_\nu \Gamma^\sigma_{\mu\sigma} - \partial_\sigma \Gamma^\sigma_{\mu\nu} + \Gamma^\rho_{\mu\sigma}\Gamma^\sigma_{\rho\nu} - \Gamma^\rho_{\mu\nu}\Gamma^\sigma_{\rho\sigma},$$

and from Problem 2.7 we have

$$\begin{aligned}
&\Gamma^0_{01} = A'/2A, && \Gamma^1_{00} = A'/2B, && \Gamma^1_{11} = B'/2B, \\
&\Gamma^1_{22} = -r/B, && \Gamma^1_{33} = -(r\sin^2\theta)/B, && \Gamma^2_{12} = 1/r, \\
&\Gamma^2_{33} = -\sin\theta\cos\theta, && \Gamma^3_{13} = 1/r, && \Gamma^3_{23} = \cot\theta,
\end{aligned}$$

all other connection coefficients being zero. Here we have labeled the coordinates according to $x^0 \equiv t$, $x^1 \equiv r$, $x^2 \equiv \theta$, $x^3 \equiv \phi$, and a prime denotes differentiation with respect to r. Tedious substitution then shows that $R_{\mu\nu} = 0$ gives (see Exercise 3.7.1):

$$R_{00} = -\frac{A''}{2B} + \frac{A'}{4B}\left(\frac{A'}{A} + \frac{B'}{B}\right) - \frac{A'}{rB} = 0, \tag{3.54}$$

$$R_{11} = \frac{A''}{2A} - \frac{A'}{4A}\left(\frac{A'}{A} + \frac{B'}{B}\right) - \frac{B'}{rB} = 0, \tag{3.55}$$

$$R_{22} = \frac{1}{B} - 1 + \frac{r}{2B}\left(\frac{A'}{A} - \frac{B'}{B}\right) = 0, \tag{3.56}$$

$$R_{33} = R_{22}\sin^2\theta = 0. \tag{3.57}$$

Fortunately, $R_{\mu\nu} = 0$ identically for $\mu \neq \nu$.

Of these four equations, only the first three are useful. Adding B/A times equation (3.54) to equation (3.55) gives (after some manipulation)

$$A'B + AB' = 0,$$

which implies that $AB =$ constant. We can identify this constant as c^2 from the boundary condition (3.53), so

$$AB = c^2 \quad \text{and} \quad B = c^2/A.$$

Substitution in equation (3.56) then gives $A + rA' = c^2$, which is equivalent to

$$d(rA)/dr = c^2.$$

Integrating, we have

$$rA = c^2(r + k),$$

where k is constant, so

$$A(r) = c^2(1 + k/r) \quad \text{and} \quad B(r) = (1 + k/r)^{-1}.$$

In solving for A and B we have used only the sum of equations (3.55) and (3.56), but not the equations separately. However, it is a simple matter to check that, with these forms for A and B, the equations are satisfied separately. Thus we have solved the field equations, and obtained Schwarzschild's solution in the form

$$c^2 d\tau^2 = c^2(1 + k/r)dt^2 - (1 + k/r)^{-1}dr^2 - r^2 d\theta^2 - r^2 \sin^2\theta \, d\phi^2,$$

where k is a constant, which we now proceed to identify. It clearly must in some way represent the mass of the object producing the gravitational field.

In the region of spacetime where k/r is small (i.e., in the asymptotic region) the line element differs but little from that of flat spacetime in spherical coordinates, so here r is approximately radial distance, the approximation getting better as $r \to \infty$. Moreover, if we put

$$x^0 \equiv ct, \quad x^1 \equiv r\sin\theta\cos\phi, \quad x^2 \equiv r\sin\theta\sin\phi, \quad x^3 \equiv r\cos\theta, \tag{3.58}$$

we obtain a metric tensor of the form $g_{\mu\nu} = \eta_{\mu\nu} + h_{\mu\nu}$ (see Exercise 3.7.2), where in the asymptotic region the $h_{\mu\nu}$ are small, and $h_{00} = k/r$. But in this region, where r is approximately radial distance, the corresponding Newtonian potential is $V = -MG/r$, where M is the mass of the body producing the field, and G is the gravitational constant. Since $h_{00} \equiv 2V/c^2$, we conclude that $k = -2MG/c^2$, and Schwarzschild's solution for the empty spacetime outside a spherical body of mass M is

$$c^2 d\tau^2 = c^2(1 - 2MG/c^2 r)dt^2 - (1 - 2MG/c^2 r)^{-1}dr^2 - r^2 d\theta^2 - r^2 \sin^2\theta \, d\phi^2. \tag{3.59}$$

That is,

$$[g_{\mu\nu}] = \begin{bmatrix} 1 - 2MG/c^2 r & 0 & 0 & 0 \\ 0 & -(1 - 2MG/c^2 r)^{-1} & 0 & 0 \\ 0 & 0 & -r^2 & 0 \\ 0 & 0 & 0 & -r^2\sin^2\theta \end{bmatrix}.$$

This solution is the basis of our discussions in the next chapter.

Exercises 3.7

1. Check the expressions given for $R_{\mu\nu}$ in equations (3.54)–(3.57), and that $R_{\mu\nu} = 0$ for $\mu \neq \nu$.

2. If in Schwarzschild's solution we introduce coordinates x^μ defined by equation (3.58), what form does $g_{\mu\nu}$ take?

Problems 3

1. Show that in a two-dimensional Riemannian manifold all components of R_{ABCD} are either zero or $\pm R_{1212}$.
   In terms of the usual polar angles (see Exercise 1.6.2) the metric tensor field of a sphere of radius a is given by
$$[g_{AB}] = \begin{bmatrix} a^2 & 0 \\ 0 & a^2 \sin^2\theta \end{bmatrix}.$$
   Show that $R_{1212} = a^2 \sin^2\theta$, and hence deduce that
$$[R_{AB}] = \begin{bmatrix} -1 & 0 \\ 0 & -\sin^2\theta \end{bmatrix}$$
   and $R = -2/a^2$.

2. In a certain N-dimensional Riemannian manifold the covariant curvature tensor may be expressed as
$$R_{abcd} = g_{ac}S_{bd} + g_{bd}S_{ac} - g_{ad}S_{bc} - g_{bc}S_{ad},$$
   where S_{ab} is a type $(0,2)$ tensor. Show that, provided $N > 2$, $S_{ab} = S_{ba}$, and that, provided $N > 3$, $S_{ab;c} = S_{ac;b}$.

3. *Dust* is a fluid without internal stress or pressure, so its stress tensor is $T^{\mu\nu} \equiv \rho u^\mu u^\nu$. Show that $T^{\mu\nu}{}_{;\mu} = 0$ implies that the dust particles follow geodesics.

4. Let
$$E^{\mu\nu} \equiv -\mu_0^{-1}[F^{\rho\mu}F_\rho{}^\nu - \tfrac{1}{4}g^{\mu\nu}(F_{\rho\sigma}F^{\rho\sigma})],$$
   where $F^{\mu\nu}$ is the free-space electromagnetic field tensor. Show that, by virtue of Maxwell's equations,
$$E^{\mu\nu}{}_{;\mu} = F^\nu{}_\rho j^\rho,$$
   where j^ρ is the 4-current density.
   If the stress tensor for a charged unstressed fluid in free space is defined to be

$$T^{\mu\nu} \equiv \mu u^\mu u^\nu + E^{\mu\nu},$$

where μ is its proper density (rather than ρ, to avoid confusion with charge density) and u^μ its world velocity, show that $T^{\mu\nu}{}_{;\mu} = 0$ by virtue of the continuity equation (for matter) and the equation of motion of the fluid. (See Sec. A.8 for the relevant definitions and equations, but adapt them to curved spacetime.)

5. Let O, A, B, C be four points in a manifold with coordinates x^a, $x^a + \xi^a$, $x^a + \eta^a$, $x^a + \xi^a + \eta^a$, respectively, where ξ^a and η^a are small, so that OACB is a small "parallelogram," as shown in Figure 3.4. Working to second order of small quantities, obtain an expression for the vector $\lambda^a_{(\text{via A})}$ obtained by transporting a vector λ^a_{O} at O to C along the edges OA and AC, and the corresponding expression $\lambda^a_{(\text{via B})}$ for transporting it along OB and BC.
   Show that the difference $\Delta\lambda^a \equiv \lambda^a_{(\text{via A})} - \lambda^a_{(\text{via B})}$ is (to second order) given by
   $$\Delta\lambda^a = -\tfrac{1}{2}(R^a{}_{bcd})_{\text{O}}\lambda^b_{\text{O}}(\xi^c\eta^d - \xi^d\eta^c).$$
   (The quantity $\xi^c\eta^d - \xi^d\eta^c$ is related to the area of the parallelogram and this is essentially the same result as equation (3.30).)

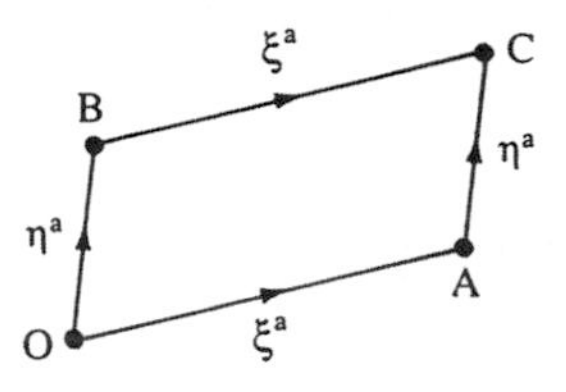

Fig. 3.4. A small parallelogram.

6. In Section 3.7 we remarked that the field equations of general relativity were nonlinear. Explain why this is not surprising.
   Does the principle of superposition hold for solutions of the field equations?
   If not, why not?

7. Show that the Schwarzschild line element (3.59) may be put into the *isotropic form*

$$c^2 d\tau^2 = c^2\left(1 - \frac{GM}{2\rho c^2}\right)^2\left(1 + \frac{GM}{2\rho c^2}\right)^{-2} dt^2 - \left(1 + \frac{GM}{2\rho c^2}\right)^4 (d\rho^2 + \rho^2 d\theta^2 + \rho^2 \sin^2\theta d\phi^2),$$

where the new coordinate ρ is defined by

$$r \equiv \rho \left(1 + \frac{GM}{2\rho c^2} \right)^2 .$$

4

Physics in the vicinity of a massive object

4.0 Introduction

In Chapter 3 we obtained the static spherically symmetric solution of Schwarzschild, and identified it as representing the gravitational field surrounding a spherically symmetric body of mass M situated in an otherwise empty spacetime. This solution is asymptotically flat, and in no way incorporates the gravitational effects of distant matter in the Universe. Nevertheless, it seems reasonable to adopt it as a model for the gravitational field in the vicinity of a spherical massive object such as a star, where the star's mass is the principal contributor to the gravitational field.

Suppose, somehow, that we are watching the trajectories of laser beams and particles in the vicinity of a star, all of these trajectories being displayed on a large television screen with the star a rather small dot in the middle. If there is a "mass-control" knob which controls the mass M of the star, we are really asking in this chapter what happens when we turn the knob so as to increase M. With M turned right down to zero, the Schwarzschild line element reduces to that of flat spacetime in spherical coordinates. The coordinates t and r then have simple physical meanings: t is the time as measured by clocks which are stationary in the reference system employed, and r is the radial distance from the origin. Turning M up introduces curvature, so that spacetime is no longer flat, and there is no reason to assume that the coordinates have the simple physical meanings they had in flat spacetime. The relationship between coordinates and physically observable quantities is investigated in Section 4.1.

The Schwarzschild solution is the basis for four of the tests of general relativity listed in the Introduction, namely perihelion advance, the bending of light, time delay in radar sounding, and the geodesic effect.[1] The third of these may be discussed without a detailed knowledge of the geodesics, and this we do in Section 4.2. The question of perihelion advance and the bending

[1]They could therefore be the tests of any other theory of gravitation which yielded the Schwarzschild solution. See also Biswas, 1994.

of light does require some knowledge of the geodesics, and these matters are discussed in Sections 4.4 to 4.6.

Spectral shift is more a test of the principle of equivalence than of general relativity, but inasmuch as the latter is based on the former, it does yield a test of the general theory, and it is appropriate to discuss it in the context of the Schwarzschild solution. This we do in Section 4.3.

The fifth test mentioned in the Introduction is presently being measured. Satellite experiments began in April 2004, and (at the time of writing) are still under way. We consider the theory behind this test in Section 4.7.

Before embarking on our detailed discussion, let us say something about the ranges of the coordinates appearing in the Schwarzschild solution. Inasmuch as the metric tensor components $g_{\mu\nu}$ do not depend on t, the solution is static, and we can take $-\infty < t < \infty$. The coordinates θ and ϕ pick out radial directions as in spherical coordinates in Euclidean space and so have the ranges $0 \leq \theta \leq \pi$, $0 \leq \phi < 2\pi$. However, no trouble will be caused if we let ϕ extend beyond the quoted range, provided we identify the event with coordinates (t, r, θ, ϕ_1) with that with coordinates (t, r, θ, ϕ_2) whenever ϕ_1 and ϕ_2 differ by a multiple of 2π. The radial coordinate r can decrease from infinity until it reaches either the value r_B corresponding to the boundary of the object, or the value $2GM/c^2$, if r_B is not reached first. The reason for the first lower bound is that the solution we have obtained is the *exterior solution*, valid only where the empty spacetime field equations hold. The reason for the second one is that as r tends to $2GM/c^2$, the component g_{11} of the metric tensor tends to infinity (see the line element (3.59)). So the range of r is $r_B < r < \infty$ or $2GM/c^2 < r < \infty$, as appropriate. Should r decrease to $2GM/c^2$ without r_B being reached, then the object is a *black hole*, and we discuss this situation in Section 4.8. In order to be able to step over the threshold at $r = 2GM/c^2$, we must introduce a coordinate system different from that used to derive Schwarzschild's solution.

The chapter finishes with a brief consideration of some other coordinate systems used in connection with the Schwarzschild solution, and a look at the more general case of rotating objects.

4.1 Length and time

The Schwarzschild spacetime has the line element

$$\boxed{c^2 d\tau^2 = (1 - 2m/r)c^2 dt^2 - (1 - 2m/r)^{-1} dr^2 - r^2 d\theta^2 - r^2 \sin^2\theta \, d\phi^2,} \tag{4.1}$$

where for convenience we have put $m \equiv GM/c^2$. If we take a slice given by $t =$ constant we obtain a three-dimensional manifold with the line element

$$ds^2 = (1 - 2m/r)^{-1} dr^2 + r^2 d\theta^2 + r^2 \sin^2\theta \, d\phi^2, \tag{4.2}$$

obtained by putting $dt = 0$ in equation (4.1). Putting

$$ds^2 = \tilde{g}_{ij} dx^i dx^j \quad (i, j = 1, 2, 3, \ x^1 \equiv r, \ x^2 \equiv \theta, \ x^3 \equiv \phi),$$

so that $\tilde{g}_{ij} \equiv -g_{ij}$, we see that $\tilde{g}_{ij}$ is a positive-definite metric tensor field on this 3-manifold, so the slice is a *space* rather than a spacetime. Moreover, no $\tilde{g}_{ij}$ depends on t, so the spaces given by $t =$ constant have an enduring permanence which allows us to refer to events with the same r, θ, ϕ coordinates, but different t coordinates, as occurring at the *same point* in space. We may also speak of *fixed points* in space. This splitting of spacetime into space and time is possible in any static spacetime, but is not a feature of spacetimes in general, and it should be borne in mind that because of this there are fewer problems of definition and identification in static spacetimes than in nonstatic ones.

If we turn M (or m) down to zero, then the line element (4.1) becomes that of flat spacetime in spherical coordinates, while the line element (4.2) becomes that of Euclidean space in spherical coordinates (see Example 1.3.1). Turning M up introduces distortion into both spacetime and space, so that neither is flat. This distortion is effectively measured by the dimensionless quantity $2m/r$ occurring in the two line elements, and is greatest when r is least, that is, when $r = r_B$, the value of r at the boundary of the object, assuming it is not a black hole. For the Earth, $2m/r_B$ is about 10^{-9}, for the Sun it is about 10^{-6}, but for a proton it is as low as 10^{-36}. For white dwarfs, however, it is not negligible, and for typical neutron stars it can be as much as 10%–15%.

In the flat spacetime given by $m = 0$, the coordinate r is simply the distance from the origin, but if we turn m up, things are not so simple, for r then has a positive lower bound (see previous section) and our origin has disappeared. What then does r represent? If we take the sphere in space given by $r =$ constant, its line element is

$$dL^2 = r^2(d\theta^2 + \sin^2\theta \, d\phi^2), \tag{4.3}$$

obtained by putting $dr = 0$ in the line element (4.2). It follows that this sphere has the two-dimensional geometry of a sphere of radius r embedded in Euclidean space (see Exercise 1.6.2), and just as in the flat space, infinitesimal tangential distances are given by

$$dL \equiv r(d\theta^2 + \sin^2\theta \, d\phi^2)^{1/2}. \tag{4.4}$$

But what about radial distances given by θ and ϕ constant? The line element shows that for these the infinitesimal radial distance is

$$\boxed{dR \equiv (1 - 2m/r)^{-1/2} dr,} \tag{4.5}$$

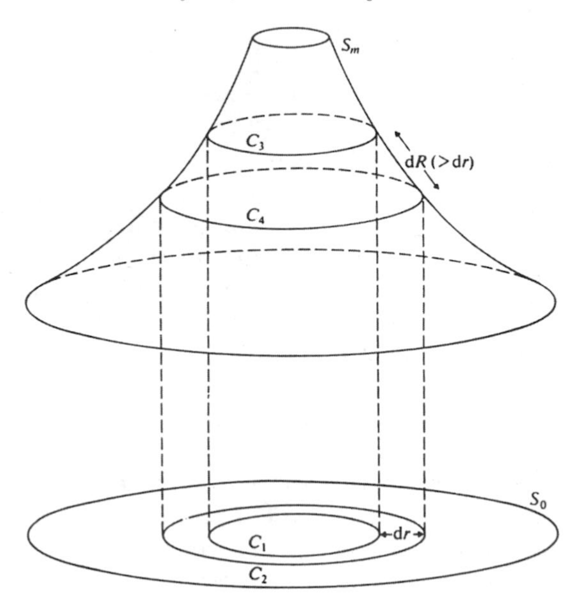

Fig. 4.1. Radial distance in the Schwarzschild geometry. Here $dR = f(r)dr$, where $f(r) = (1 - 2m/r)^{-1/2}$. The curve gives the value of $f(r)$, which tends to infinity as $r \to 2m$ and to unity as $r \to \infty$.

so $dR > dr$ and r *no longer measures radial distance.*[2] The apparent incompatibility of the distances (4.4) and (4.5) is explained by the curvature of space. In Figure 4.1, the flat disk S_0 represents a portion of flat space (m turned down to zero), while the curved surface S_m represents a portion of the curved space (m turned up). The circles C_1 and C_3 represent spheres having the geometry of a sphere of radius r in Euclidean space, while C_2 and C_4 represent neighboring spheres having the geometry of a sphere of radius $r + dr$ in Euclidean space. However, it is only in the flat space represented by S_0 that the measured radial distance between the spheres is dr. In the curved space represented by S_m the measured distance is dR given by equation (4.5), and this exceeds dr. If we were to measure the circumference of a great circle of the sphere $r =$ constant using small measuring rods, then the same number of rods would be needed in flat space as in the curved space. On the other hand, if we were to measure the radial distance between points with radial coordinates r_1 and r_2, then more rods would be needed in the curved space than in the flat space (see Example 4.1.2, at the end of this section, and Fig. 4.2).

[2]We emphasize that coordinates are nothing more than "street numbers": there is no reason to believe that the measured distance between 36th St. and 37th St. is equal to the measured distance between 40th St. and 41st St. For the measured distance, we need to integrate ds, where ds is given by the line element, as explained in Sec. 1.9.

Something similar happens with measuring radii and circumferences on the surface of the Earth, as the following example shows.

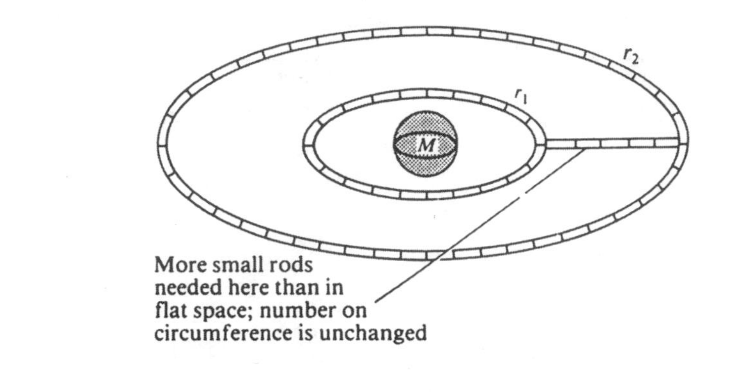

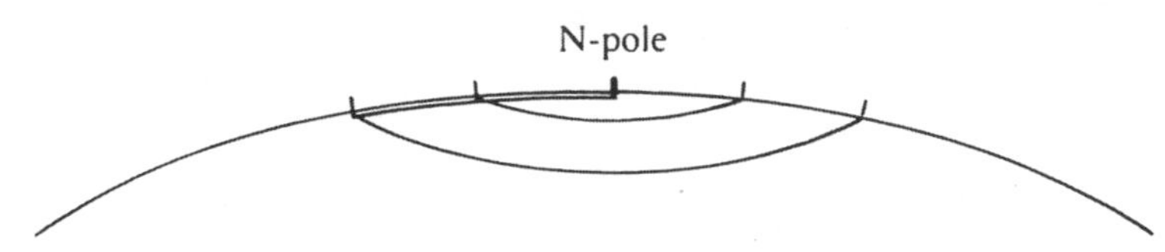

Fig. 4.2. Measuring distances in the Schwarzschild geometry (upper figure) and on the surface of the Earth (lower figure). In each case, the circumference of the circle is less than $2\pi R$.

Example 4.1.1

If we mark out a circle of radius 1 m, by joining together all points on the surface of the Earth that are a distance 1 m from the North Pole, then the circumference of the resulting circle is 2π m. However, if we mark out the Equator in a similar way by joining together all points that are a distance R from the North Pole (where R is the distance from the North Pole to the Equator measured over the surface of the Earth), then the circumference of the resulting circle is not $2\pi R$: fewer measuring rods are needed to cover the circumference of this circle on the Earth than are needed to cover a circle on a flat surface constructed in the same way.

Let us now turn our attention to time. One of the basic assumptions taken over from special relativity is that clocks record proper time intervals along their world lines. Infinitesimal proper time intervals are given by the line element (4.1), and for a clock at a fixed distance in space (r, θ, ϕ constant) this gives

$$\boxed{d\tau = (1 - 2m/r)^{1/2} dt.} \tag{4.6}$$

So in flat spacetime (m turned down to zero) $d\tau = dt$, and such a clock records the coordinate time t. However, in the curved spacetime (m turned up) $d\tau < dt$, and fixed clocks do not record coordinate time.

Special relativity v constant	Schwarzschild r variable
$dl = dl_0(1 - v^2/c^2)^{1/2}$	$dr = dR(1 - 2m/r)^{1/2}$
$dt = d\tau(1 - v^2/c^2)^{-1/2}$	$dt = d\tau(1 - 2m/r)^{-1/2}$

Table 4.1. Comparison of length and time.

It is tempting to compare the relations (4.5) and (4.6) with similar formulae from special relativity (see Table 4.1). However, there are important differences. The square root in the Schwarzschild solution involves the coordinate r and therefore depends on position, whereas that in the special-relativistic case is constant. Moreover, if we used a different coordinate system for describing the Schwarzschild solution, for example, isotropic coordinates (see Problem 3.7), then the expressions would have different forms altogether.

One final point to note is that as $r \to \infty$, $dR \to dr$ in equation (4.5) and $d\tau \to dt$ in equation (4.6), so asymptotically the coordinate distance dr coincides with the actual distance dR, and the coordinate time dt with the proper time $d\tau$.

Examples 4.1.2

(a) If a stick of length 1 m lies radially in the field of a star where m/r is 10^{-2}, what coordinate distance does it take up?
Answer. From equation (4.5), the coordinate distance is

$$\Delta r = (1 - 2m/r)^{1/2}\Delta R = (1 - 2 \times 10^{-2})^{1/2}\,\text{m} \approx 0.99\,\text{m}.$$

(b) A long stick is lying radially in the field of a spherical object of mass M. If the r coordinates of its ends are r_1 and r_2 ($r_1 < r_2$), what is its length?
Answer. Since the stick is long, we must integrate the length differential dR. This gives the length as

$$\int_{r_1}^{r_2} (1 - 2GM/rc^2)^{-1/2} dr = \Big[r^{1/2}(r - 2GM/c^2)^{1/2} + (2GM/c^2)\ln\{r^{1/2} + (r - 2GM/c^2)^{1/2}\}\Big]_{r_1}^{r_2}. \quad (4.7)$$

Note that when $GM/rc^2 \ll 1$ this reduces to $r_2 - r_1$.

Exercise 4.1

1. Check the integral (4.7).

4.2 Radar sounding

Suppose that an observer is at a fixed point in space in the field of a massive object, and that directly between him and this object there is a small body. We can imagine the observer sending radar pulses in a radial direction towards the body, these pulses being reflected by it and subsequently received by the observer at some later time. Let us calculate the time lapse between transmission and subsequent reception of a radar pulse by the observer. If the

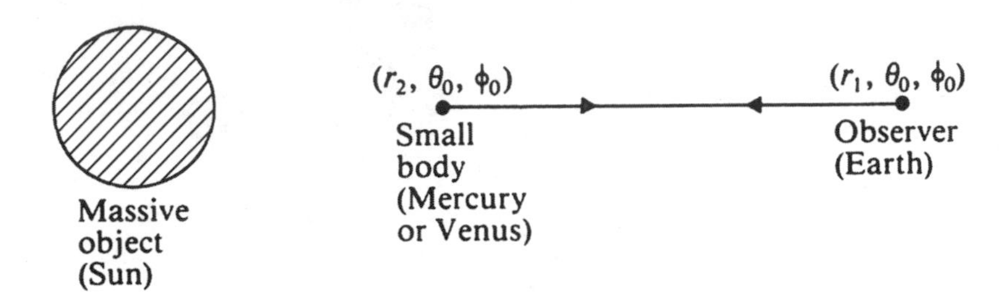

Fig. 4.3. Radar sounding.

spatial coordinates of the observer are (r_1, θ_0, ϕ_0), then those of the body are (r_2, θ_0, ϕ_0), with $r_2 < r_1$ (see Fig. 4.3). The radar pulses travel in a radial direction with the speed of light, so putting $d\tau = 0$ and $d\theta = d\phi = 0$ in the line element (4.1), we have

$$(1 - 2m/r)c^2 dt^2 = (1 - 2m/r)^{-1} dr^2,$$

which gives

$$dr/dt = \pm c(1 - 2m/r).$$

This expression gives the *coordinate speed of light in the radial direction.* The coordinate time for the whole trip is therefore

$$\begin{aligned} \Delta t &= -\frac{1}{c}\int_{r_1}^{r_2} \frac{dr}{1 - 2m/r} + \frac{1}{c}\int_{r_2}^{r_1} \frac{dr}{1 - 2m/r} \\ &= \frac{2}{c}\int_{r_2}^{r_1} \frac{dr}{1 - 2m/r}. \end{aligned} \tag{4.8}$$

However, we require the proper time lapse as measured by the observer at r_1. (The observer's clock records *proper* time.) From equation (4.6) this is

$$\begin{aligned}\Delta\tau &= \left(1 - \frac{2m}{r_1}\right)^{1/2} \Delta t = \frac{2}{c}\left(1 - \frac{2m}{r_1}\right)^{1/2} \int_{r_2}^{r_1} \frac{dr}{1 - 2m/r} \\ &= \frac{2}{c}\left(1 - \frac{2m}{r_1}\right)^{1/2} \left(r_1 - r_2 + 2m \ln \frac{r_1 - 2m}{r_2 - 2m}\right).\end{aligned} \tag{4.9}$$

The distance traveled by the radar pulse is twice the integral (4.7), so on the basis of the *classical theory* one would expect a round-trip time of

$$\Delta\tilde{\tau} = (2 \times \text{integral (4.7)})/c,$$

and $\Delta\tau \neq \Delta\tilde{\tau}$. The difference forms the basis of the so-called fourth test of general relativity, in which the massive object is the Sun, the observer is on Earth, and the small body is either Mercury or Venus. Of course, the Earth is not at a fixed point in space, but we neglect its motion during the travel time of a pulse.

With M equal to the mass of the Sun, and r_1 and r_2 the orbital values of r for the Earth and the other planet involved, $2m/r$ is small for $r_2 < r < r_1$, and this leads to the approximations:

$$\begin{aligned}\Delta\tau &\approx (2/c)\left[r_1 - r_2 - m(r_1 - r_2)/r_1 + 2m \ln(r_1/r_2)\right], \\ \Delta\tilde{\tau} &\approx (2/c)\left[r_1 - r_2 + m \ln(r_1/r_2)\right].\end{aligned} \tag{4.10}$$

Hence there is a general-relativity-induced delay

$$\Delta\tau - \Delta\tilde{\tau} \approx \frac{2GM}{c^3}\left(\ln \frac{r_1}{r_2} - \frac{r_1 - r_2}{r_1}\right). \tag{4.11}$$

For inferior conjunction, with the planet between the Earth and the Sun, this time delay is too small to measure. However, it is increased considerably if they are in superior conjunction, and an experiment was suggested by Shapiro in 1964 which involved radar sounding of Mercury and Venus as they passed behind the Sun.[3] The analysis above will not cope with this situation, where the Sun prevents direct radar sounding in the radial direction.

In using a time-delay formula such as formula (4.11), or its modification for nonradial motion, one should ask oneself certain questions. Can the Earth's motion in its orbit be ignored? Can the Earth's own gravitational field be ignored? Can accepted planetary distances be used for r_1 and r_2, which are after all coordinate values and not distances (see Sec. 4.1)? What is the effect of dispersion by the solar wind? When such considerations have been taken into account, one may go ahead and perform one's experiment to check the theoretical with the observed time delay. Recent tests using Mercury and Venus have yielded agreement to well within the experimental uncertainty of 20% in 1968, and 5% in 1971, while tests using the spacecrafts *Mariner 6* and

[3]See Shapiro, 1964.

7 have yielded agreement to well within the experimental uncertainty of 3% in 1975.[4]

Exercise 4.2

1. Check the approximations (4.10).

4.3 Spectral shift

Suppose that a signal is sent from an emitter at a fixed point (r_E, θ_E, ϕ_E), that it travels along a null geodesic and is received by a receiver at a fixed point (r_R, θ_R, ϕ_R). If t_E is the coordinate time of emission and t_R the coordinate time of reception, then the signal passes from the event with coordinates $(t_E, r_E, \theta_E, \phi_E)$ to the event with coordinates $(t_R, r_R, \theta_R, \phi_R)$ (see Fig. 4.4). Let u be an affine parameter along the null geodesic with $u = u_E$ at the event of emission and $u = u_R$ at the event of reception. Since the geodesic is null,

$$(1 - 2m/r)c^2(dt/du)^2 = (1 - 2m/r)^{-1}(dr/du)^2 + r^2(d\theta/du)^2 + r^2\sin^2\theta(d\phi/du)^2,$$

so

$$\frac{dt}{du} = \frac{1}{c}\left[\left(1 - \frac{2m}{r}\right)^{-1} \tilde{g}_{ij}\frac{dx^i}{du}\frac{dx^j}{du}\right]^{1/2}, \qquad (4.12)$$

where $\tilde{g}_{ij} = -g_{ij}$. On integrating we have

$$t_R - t_E = \frac{1}{c}\int_{u_E}^{u_R}\left[\left(1 - \frac{2m}{r}\right)^{-1} \tilde{g}_{ij}\frac{dx^i}{du}\frac{dx^j}{du}\right]^{1/2} du.$$

The integral on the right-hand side depends only on the path through space, so with a spatially fixed emitter and a spatially fixed receiver, $t_R - t_E$ is the same for all signals sent. So for two signals we have

$$t_R^{(1)} - t_E^{(1)} = t_R^{(2)} - t_E^{(2)},$$

giving

$$\Delta t_R = t_R^{(2)} - t_R^{(1)} = t_E^{(2)} - t_E^{(1)} = \Delta t_E. \qquad (4.13)$$

That is, the coordinate time difference at the point of emission equals the coordinate time difference at the point of reception. However, the clock of an observer situated at the point of emission records proper time and not coordinate time, the two being related by a finite version of equation (4.6). This gives a proper time interval

[4] See Shapiro, 1968; Shapiro et al., 1971; and Anderson et al., 1975.

$$\Delta\tau_E = (1 - 2m/r_E)^{1/2}\Delta t_E,$$

and similarly

$$\Delta\tau_R = (1 - 2m/r_R)^{1/2}\Delta t_R.$$

Since $\Delta t_R = \Delta t_E$, we have

$$\frac{\Delta\tau_R}{\Delta\tau_E} = \left[\frac{1 - 2m/r_R}{1 - 2m/r_E}\right]^{1/2}. \tag{4.14}$$

Equation (4.14) is the basis of the gravitational spectral-shift formula, which we shall now derive.

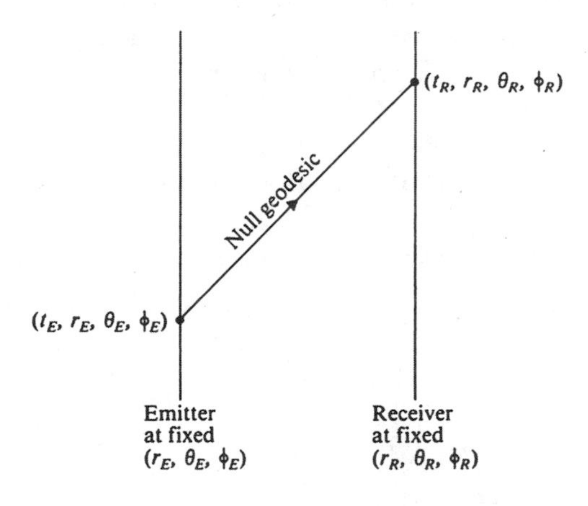

Fig. 4.4. Spacetime diagram illustrating emission and reception of a signal.

Suppose the emitter is a pulsating atom, and that in the proper time interval $\Delta\tau_E$ it emits n pulses. An observer situated at the emitter will assign to the atom a frequency of pulsation $\nu_E \equiv n/\Delta\tau_E$, and this is the *proper frequency* of the pulsating atom. An observer situated at the receiver will see these n pulses in a proper time interval $\Delta\tau_R$ (see Fig. 4.5), and therefore assign a frequency $\nu_R \equiv n/\Delta\tau_R$ to the pulsating atom. Since $\Delta\tau_R \neq \Delta\tau_E$ the observed frequency differs from the proper frequency. In fact, equation (4.14) gives

$$\boxed{\frac{\nu_R}{\nu_E} = \left[\frac{1 - 2m/r_E}{1 - 2m/r_R}\right]^{1/2} = \left[\frac{1 - 2GM/r_E c^2}{1 - 2GM/r_R c^2}\right]^{1/2},} \tag{4.15}$$

on putting $m \equiv GM/c^2$. If $r_E c^2 \gg 2GM$ and $r_R c^2 \gg 2GM$, then this reduces to

$$\frac{\nu_R}{\nu_E} \approx 1 + \frac{GM}{c^2}\left(\frac{1}{r_R} - \frac{1}{r_E}\right). \tag{4.16}$$

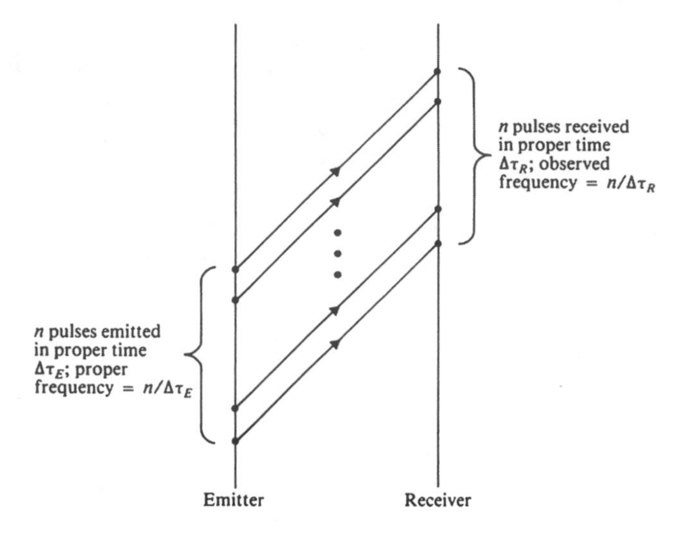

Fig. 4.5. Proper and observed frequencies.

From this we can obtain the fractional shift

$$\frac{\Delta\nu}{\nu_E} \equiv \frac{\nu_R - \nu_E}{\nu_E} \approx \frac{GM}{c^2}\left(\frac{1}{r_R} - \frac{1}{r_E}\right). \tag{4.17}$$

If the emitter is nearer the massive object than the receiver is, then $1/r_R < 1/r_E$, and the shift is towards the red, but if the receiver is nearer the massive object, then it is towards the blue.

There are two points relating to the formulae of spectral shift which are worth noting. The first is that the formula (4.15) generalizes to any static spacetime; that is, one for which there exists a timelike coordinate t which gives a splitting of the line element into the form

$$c^2 d\tau^2 = g_{00}(x^k)dt^2 + g_{ij}(x^k)dx^i dx^j.$$

In such a spacetime it makes sense to talk of fixed points in space, and following an argument analogous to that above leads to

$$\nu_R/\nu_E = \left[g_{00}(x_E^k)/g_{00}(x_R^k)\right]^{1/2}, \tag{4.18}$$

where x_E^k are the spatial coordinates of the emitter and x_R^k are those of the receiver.

The second point is that the version (4.17) may be derived using an eclectic argument not based on general relativity. Suppose for the sake of argument that the emitter is nearer the massive object than the receiver is. Then in traveling from the emitter to the receiver a photon suffers a loss in "intrinsic" energy equal to its gain in gravitational potential energy. The loss in intrinsic energy is $h(\nu_E - \nu_R)$, while the gain in gravitational potential energy is

$$\frac{h\nu_E GM}{c^2}\left(\frac{1}{r_E}-\frac{1}{r_R}\right),$$

on assigning the mass $h\nu_E/c^2$ to the photon. Equating these leads to the fractional-shift formula (4.17). This formula assumes that the photon's energy has both inertial and gravitational mass, and depends in an essential way on the equivalence principle.

Terrestrial experiments confirming the formula (4.17) were performed in 1960 by Pound and Rebka using a vertical separation of 22.5 m in the Jefferson Physics Laboratory at Harvard.[5] The formula should also be amenable to testing by observing the spectra of stars. For an observer on Earth m/r_R is negligible, and the effect depends essentially on m/r_E. Since the observed spectrum is that of atoms on the surface of the star, the effect is greatest for dense objects, such as white dwarfs, for which m/r_E is large. However, data concerning stellar masses and radii are not usually accurate enough for such observations to compete with terrestrial ones. Moreover, the random motion of the radiating atoms produces Doppler shifts which broaden the spectral lines, making it difficult to obtain an accurate value for the gravitational shift.

The following example makes use of the spectral-shift formula (4.15).

Example 4.3.1

The wavelength of a helium–neon laser is measured inside a *Skylab* freely floating far out in deep space, and is found to be 632.8 nm. What wavelength would an experimenter measure (see Fig. 4.6) if:

(a) he and the laser fell freely together towards a neutron star?
(b) he remained in the freely floating *Skylab* while the laser transmitted radially from the surface of the neutron star of mass 10^{30} kg and radius $r_B = 10^4$ m?
(c) he were beside the laser, both on the surface of the neutron star?
(d) he were on the surface of the neutron star while the laser was back in the distant *Skylab*?

Answer.

(a) Since the observer is at rest relative to the laser he observes its proper wavelength as determined in the *Skylab*, namely 632.8 nm.
(b) The wavelength version of formula (4.15) is

$$\frac{\lambda_R}{\lambda_E}=\left[\frac{1-2GM/r_Rc^2}{1-2GM/r_Ec^2}\right]^{1/2}, \tag{4.19}$$

and if we assume that the *Skylab* is so distant that we may take $1/r_R = 0$, and ignore its motion in space, then this gives an observed wavelength of

$$\lambda_R \approx \lambda_E(1-2MG/r_Ec^2)^{-1/2} = 685.6\,\text{nm},$$

[5]See Pound and Rebka, 1960.

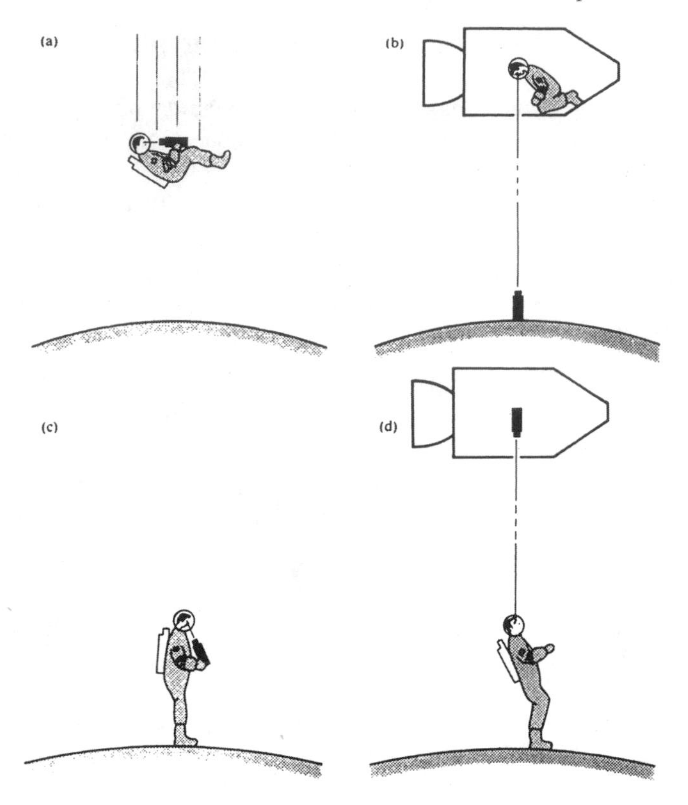

Fig. 4.6. Observer and laser.

on putting $\lambda_E = 632.8\,\text{nm}$, $G = 6.67 \times 10^{-11}\,\text{N}\,\text{m}^2\,\text{kg}^{-2}$, $M = 10^{30}\,\text{kg}$, $r_E = r_B = 10^4\,\text{m}$, and $c = 3 \times 10^8\,\text{m}\,\text{s}^{-1}$.

(c) Here $r_E = r_B$, so formula (4.19) gives a measured wavelength of 632.8 nm.

(d) Here we take $1/r_E = 0$ in formula (4.19) giving

$$\lambda_R \approx \lambda_E(1 - 2MG/r_R c^2)^{1/2},$$

where $\lambda_E = 632.8\,\text{nm}$ and $r_R = r_B = 10^4\,\text{m}$. This gives a measured wavelength of approximately 584 nm.

Exercise 4.3

1. Find the fractional shift in frequency, as measured on Earth, for light from a star of mass 10^{30} kg, assuming that the photons come from just above the star's atmosphere where $r_B = 1000\,\text{km}$.

4.4 General particle motion (including photons)

The paths of particles with mass moving in the vicinity of a spherical massive object are given by the timelike geodesics of spacetime, while the paths of photons are given by the null geodesics. Our plan for this section is to consider first the timelike geodesics, and then to see what modifications are needed for null geodesics.

For a timelike geodesic we may use its proper time τ as an affine parameter. From Section 2.1 we see that the four geodesic equations are given by

$$\frac{d}{d\tau}\left(\frac{\partial L}{\partial \dot{x}^\mu}\right) - \frac{\partial L}{\partial x^\mu} = 0, \tag{4.20}$$

where

$$\begin{aligned} L(\dot{x}^\sigma, x^\sigma) &\equiv \tfrac{1}{2} g_{\mu\nu}\dot{x}^\mu \dot{x}^\nu \\ &= \tfrac{1}{2}\left(c^2(1-2m/r)\dot{t}^2 - (1-2m/r)^{-1}\dot{r}^2 - r^2(\dot{\theta}^2 + \sin^2\theta\,\dot{\phi}^2)\right). \end{aligned}$$

Here dots denote derivatives with respect to τ, the coordinates are $x^0 \equiv t$, $x^1 \equiv r$, $x^2 \equiv \theta$, $x^3 \equiv \phi$, and we have again put $m = GM/c^2$.

Because of the spherical symmetry, there is no loss of generality in confining our attention to particles moving in the "equatorial plane" given by $\theta = \pi/2$. With this value for θ, the third ($\mu = 2$) of equations (4.20) is satisfied, and the second of these ($\mu = 1$) reduces to

$$\boxed{\left(1-\frac{2m}{r}\right)^{-1}\ddot{r} + \frac{mc^2}{r^2}\dot{t}^2 - \left(1-\frac{2m}{r}\right)^{-2}\frac{m}{r^2}\dot{r}^2 - r\dot{\phi}^2 = 0.} \tag{4.21}$$

Since t and ϕ are cyclic coordinates, we have immediate integrals of the two remaining equations (see Sec. 2.1):

$$\partial L/\partial \dot{t} = \text{const}, \quad \partial L/\partial \dot{\phi} = \text{const}.$$

With $\theta = \pi/2$ these are (for $\partial L/\partial \dot{t}$ and $\partial L/\partial \dot{\phi}$, respectively)

$$\boxed{(1-2m/r)\dot{t} = k,} \tag{4.22}$$

$$\boxed{r^2\dot{\phi} = h,} \tag{4.23}$$

where k and h are integration constants. We also have the relation (2.69) which defines τ. With $\theta = \pi/2$ this becomes

$$c^2(1-2m/r)\dot{t}^2 - (1-2m/r)^{-1}\dot{r}^2 - r^2\dot{\phi}^2 = c^2, \tag{4.24}$$

and may be used in place of the rather complicated equation (4.21).

Equation (4.22) gives the relation between the coordinate time t and the proper time τ; equation (4.23) is clearly analogous to the equation of conservation of angular momentum; as we shall see, equation (4.24) yields an equation analogous to that expressing conservation of energy.

Equation (4.24) gives

$$c^2(1-2m/r)\dot{t}^2/\dot{\phi}^2 - (1-2m/r)^{-1}(dr/d\phi)^2 - r^2 = c^2/\dot{\phi}^2,$$

and substituting for $\dot{\phi}$ and $\dot{t}$ from equations (4.22) and (4.23) gives

$$(dr/d\phi)^2 + r^2(1+c^2r^2/h^2)(1-2m/r) - c^2k^2r^4/h^2 = 0.$$

If we put $u \equiv 1/r$ and $m = GM/c^2$ this reduces to

$$\boxed{\left(\frac{du}{d\phi}\right)^2 + u^2 = E + \frac{2GM}{h^2}u + \frac{2GM}{c^2}u^3,} \tag{4.25}$$

where $E \equiv c^2(k^2-1)/h^2$. Comparing this with the analogous Newtonian equation (4.41) we see that it corresponds to an energy equation. Comparison also shows that the last term on the right is, in a sense, a relativistic correction, and this is the point of view that we shall adopt when discussing the advance of the perihelion in planetary orbits in the next section. In theory equation (4.25) may be integrated to give u, and hence r, as a function of ϕ, to obtain the particle paths in the equatorial plane. Except in special cases, this integration is impossible in practice, and we resort to approximation methods when discussing planetary motion.

Two interesting special cases may be examined in detail, namely vertical free-fall and motion in a circle.

Vertical free-fall. For vertical free-fall, ϕ is constant, which implies that equation (4.23) is satisfied with $h = 0$. In deriving equation (4.25) we assumed that $\dot{\phi}$ and h were nonzero, so that equation cannot be used. However, it was based on equation (4.24), which, with $\dot{\phi} = 0$ and the expression for $\dot{t}$ given by equation (4.22) substituted, reduces to

$$\dot{r}^2 - c^2k^2 + c^2(1-2m/r) = 0. \tag{4.26}$$

This equation enables us to give a meaning to the integration constant k, for if the particle is at rest ($\dot{r} = 0$) when $r = r_0$, then $k^2 = 1 - 2m/r_0$. Since τ increases with t, equation (4.22) shows that k is the positive square root[6] of $1 - 2m/r_0$. Hence k is not a universal constant, but depends on the geodesic in question. In particular, if $\dot{r} \to 0$ as $r \to \infty$, then $k = 1$.

Differentiating equation (4.26) gives

[6] We are assuming that t has been chosen to increase into the future.

$$2\dot{r}\ddot{r} + (2mc^2/r^2)\dot{r} = 0,$$

which can be written as

$$\ddot{r} + GM/r^2 = 0. \tag{4.27}$$

This equation has exactly the same form as its Newtonian counterpart. However, it should be remembered that in equation (4.27) the coordinate r is not the vertical distance, and dots are derivatives with respect to proper time, whereas in the Newtonian version r would be vertical distance and dots would be derivatives with respect to the universal time.

Putting $k^2 = 1 - 2m/r_0$ and $m = MG/c^2$ in equation (4.26), we get

$$\tfrac{1}{2}\dot{r}^2 = MG\left(\frac{1}{r} - \frac{1}{r_0}\right). \tag{4.28}$$

Since the left-hand side is positive, this only makes sense if $r < r_0$. It has exactly the same form as the Newtonian equation expressing the fact that a particle (of unit mass) falling from rest at $r = r_0$ gains a kinetic energy equal to the loss in gravitational potential energy. However, the different meanings of r and the dot mentioned above should be borne in mind.

Equation (4.28) allows us to calculate the proper time experienced by the particle in falling from rest at $r = r_0$. If $\tau = 0$ when $r = r_0$, then this time is

$$\tau = \frac{1}{\sqrt{2MG}} \int_r^{r_0} \left(\frac{r_0 r}{r_0 - r}\right)^{1/2} dr, \tag{4.29}$$

where, because $\dot{r} < 0$, we have taken the negative square root when solving equation (4.28) for $dr/d\tau$. The lower limit of integration may be taken down to $2GM/c^2$ (i.e., $2m$) unless the boundary of the massive object is reached first, and as $r \longrightarrow 2m$ the integral clearly remains finite (see Exercise 4.8.2). However, if one calculates the coordinate time t for falling to $r = 2m \equiv 2GM/c^2$, then one finds it to be infinite. Using $\dfrac{dt}{dr} = \dfrac{dt}{d\tau}\dfrac{d\tau}{dr}$ with

$$\frac{dt}{d\tau} = \frac{k}{1 - 2m/r} = \frac{(1 - 2m/r_0)^{1/2}}{1 - 2m/r}$$

from equation (4.22), and

$$\frac{d\tau}{dr} = -\frac{1}{c\sqrt{2m}}\left(\frac{r_0 r}{r_0 - r}\right)^{1/2}$$

from equation (4.28) (with $MG = mc^2$), gives

$$\frac{dt}{dr} = -\frac{1}{c\sqrt{2m}} \frac{r^{3/2}(r_0 - 2m)^{1/2}}{(r - 2m)(r_0 - r)^{1/2}}, \tag{4.30}$$

so the coordinate time to fall from $r = r_0$ to $r = 2m + \varepsilon$ $(\varepsilon > 0)$ is

$$t_\varepsilon = \left(\frac{r_0 - 2m}{2mc^2}\right)^{1/2} \int_{2m+\varepsilon}^{r_0} \frac{r^{3/2} dr}{(r-2m)(r_0-r)^{1/2}}.$$

With $2m + \varepsilon < r < r_0$ we have $r > 2m$ and $r_0 - r < r_0$, so

$$t_\varepsilon > \left(\frac{r_0 - 2m}{2mc^2}\right)^{1/2} \frac{(2m)^{3/2}}{r_0^{1/2}} \int_{2m+\varepsilon}^{r_0} \frac{dr}{r-2m}.$$

But

$$\int_{2m+\varepsilon}^{r_0} \frac{dr}{r-2m} = \ln \frac{r_0 - 2m}{\varepsilon} \to \infty \text{ as } \varepsilon \to 0,$$

showing that $t_\varepsilon \to \infty$ also. Hence the coordinate time taken to fall to $r = 2m$ is infinite, as asserted.

The way in which the coordinate time t depends on r for a radially falling particle becomes more comprehensible if we compare its *coordinate speed* $v(r)$, defined by $v(r) \equiv |dr/dt|$, with that of a particle falling according to the classical Newtonian theory with speed $\tilde{v}(r)$. For simplicity, let us consider a particle falling from rest at infinity. Letting $r_0 \to \infty$ in equation (4.30) gives

$$v(r) = (2mc^2)^{1/2}(r-2m)/r^{3/2},$$

whereas the corresponding classical expression is (with $MG = mc^2$)

$$\tilde{v}(r) = (2mc^2)^{1/2}/r^{1/2}.$$

A short calculation shows that as r decreases from infinity $v(r)$ increases until it reaches a maximum value of $2c/3\sqrt{3}$ at $r = 6m$, after which $v(r)$ decreases, and $v(r) \to 0$ as $r \to 2m$. On the other hand, as r decreases, $\tilde{v}(r)$ increases, and $\tilde{v}(r) \to \infty$ as $r \to 0$. The graphs of $v(r)$ and $\tilde{v}(r)$ are given in Figure 4.7.

Motion in a circle. For circular motion in the equatorial plane we have $r =$ constant, and $\dot{r} = \ddot{r} = 0$. Equation (4.21) then reduces to

$$mc^2\dot{t}^2 = r^3\dot{\phi}^2, \tag{4.31}$$

giving

$$(d\phi/dt)^2 = GM/r^3, \tag{4.32}$$

on putting $mc^2 = GM$. Hence the change in coordinate time t for one complete revolution is

$$\Delta t = 2\pi(r^3/GM)^{1/2}. \tag{4.33}$$

This expression is exactly the same as the Newtonian expression for the period of a circular orbit of radius r, that is, Kepler's third law. Although we cannot say that r is the radius of the orbit in the relativistic case, we see that the spatial distance traveled in one complete revolution is $2\pi r$, just as in the Newtonian case.

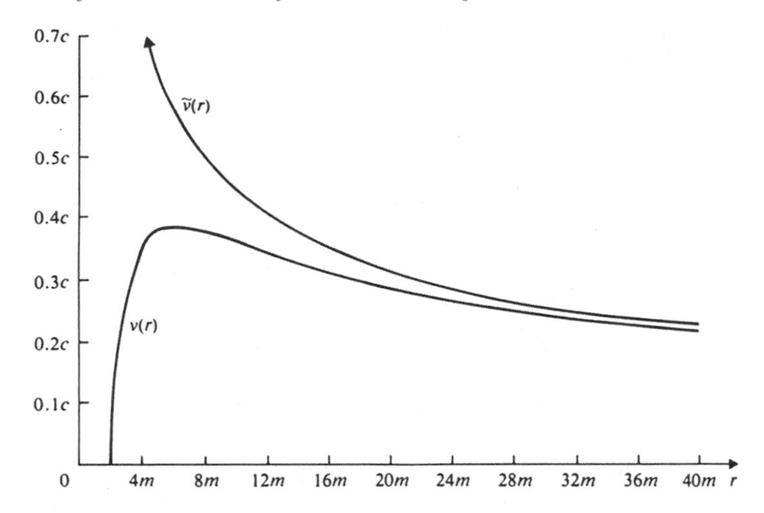

Fig. 4.7. Comparison of the coordinate speed $v(r)$ with the Newtonian speed $\tilde{v}(r)$ for a particle falling from rest at infinity.

Figure 4.8 is a spacetime diagram illustrating one complete revolution as viewed by an observer fixed at the point where $r = r_0$. B_1 is the event of the observer's viewing the start of the orbit at A_1, while B_2 is that of his viewing its completion at A_2. The coordinate time between A_1 and A_2 is Δt as given by equation (4.33), and we know from the argument used in deriving the spectral-shift formula (see Sec. 4.3) that Δt is also the coordinate time between B_1 and B_2. So the proper time $\Delta\tau_0$ which the observer measures for the orbital period is (from equation (4.6))

$$\Delta\tau_0 = (1 - 2m/r_0)^{1/2}\Delta t. \tag{4.34}$$

As $r_0 \to \infty$, $\Delta\tau_0 \to \Delta t$, so Δt is the orbital period as measured by an observer at infinity. So Δt turns out to be directly observable, and this suggests an indirect means of measuring the coordinate r, by measuring the orbital period of a test particle in a circular orbit given by the value of r. However, this depends on a knowledge of M, which must be known independently, that is, not found by methods involving orbital periods.

Equation (4.34) shows that for a fixed observer the period $\Delta\tau_0$ assigned to the orbit depends on his position. It is natural to ask what period $\Delta\tau$ an observer travelling with the orbiting particle would assign to the orbit. The relationship between t and τ is given by equation (4.22), so the answer to the question depends on the value of the integration constant k. From equations (4.31) and (4.22) we have

$$\dot{t}^2 = \frac{r^2k^2}{(r-2m)^2} \quad \text{and} \quad \dot{\phi}^2 = \frac{mc^2k^2}{r(r-2m)^2}, \tag{4.35}$$

and substitution in equation (4.24) (with $\dot{r} = 0$) gives

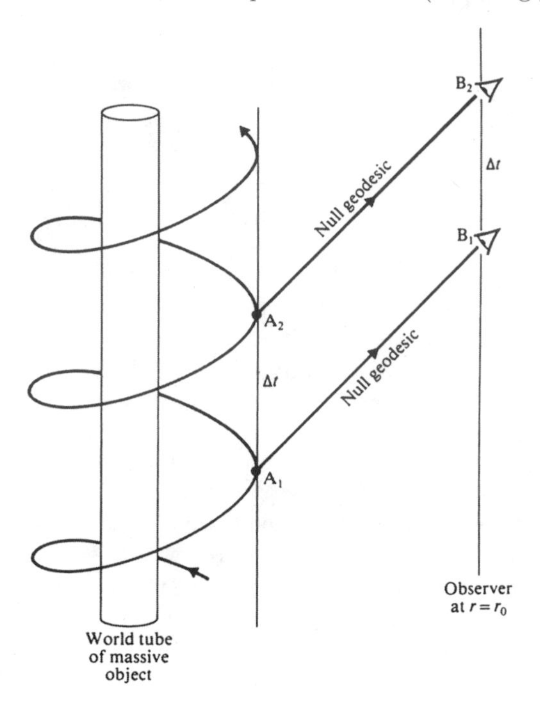

Fig. 4.8. Spacetime diagram illustrating a circular orbit as viewed by a fixed observer.

$$k^2 = \frac{(r-2m)^2}{r(r-3m)}. \tag{4.36}$$

From equation (4.22), the orbiting observer assigns a period $\Delta\tau$ to the orbit given by

$$\begin{aligned}\Delta\tau &= (1-2m/r)k^{-1}\Delta t \\ &= \left(\frac{r-3m}{r}\right)^{1/2}\Delta t = 2\pi\left[\frac{r^3}{GM}\left(1-\frac{3MG}{rc^2}\right)\right]^{1/2}.\end{aligned} \tag{4.37}$$

Since $k^2 > 0$, equation (4.36) implies that *circular orbits are impossible, unless* $r > 3m$. In the limit, as $r \to 3m$, $\Delta\tau \to 0$, suggesting that photons can orbit at $r = 3m$, and we shall see later in this section that this is indeed the case.

Imagine now a situation where we have two astronauts in a spacecraft which is in a circular orbit at a value of r greater than $3m$. Suppose one of them leaves the craft, uses his rocket-pack to maintain a hovering position at a fixed point in space, and then rejoins the craft after it has completed one orbit. According to equation (4.34), the hovering astronaut measures the time of absence as

$$\Delta\tau_{\text{hov}} = (1-2m/r)^{1/2}\Delta t,$$

while the orbiting astronaut measures it as

$$\Delta\tau_{\text{orb}} = (1 - 2m/r)k^{-1}\Delta t,$$

so

$$\frac{\Delta\tau_{\text{hov}}}{\Delta\tau_{\text{orb}}} = \frac{k}{(1 - 2m/r)^{1/2}} = \left(\frac{r - 2m}{r - 3m}\right)^{1/2} > 1.$$

This shows that if the two astronauts were the same age at the time one of them left the spacecraft for a period of powered flight, then on his return he is older than his companion who remained in the freely falling spacecraft. This result contrasts with the twin paradox of special relativity where the twin undertaking a powered excursion returns to find himself younger.[7]

Photons. Let us now look at the null geodesics which give the paths of photons (and any other particles having rest mass equal to zero). We cannot use proper time τ as a parameter, so let w be any affine parameter along the geodesic, and let dots now denote derivatives with respect to w. For photons moving in the equatorial plane, equations (4.21) to (4.23) remain the same, but the right-hand side of equation (4.24) must be replaced by zero:

$$c^2(1 - 2m/r)\dot{t}^2 - (1 - 2m/r)^{-1}\dot{r}^2 - r^2\dot{\phi}^2 = 0. \tag{4.38}$$

This leads to a modified form of equation (4.25):

$$\boxed{\left(\frac{du}{d\phi}\right)^2 + u^2 = F + \frac{2GM}{c^2}u^3,} \tag{4.39}$$

where $F \equiv c^2k^2/h^2$. We make use of this equation when discussing the bending of light in Section 4.6. To complete the present section we shall discuss two consequences of the null geodesic equations.

The first is the possibility of having photons in a circular orbit. With $\dot{r} = \ddot{r} = 0$, equation (4.21) gives $\dot{\phi}^2/\dot{t}^2 = mc^2/r^3$, while equation (4.38) gives $\dot{\phi}^2/\dot{t}^2 = c^2(1 - 2m/r)/r^2$. Equating these gives $r = 3m$ as the only possible value of r for which photons can go into orbit.

The second consequence is that by investigating the radial null geodesics we can discover what sort of picture a fixed observer gets of any particle falling into a black hole. Suppose the observer is fixed at $r = r_0$, and he drops a particle from rest. According to the discussion above it takes an infinite coordinate time to fall to $r = 2m$, and we see from equation (4.30) that $dr/dt \to 0$ as $r \to 2m$ (see also Fig. 4.7). So in an r, t diagram the path of the falling particle is asymptotic to $r = 2m$, as shown in Figure 4.9. However, as we have seen, the proper time to fall to $r = 2m$ as measured by an observer

[7]This result apparently contradicts the dictum that a timelike geodesic maximizes proper time.

falling with the particle is finite. Moreover, as $r \to 2m$ equation (4.28) shows that $dr/d\tau \to -c(1 - 2m/r_0)^{1/2}$, so the particle has not run out of steam by the time it gets down to $r = 2m$ and presumably passes beyond the threshold. (So in some respects our r, t diagram is misleading.)

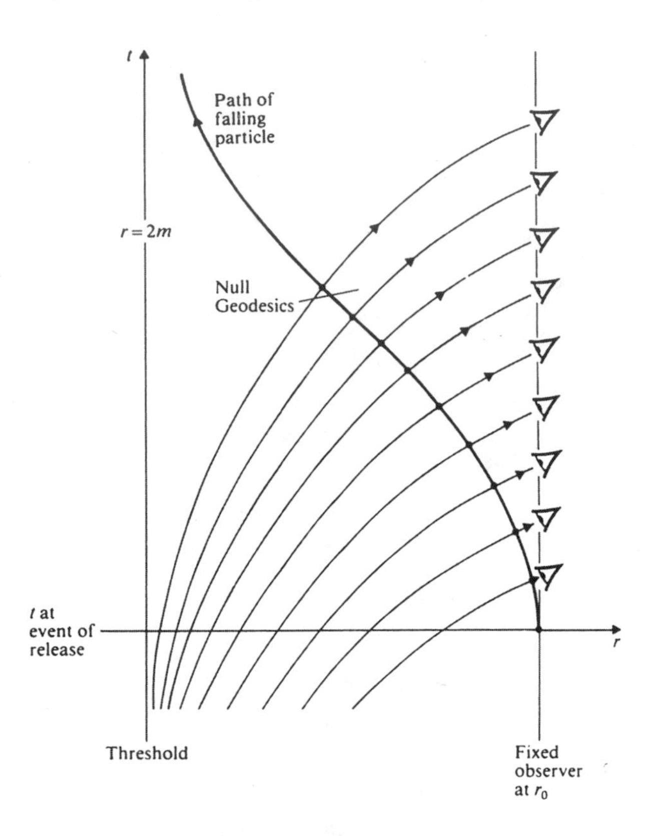

Fig. 4.9. r, t diagram illustrating the observation of a falling particle by a fixed observer.

What the observer sees is governed by the outgoing radial null geodesics issuing from the falling particle. For such a geodesic, equation (4.38) gives

$$c^2(1 - 2m/r)\dot{t}^2 - (1 - 2m/r)^{-1}\dot{r}^2 = 0,$$

so

$$dr/dt = c(1 - 2m/r). \tag{4.40}$$

As $r \to 2m$, $dr/dt \to 0$ and an outgoing radial null geodesic is also asymptotic to $r = 2m$ in our r, t diagram. The outgoing null geodesics are therefore as shown in Figure 4.9. If we follow these back from the eye of the fixed observer, then we discover that he always sees the falling particle *before* it gets to $r = 2m$, as we asserted above.

Our present coordinate system is inadequate for discussing what happens at and beyond $r = 2m$, so care should be exercised in interpreting diagrams such as Figure 4.9. When we discuss black holes in Section 4.8 we make use of another coordinate system that allows us to step over the threshold at $r = 2m$, and to give an improved version of Figure 4.9.

Exercises 4.4

1. Obtain the second and third of equations (4.20) (i.e., for $\mu = 1$ and 2), and hence show that $\theta = \pi/2$ satisfies the third equation, and that with $\theta = \pi/2$ the second reduces to equation (4.21).

2. Check equation (4.25) for timelike geodesics, and the corresponding equation (4.39) for null geodesics.

3. An observer stationed where $r = r_0$ watches a light signal emitted from a point where $r = r_1$. It travels radially inwards and is reflected by a fixed mirror at $r = r_2$, so that it travels back to its point of origin at $r = r_1$. How long does the round-trip take according to the observer at $r = r_0$? (Assume that $2m < r_2 < r_1 < r_0$.)

4.5 Perihelion advance

For a particle moving in the equatorial plane under the Newtonian gravitational attraction of a spherical object of mass M situated at the origin, classical angular momentum and energy considerations lead to the equation

$$(du/d\phi)^2 + u^2 = E + 2GMu/h^2, \tag{4.41}$$

where $u \equiv 1/r$, E is a constant related to the energy of the orbit, and h is the angular momentum per unit mass given by $r^2 d\phi/dt = h$ (see Problem 4.6). The solution of this equation is well known from mechanics as

$$u = (GM/h^2)[1 + e\cos(\phi - \phi_0)], \tag{4.42}$$

where ϕ_0 is a constant of integration, and $e^2 \equiv 1 + Eh^4/G^2M^2$. Equation (4.42) is that of a conic section with eccentricity e.

The general-relativistic analogue of equation (4.41) is equation (4.25), and we expect the extra term (equal to $2GMu^3/c^2$) to perturb the Newtonian orbit in some way. If we take the Schwarzschild solution as a model for the solar system, treating the planets as particles, then this extra term makes its presence felt by an advance of the perihelion (i.e., the point of closest approach to the Sun) in each circuit of a planet about the Sun. In deriving this result we make use of an argument due to Møller.[8]

[8]See, Møller, 1972, §12.2.

Aphelion and perihelion occur where $du/d\phi = 0$, that is, at values of u satisfying

$$\frac{2GM}{c^2}u^3 - u^2 + \frac{2GM}{h^2}u + E = 0.$$

This is a cubic equation with three roots, u_1, u_2, u_3 say. Suppose that u_1 gives the aphelion and u_2 the perihelion, so $u_1 \le u \le u_2$. Let us introduce a new variable $\bar{u} \equiv u/u_0$, where $u_0 \equiv (u_1 + u_2)/2$. This enables us to write the cubic equation above as

$$\varepsilon\bar{u}^3 - \bar{u}^2 + \frac{2GM}{h^2 u_0}\bar{u} + \frac{E}{u_0^2} = 0,$$

where $\varepsilon \equiv 2GMu_0/c^2$. The quantity ε and the variable $\bar{u}$ are dimensionless, and for planetary orbits in the solar system ε is extremely small (about 5.1×10^{-8} for Mercury and 2×10^{-8} for the Earth). We shall therefore work to first order in ε, neglecting its square and higher powers.[9]

The three roots of the cubic in $\bar{u}$ are $\bar{u}_1 \equiv u_1/u_0$, $\bar{u}_2 \equiv u_2/u_0$, and $\bar{u}_3 \equiv u_3/u_0$. The equation (4.25) is equivalent to

$$(d\bar{u}/d\phi)^2 = \varepsilon(\bar{u} - \bar{u}_1)(\bar{u}_2 - \bar{u})(\bar{u}_3 - \bar{u}), \tag{4.43}$$

on writing the cubic expression in terms of its factors. But

$$\varepsilon\bar{u}_3 = 1 - \varepsilon(\bar{u}_1 + \bar{u}_2) = 1 - 2\varepsilon,$$

since the sum of the roots[10] is $1/\varepsilon$ and $\bar{u}_1 + \bar{u}_2 = 2$, giving

$$(d\bar{u}/d\phi)^2 = (\bar{u} - \bar{u}_1)(\bar{u}_2 - \bar{u})(1 - \varepsilon(2 + \bar{u})).$$

So to first order in ε,

$$\frac{d\phi}{d\bar{u}} = \frac{1 + \frac{1}{2}\varepsilon(2 + \bar{u})}{[(\bar{u} - \bar{u}_1)(\bar{u}_2 - \bar{u})]^{1/2}}.$$

Putting $\beta \equiv \frac{1}{2}(\bar{u}_2 - \bar{u}_1)$ allows us to write the above as

$$\frac{d\phi}{d\bar{u}} = \frac{\frac{1}{2}\varepsilon(\bar{u} - 1) + 1 + \frac{3}{2}\varepsilon}{[\beta^2 - (\bar{u} - 1)^2]^{1/2}}.$$

This form for $d\phi/du$ allows us to integrate it to find the angle $\Delta\phi$ between an aphelion and the next perihelion. Using $\Delta\phi = \int_{\bar{u}_1}^{\bar{u}_2}(d\phi/d\bar{u})d\bar{u}$, we get

---

[9]The intention here is to make clear what is small and what is not by using dimensionless quantities and variables that do not depend on the units used. The constant u_0 is taken as a *characteristic value* for u and is used to define the variable $\bar{u}$, whose value for nearly circular orbits is then not too different from unity. We are effectively *scaling* u, so that the problem is formulated in a dimensionless way. (The angular variable ϕ is already dimensionless.) See, for example Logan, 1987, §1.3.

[10]If $a(x-x_1)(x-x_2)(x-x_3) \equiv ax^3+bx^2+cx+d$, then expanding the left-hand side and comparing coefficients of x^2 gives $-a(x_1+x_2+x_3) = b$, so $x_1+x_2+x_3 = -b/a$.

$$\begin{aligned}\Delta\phi &= \left[-\tfrac{1}{2}\varepsilon\left(\beta^2-(\bar{u}-1)^2\right)^{1/2}+\left(1+\tfrac{3}{2}\varepsilon\right)\arcsin\frac{\bar{u}-1}{\beta}\right]_{\bar{u}_1}^{\bar{u}_2}\\ &= (1+\tfrac{3}{2}\varepsilon)\pi.\end{aligned} \tag{4.44}$$

Doubling $\Delta\phi$ gives the angle between successive perihelions, and shows that in each circuit this is advanced by

$$3\varepsilon\pi = \frac{3GM\pi}{c^2}(u_1+u_2) = \frac{3GM\pi}{c^2}\left(\frac{1}{r_1}+\frac{1}{r_2}\right), \tag{4.45}$$

where r_1 and r_2 are the values of r at aphelion and perihelion.

Although the quantity (4.45) is incredibly small, the effect is cumulative, and eventually becomes susceptible to observation. It is greatest for the planet Mercury, which is the one closest to the Sun, and amounts to $43''$ per century. There is excellent agreement between the theoretical and observed values, but the comparison is not as straightforward as it might seem. In deriving the quantity (4.45) we assumed that planets behaved like particles, and ignored their gravitational influence on each other. In fact this influence cannot be ignored, and the effect of the other planets on Mercury causes a perturbation of its orbit. However, after taking this into account (using Newtonian methods) there remains an anomalous advance of the perihelion not explicable in Newtonian terms, and it is this which is accounted for by general relativity. The predicted advances for Mercury, Venus, and Earth, together with the observed anomalous advances, are given in the Introduction.

Although we are here discussing tiny effects in our solar system, it is interesting to note that between 1975 and 1989 the eccentric orbit of the famous neutron-star pair[11] PSR 1913+16 has been observed to advance over $60°$. However, as mentioned in the Introduction, our analysis above does *not* treat the motion of two massive stars, and so the results cannot be carried over without modification. Another binary pulsar, PSR 1855+09, is also being studied for large general-relativistic effects.

Exercise 4.5

1. Check the calculations leading to the result (4.44).

4.6 Bending of light

We have already noted that a massive object can have a considerable effect on light: photons can orbit at $r = 3m$. However, we do not expect to be able to observe this extreme effect in nature, principally because we do not expect to

[11] The pulsar was discovered in 1975 by Hulse and Taylor. See Hulse and Taylor, 1975, and Weisberg and Taylor, 1984.

find many objects with $r_B < 3m$. More modest deflections of light passing a massive object can be observed, and in this section we give the theory behind the observations.

The path of a photon travelling in the equatorial plane is given by equation (4.39). With M turned right down to zero, this becomes

$$(du/d\phi)^2 + u^2 = F, \tag{4.46}$$

a particular solution of which is

$$u = u_0 \sin\phi \quad \text{or} \quad r_0 = r\sin\phi, \tag{4.47}$$

where $u_0^2 \equiv 1/r_0^2 = F$. This solution represents the straight-line path taken by a photon originating from infinity in the direction $\phi = 0$, and going off to infinity in the direction $\phi = \pi$. The point on the path nearest to the origin O is at a distance r_0 from it, and is given by $\phi = \pi/2$ (see Fig. 4.10). On turning M up, we expect this path to be modified in some way.

Turning M up means replacing equation (4.46) by equation (4.39). Taking a similar approach to that in the previous section, let us introduce the dimensionless variable $\bar{u} \equiv u/u_0$, where (as in the case where $M = 0$) u_0 is the value of u at the point of closest approach, and set $\varepsilon \equiv 2GMu_0/c^2$. In many situations, ε is extremely small, in particular that of a photon from a distant star reaching the Earth after grazing the Sun, for which the value is about 4.2×10^{-6}. As in the previous section, we shall work to first order in ε and neglect its square and higher powers.

We can now write equation (4.39) in the equivalent form

$$\left(\frac{d\bar{u}}{d\phi}\right)^2 + \bar{u}^2 = \frac{F}{u_0^2} + \varepsilon\bar{u}^3. \tag{4.48}$$

At the point of closest approach, $d\bar{u}/d\phi = 0$ and $\bar{u} = 1$, so $F/u_0^2 = 1 - \varepsilon$ and equation (4.48) becomes

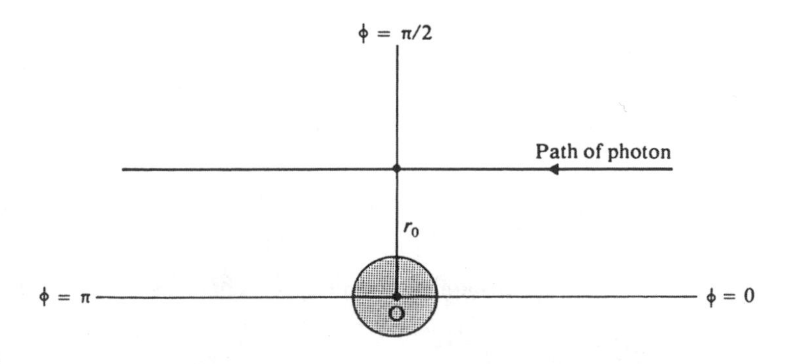

Fig. 4.10. Photon path in the equatorial plane of flat spacetime ($M = 0$).

$$(d\bar{u}/d\phi)^2 + \bar{u}^2 = 1 - \varepsilon + \varepsilon\bar{u}^3. \tag{4.49}$$

This equation should have a solution close to $\bar{u} = \sin\phi$, which is the form the flat-spacetime solution (4.47) takes when expressed in terms of $\bar{u}$. Let this be

$$\bar{u} = \sin\phi + \varepsilon v,$$

where v is some function of ϕ to be determined. Substitution in equation (4.49), and working to first order in ε, gives

$$2(dv/d\phi)\cos\phi + 2v\sin\phi = \sin^3\phi - 1,$$

which can be rewritten as

$$d(v\sec\phi)/d\phi = \tfrac{1}{2}(\sec\phi\tan\phi - \sin\phi - \sec^2\phi)$$

after some manipulation. Integrating gives

$$v = \tfrac{1}{2}(1 + \cos^2\phi - \sin\phi) + A\cos\phi,$$

where A is a constant of integration. Let us fix A by requiring that the photon originates from infinity in the direction $\phi = 0$, as in the flat-spacetime case. Then $v = 0$ when $\phi = 0$, so $A = -1$, and

$$\bar{u} = (1 - \tfrac{1}{2}\varepsilon)\sin\phi + \tfrac{1}{2}\varepsilon(1 - \cos\phi)^2 \tag{4.50}$$

is the equation of the path of the photon, to first order in ε.

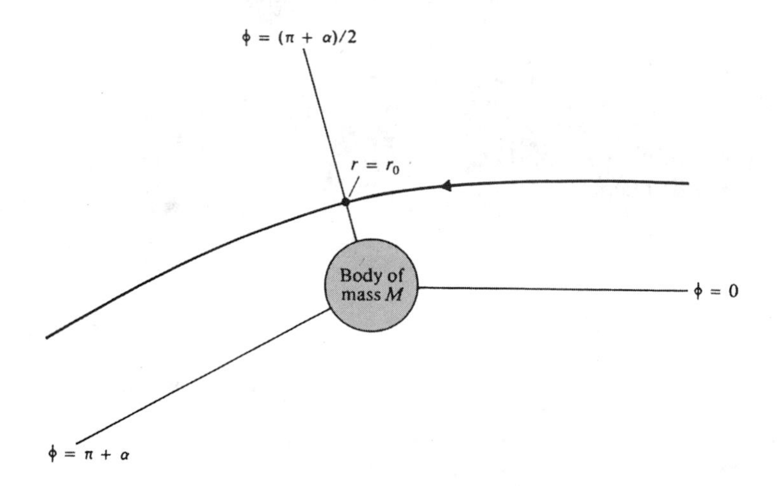

Fig. 4.11. Photon path in the equatorial plane of a massive body ($M > 0$).

We no longer expect the photon to go off to infinity in the direction π, but in a direction $\pi + \alpha$, where α is small. Putting $\bar{u} = 0$ and $\phi = \pi + \alpha$ in equation (4.50), and ignoring squares and higher powers of α, and also $\varepsilon\alpha$, gives

$$0 = -\alpha + 2\varepsilon,$$

so $\alpha = 2\varepsilon$. (We have used $\sin(\pi + \alpha) \approx -\alpha$ and $\cos(\pi + \alpha) \approx -1$.) So in its flight past the massive object the photon is deflected through an angle

$$\alpha = 2\varepsilon = 4GM/r_0c^2 \tag{4.51}$$

(see Fig. 4.11).

The deflection increases as the impact parameter r_0 decreases. In the case of light passing through the gravitational field of the Sun, the smallest that r_0 can be is its value at the Sun's surface. If we take for r_0 the accepted value of the Sun's radius (a good enough approximation[12]), then the formula (4.51) gives a value of $1.75''$ for the total deflection of light originating and terminating at infinity.[13]

This theoretical result has been checked by observation, but the experiment is a difficult one. One method involves photographing the star field around the Sun during a total eclipse, and comparing the photograph with one of the same star field taken six months later. The problems facing experimenters include

(a) the marked change in conditions which occur when bright sunlight changes to the semidarkness of an eclipse;
(b) the time lapse of six months, which makes it difficult to reproduce similar conditions when taking the comparison photograph;
(c) the smallness of the effect, which pushes photography to its limits.

Another method involves the measurement of the relative positions of two radio sources (by interferometric means) as one of them passes behind the Sun.[14] Some detailed figures are given in the Introduction.

4.7 Geodesic effect

If in *flat* spacetime a spacelike vector λ^μ is transported along a timelike geodesic without changing its spatial orientation, then, in Cartesian coordinates, it satisfies $d\lambda^\mu/d\tau = 0$, where τ is the proper time along the geodesic. That is, λ^μ is parallelly transported through *spacetime* along the geodesic. (Since $D\lambda^\mu/d\tau = d\lambda^\mu/d\tau$ when Cartesian coordinates are used.) Moreover, if at some point λ^μ is orthogonal to the tangent vector $\dot{x}^\mu \equiv dx^\mu/d\tau$ to the

[12]Although r_0 is a coördinate value, the circumference of the Sun's disk is $2\pi r_0$, which is measured by optical means.

[13]In 1911, prior to general relativity, Einstein predicted a deflection equal to half this amount. See Hoffman, 1972, Chap. 8, for history, and Kilmister, 1973, Extract 3, for a translation of Einstein's paper.

[14]See, for example, Riley, 1973, where references for other experiments are given.

geodesic, then $\eta_{\mu\nu}\lambda^\mu \dot{x}^\nu = 0$, and this relationship is preserved under parallel transport. This orthogonality condition simply means that λ^μ has no temporal component in an instantaneous rest frame of an observer traveling along the geodesic. The corresponding criteria for transporting a spacelike vector λ^μ in this fashion in the curved spacetime of general relativity are, therefore,

$$d\lambda^\mu/d\tau + \Gamma^\mu_{\nu\sigma}\lambda^\nu \dot{x}^\sigma = 0 \tag{4.52}$$

and

$$g_{\mu\nu}\lambda^\mu \dot{x}^\nu = 0, \tag{4.53}$$

where $\dot{x}^\mu \equiv dx^\mu/d\tau$.

The *geodesic effect* (sometimes termed the *geodetic effect*) is a consequence of the fact that if a spacelike vector is transported without rotation along a geodesic corresponding to a circular orbit of the Schwarzschild solution, then on its return to the same point in space, after completing one revolution, its spatial orientation has changed. To see this, we must integrate the system of equations (4.52) using expressions for $\Gamma^\mu_{\nu\sigma}$ and $\dot{x}^\sigma$ corresponding to a circular orbit, which without loss of generality we may take to be in the equatorial plane.

For such an orbit $\dot{x}^1 = \dot{x}^2 = 0$, and most of the $\Gamma^\mu_{\nu\sigma}$ are zero. Making use of the results of Problem 2.7 (with $\theta = \pi/2$), we see that equations (4.52) reduce to

$$d\lambda^0/d\tau + \Gamma^0_{10}\lambda^1 \dot{x}^0 = 0, \tag{4.54}$$

$$d\lambda^1/d\tau + \Gamma^1_{00}\lambda^0 \dot{x}^0 + \Gamma^1_{33}\lambda^3 \dot{x}^3 = 0, \tag{4.55}$$

$$d\lambda^2/d\tau = 0, \tag{4.56}$$

$$d\lambda^3/d\tau + \Gamma^3_{13}\lambda^1 \dot{x}^3 = 0, \tag{4.57}$$

where

$$\Gamma^0_{10} = \frac{m}{r^2}\left(1 - \frac{2m}{r}\right)^{-1}, \quad \Gamma^1_{00} = \frac{mc^2}{r^2}\left(1 - \frac{2m}{r}\right),$$

$$\Gamma^1_{33} = -r\left(1 - \frac{2m}{r}\right), \quad \Gamma^3_{13} = \frac{1}{r}.$$

Let us put $\dot{x}^\mu = (a, 0, 0, \Omega a)$, where $a \equiv \dot{t}$ and $\Omega \equiv d\phi/dt$, so that Ω is the angular *coordinate* speed around the circular orbit. Equations (4.35), (4.36), and (4.31) show that

$$a = \left(\frac{r}{r - 3m}\right)^{1/2} \quad \text{and} \quad \Omega = c\left(\frac{m}{r^3}\right)^{1/2}$$

on assuming that ϕ increases with t. Both a and Ω are constants. The orthogonality condition (4.53) reduces to

$$c^2(1 - 2m/r)\lambda^0\dot{x}^0 - r^2\lambda^3\dot{x}^3 = 0,$$

and allows us to express λ^0 in terms of λ^3:

$$\lambda^0 = \left[\Omega r^2/c^2(1 - 2m/r)\right]\lambda^3.$$

A short calculation then shows that equation (4.54) is equivalent to equation (4.57), and the system of equations (4.54)–(4.57) reduces to

$$\begin{aligned} d\lambda^1/d\tau - (r\Omega/a)\lambda^3 &= 0, \\ d\lambda^2/d\tau &= 0, \\ d\lambda^3/d\tau + (a\Omega/r)\lambda^1 &= 0. \end{aligned} \tag{4.58}$$

The general solution of these is

$$\begin{aligned} \lambda^1 &= (A/a)\cos(\phi_0 - \Omega\tau), \\ \lambda^2 &= B, \\ \lambda^3 &= (A/r)\sin(\phi_0 - \Omega\tau), \end{aligned} \tag{4.59}$$

where A, B, and ϕ_0 are constants of integration. This shows that the spatial part $\boldsymbol{\lambda}$ of $\lambda^\mu \equiv (\lambda^0, \boldsymbol{\lambda})$ rotates relative to the radial direction with angular *proper* speed Ω in the negative ϕ direction. However, the radial direction itself rotates with angular *coordinate* speed Ω in the positive ϕ direction, and it is the difference between angular *proper* speed and angular *coordinate* speed which gives rise to the geodesic effect.

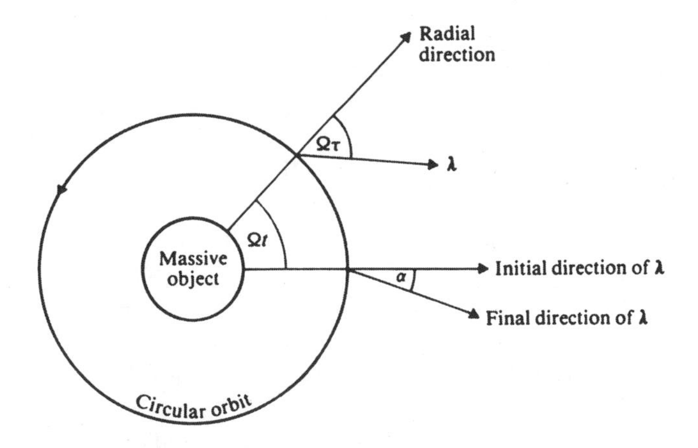

Fig. 4.12. Geodesic effect. Here the initial direction ($t = \tau = 0$) is radial.

If we take the initial direction of $\boldsymbol{\lambda}$ to be radial ($B = \phi_0 = 0$ in equation (4.59)), and choose the origin of t so that $t = 0$ when $\tau = 0$, then the situation is indicated in Figure 4.12. One revolution is completed in a coordinate time of $2\pi/\Omega$, and hence, from equation (4.37), in a proper time of $\left(\frac{r-3m}{r}\right)^{1/2}\frac{2\pi}{\Omega}$. The final direction of $\boldsymbol{\lambda}$ is therefore $2\pi - \alpha$, where

$$\alpha = 2\pi\left[1 - (1 - 3m/r)^{1/2}\right].$$

So $\alpha \approx 3\pi m/r$ for small m/r.

The axis of an orbiting gyroscope furnishes us with a spacelike vector which is transported without rotation, so the geodesic effect is, perhaps, susceptible to observation by means of gyroscopes in orbiting satellites.[15] The smaller the value of r, the greater is the effect; though small, the effect is cumulative, and for a satellite in near-Earth orbit amounts to about $8''$ per year, which should be measurable. An experiment involving a terrestrial satellite in a low circular orbit was planned for the Space Shuttle in 1993–94, but was delayed. The experiment was finally launched in April 2004, where four ping-pong-ball-sized spheres, spinning at 10,000 rpm, and using the binary star I.M. Pegasus as a reference, were put into polar orbit at a height of about 400 miles above the Earth's surface. The experiment, called Gravity Probe B, will measure both geodesic precession as well as the much smaller (and mutually perpendicular) Lense-Thirring effect of Section 4.10.

Exercise 4.7

1. What does the geodesic effect amount to for the axis of the Earth in its orbit round the Sun?
   (Take $M_\odot = 2 \times 10^{30}\,\mathrm{kg}$, $r = 1.5 \times 10^{11}\,\mathrm{m}$,
   $G = 6.67 \times 10^{-11}\,\mathrm{N\,m^2\,kg^{-2}}$, and $c = 3 \times 10^8\,\mathrm{m\,s^{-1}}$.)

4.8 Black holes

So far, our discussion of the Schwarzschild solution has been in terms of the coordinates (t, r, θ, ϕ), and we pointed out in Section 4.0 that the lower bound on r was either its value r_B at the boundary of the object, or $2m$ $(= 2GM/c^2)$, depending on which is reached first as r decreases. If $2m$ is reached first, we have a black hole, and this is the situation prevailing in this section. For an object of mass M, $2GM/c^2$ is known as its *Schwarzschild radius*.

In the limit as $r_E \to 2m$, the spectral-shift formula (4.15) produces an infinite redshift. A particle falling radially inwards appears to continue beyond the threshold at $r = 2m$, although, as we have seen, an observer viewing its fall always sees it before it passes the threshold. These two observations suggest

[15] See the paper by Everitt, Fairbank, and Hamilton in Carmeli et al., 1970.

that some odd things happen at $r = 2m$. However, the coordinates (t, r, θ, ϕ) are inadequate for discussing what happens at $r = 2m$ and beyond, so we introduce new coordinates which are valid for $r \leq 2m$.

Let us keep r, θ, ϕ, but replace t by

$$v \equiv ct + r + 2m \ln(r/2m - 1). \tag{4.60}$$

A short calculation (see Exercise 4.8.1) shows that in terms of v, r, θ, ϕ the line element is

$$c^2 d\tau^2 = (1 - 2m/r)dv^2 - 2\, dv\, dr - r^2 d\theta^2 - r^2 \sin^2\theta\, d\phi^2. \tag{4.61}$$

These new coordinates are *Eddington–Finkelstein coordinates*. They are valid for all v, for all $r > r_B$, even if $r_B < 2m$ (because none of the metric tensor components becomes infinite), and take us over the threshold at $r = 2m$.

From the line element (4.61), we see that radial null geodesics ($d\tau = 0$) are given by

$$\left(1 - \frac{2m}{r}\right)\left(\frac{dv}{dr}\right)^2 - 2\frac{dv}{dr} = 0,$$

that is, by

$$dv/dr = 0 \tag{4.62}$$

or

$$dv/dr = 2/(1 - 2m/r). \tag{4.63}$$

Differentiation of equation (4.60) gives

$$\frac{dv}{dr} = c\frac{dt}{dr} + \frac{1}{1 - 2m/r},$$

so $dv/dr = 0$ implies that $c\,dt/dr = -1/(1 - 2m/r)$, which is negative for $r > 2m$, while $dv/dr = 2/(1 - 2m/r)$ gives $c\,dt/dr = 1/(1 - 2m/r)$, which is positive for $r > 2m$. We therefore infer that equation (4.62) gives the ingoing null geodesics, while equation (4.63) gives the outgoing ones, at least in the region $r > 2m$.

Integration of equation (4.62) gives

$$v = A, \qquad A = \text{constant}, \tag{4.64}$$

while integration of equation (4.63) gives

$$v = 2r + 4m \ln|r - 2m| + B, \qquad B = \text{constant}. \tag{4.65}$$

Figure 4.13 shows a v, r diagram of radial null geodesics. In drawing this diagram we have used oblique axes, so that the ingoing null geodesics given by $v = A$ are inclined at 45°, just as they would appear in a flat spacetime diagram. We have also imagined the whole mass of the object to be concentrated at $r = 0$, and taken the ingoing null geodesics right down to $r = 0$. The

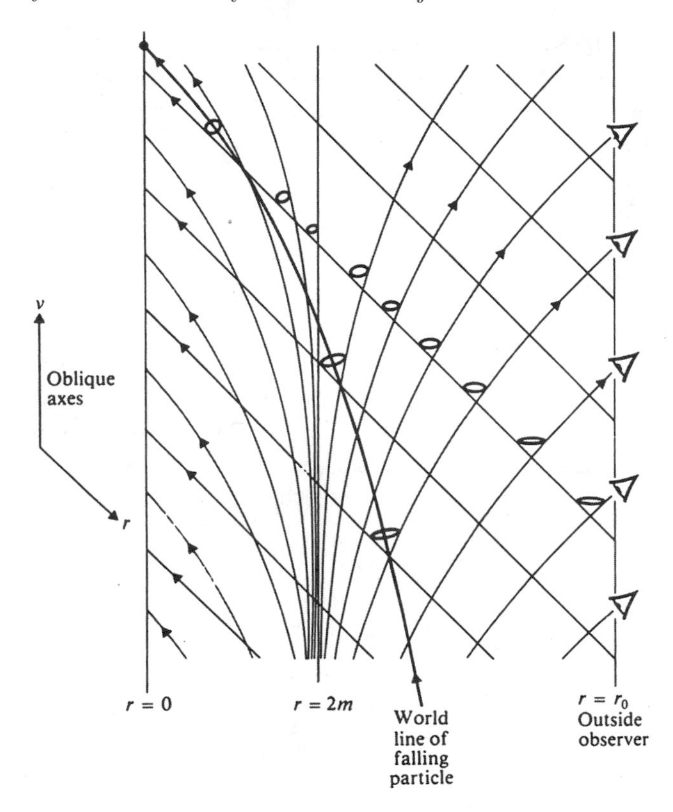

Fig. 4.13. Eddington–Finkelstein picture of ingoing and outgoing null geodesics.

outward equation (4.65) shows that $v \to -\infty$ as $r \to 2m$, so the outgoing null geodesics are asymptotic to $r = 2m$, as shown in the figure.

It may be seen from the figure that a photon starting where $r > 2m$ can travel inwards, cross the threshold at $r = 2m$, and carry on inwards, but that a photon starting where $r < 2m$ does not travel outwards. It is confined to the region $r < 2m$. Thus if a massive object had $r_B < 2m$, light could not escape from it to the region $r > 2m$. An outside observer could detect its presence through its gravitational field, but he could not see it, and it is for this reason that such an object is called *black hole.* Another way of describing the effect of a black hole on light propagation is that it causes the null cones in the tangent spaces to tilt over, and we have drawn small null cones in the figure illustrating this. The possible existence of objects from which light cannot escape was considered as early as 1798 by Laplace.[16]

[16] In classical physics the escape velocity for a particle from a star of mass M and radius r is $(2GM/r)^{1/2}$. Assigning a light corpuscle the escape velocity c yields $r = 2GM/c^2$, which is also the Schwarzschild result. See Hawking and Ellis, 1973, Appendix A, for a translation of Laplace's essay.

Since we have not changed the coordinate r, the integral (4.29), which gives the proper time for a particle to fall inwards from rest at $r = r_0$ remains the same, but we now see that it is valid for $r < 2m$. This integral may be evaluated (see Exercise 4.8.2), and remains finite as its lower limit tends to zero. For example, if $r_0 = 4m$, then the time taken to fall to $r = 2m$ is $\sqrt{2}m(\pi + 2)/c$, while that taken to fall to $r = 0$ is $2\sqrt{2}m\pi/c$ (see Exercise 4.8.3). The world line of a falling particle is also shown in Figure 4.13, and this results in an improved version of Figure 4.9.

Thus if Alice were to fall radially down a black hole (rather than a rabbit hole) clutching a clock and a lantern, then she would complete her fall within a finite time on her clock. However, an outside observer would never see her pass beyond $r = 2m$, but she would effectively disappear from view as the light from her lantern became became increasingly redshifted. Once beyond $r = 2m$ she could no longer signal to the outside observer, nor could she return to tell of her experiences.

The above considerations show that an outside observer cannot see events which occur inside the sphere $r = 2m$, and for this reason the sphere, or rather the hypersurface in spacetime which is its time development, is called an *event horizon*.

Our discussion of the properties of a black hole would be largely academic, unless there were reasons for believing that they might exist in nature. The possibility of their existence arises from the idea of gravitational collapse. If one imagined a very massive object accreting more matter by gravitational attraction, then a stage would be reached where the mutual gravitational attraction between the constituent particles was so great that the internal repulsive forces between them could no longer hold them apart. The whole object would collapse in on itself: nothing could stop this collapse, and the result would be a black hole. Quite general arguments (not based on the spherically symmetric solution of Schwarzschild) exist to show that a collapsing object leads to a singularity in spacetime.[17] If the collapse is spherically symmetric, then the singularity which is the eventual destination of the collapsing material is given by $r = 0$ in the Schwarzschild solution.

If one assumes that the general features of a collapsing object are not too far removed from those that prevail in the spherically symmetric case, then one would expect the emergence of an event horizon which would shield the object in its collapsed state from view (see Fig. 4.14). An outside observer would see the object to be always outside the event horizon. However, it would effectively disappear from view because of the increasing redshift, and a black hole in space would be the result.[18] At least two possible candidates

[17] See, for example, Misner, Thorne, and Wheeler, 1973, §34.6.

[18] It would take an infinite time to disappear. If black holes *do* exist, then this is an argument that they must have been "put in" at the beginning.

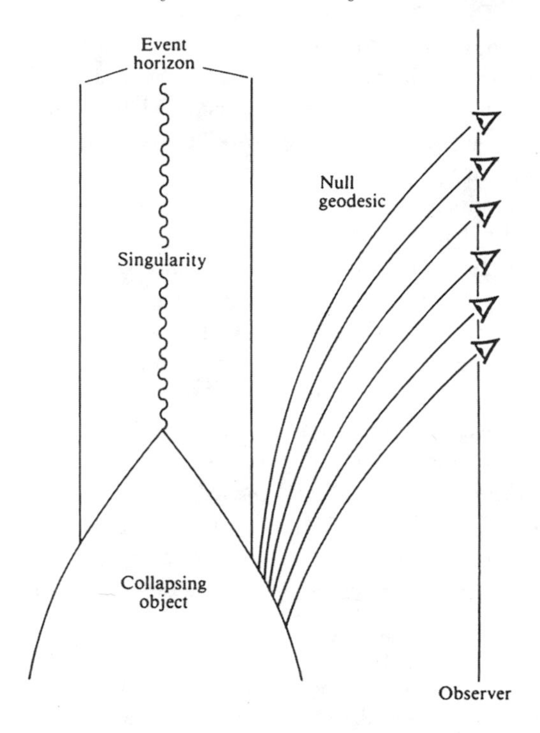

Fig. 4.14. Spacetime diagram of gravitational collapse.

for black holes have been "observed," which include the compact X-ray source Cygnus X-1 and the nucleus of the radio galaxy M87.[19]

Exercises 4.8

1. Verify the form (4.61) of the line element in Eddington–Finkelstein coordinates.

2. By making the substitution $r = r_0 \sin^2 \psi$, show that the value of the integral (4.29), giving the proper time to fall radially from rest at $r = r_0$, is

$$\tau(r_0, r) = \frac{r_0^{3/2}}{c(2m)^{1/2}} \left[\frac{\pi}{2} - \arcsin \left(\frac{r}{r_0} \right)^{1/2} + \left(\frac{r}{r_0} \right)^{1/2} \left(1 - \frac{r}{r_0} \right)^{1/2} \right].$$

3. Using the result of Exercise 4.8.2, show that $\tau(4m, 2m) = \sqrt{2}m(\pi + 2)/c$, and that $\tau(4m, r) = 2\sqrt{2}m\pi/c$ as $r \to 0$.

4. Why can Alice not return to the outside world ($r > 2m$) after falling down the black hole?

[19]See Thorne, 1974, and Young et al., 1978.

4.9 Other coordinate systems

The description of spacetime near a spherically symmetric massive object need not be in terms of the standard Schwarzschild coordinates and their corresponding line element. We have already seen the usefulness of Eddington–Finkelstein coordinates in discussing what happens beyond the event horizon at $r = 2m$.

Another example is provided by *isotropic coordinates* defined in Problem 3.7. Here the standard coordinate r is replaced by ρ, defined by

$$r \equiv \rho(1 + m/2\rho)^2, \tag{4.66}$$

and simple substitution gives the line element

$$\begin{aligned} c^2 d\tau^2 = {} & c^2(1 - m/2\rho)^2(1 + m/2\rho)^{-2} dt^2 \\ & - (1 + m/2\rho)^4(d\rho^2 + \rho^2 d\theta^2 + \rho^2 \sin^2\theta \, d\phi^2). \end{aligned} \tag{4.67}$$

These coordinates are employed in compiling the relativistic astronomical tables for the solar system used extensively throughout the world. We see that the line element has the form

$$c^2 d\tau^2 = A(\rho) dt^2 + B(\rho) d\sigma^2,$$

where $d\sigma^2$ is the line element of flat space in spherical coordinates ρ, θ, ϕ.

The particular advantage of the isotropic line element is that $d\sigma^2$ is invariant under changes of flat-space coordinates, and ρ, θ, ϕ may therefore be replaced by any other flat-space coordinates we care to use. For example, if Cartesian coordinates x, y, z (defined in terms of ρ, θ, ϕ in the usual way) are used as spatial coordinates, then

$$d\sigma^2 = dx^2 + dy^2 + dz^2$$

and

$$c^2 d\tau^2 = A(\rho) dt^2 + B(\rho)\left(dx^2 + dy^2 + dz^2\right),$$

where ρ occurring in $A(\rho)$ and $B(\rho)$ is given by $\rho^2 = x^2 + y^2 + z^2$.

Our previous results could be formulated in terms of isotropic coordinates, but the corresponding expressions are usually more complicated. For example, corresponding to equation (4.40) we would have

$$d\rho/dt = c(1 - m/2\rho)/(1 + m/2\rho)^3. \tag{4.68}$$

Kruskal (or *Kruskal–Szekeres*) *coordinates* are, like Eddington–Finkelstein coordinates, particularly useful for discussing what happens both sides of the event horizon. The r and t of the standard Schwarzschild coordinates are replaced by

$$u \equiv (r/2m - 1)^{1/2} e^{r/4m} \cosh(ct/4m),$$
$$v \equiv (r/2m - 1)^{1/2} e^{r/4m} \sinh(ct/4m).$$

This leads to the line element

$$c^2 d\tau^2 = -(32m^3/r)e^{-r/2m}(du^2 - dv^2) - r^2(d\theta^2 + \sin^2\theta \, d\phi^2), \tag{4.69}$$

where r is defined implicitly by

$$u^2 - v^2 \equiv (r/2m - 1)e^{r/2m}.$$

The particular advantage of these coordinates is that radial null geodesics are given by $u \pm v =$ constant, and are thus straight lines with 45° slopes when drawn in a u, v diagram, just as in the flat spacetime of special relativity.

We have seen in this chapter the effects on particles in a spacetime which is not flat. Note that no amount of coordinate transformation from one system to another can change the curvature of the spacetime. (The test for curvature is given in Sec. 3.2.) Another way of saying this is that we cannot transform away gravity just by turning to another coordinate system, except of course locally, but then only approximately. Globally, we cannot transform away gravity at all.

Exercise 4.9

1. Verify the form (4.69) for the line element in Kruskal coordinates.

4.10 Rotating objects; the Kerr solution[20]

In general, stars and planets spin on their axes. A Foucault pendulum at our North pole swings in a plane while the Earth spins underneath. An observer at the North Pole notices that axes marked on the snow certainly do not represent an inertial frame; rather, the pendulum dictates things, swinging in a plane which is apparently at rest with respect to the remainder of the universe.

In the seventeenth century, Newton, pondering what determined an inertial frame for his law of inertia, had asked (in essence) what would happen to the surface of water in a rotating bucket if the rest of the universe were not there[21].

In the nineteenth century the philosopher–physicist Ernst Mach, as well as Lense and Thirring in 1918,[22] had wondered whether ponderous moving masses in general might influence test particles, and in particular whether it would be possible to affect the plane of swing of a pendulum by placing it,

[20]This section may be omitted without loss of continuity.

[21]Discussed in, for example, Weinberg, 1972, §1.3.

[22]See Thirring and Lense, 1918.

for example, inside a very massive rotating cylinder. So the general question arises: does a massive *spinning* body have any sort of effect on inertial frames?

It turns out that it does. In the early 1960s R.P. Kerr found another exact solution[23] to the empty-spacetime Einstein field equations (3.40);[24] it is a generalization of the Schwarzschild solution and is now accepted as representing the gravitational field external to a rotating object in an otherwise empty spacetime. In Boyer–Lindquist coordinates,[25] in which t, r, θ, ϕ play a similar role to their counterparts in the standard form of the Schwarzschild solution, the line element has the form

$$\boxed{\begin{aligned} c^2 d\tau^2 = {} & \left(1 - \frac{2mr}{\rho^2}\right) c^2 dt^2 + \frac{4mcra\sin^2\theta}{\rho^2}\, dt\, d\phi - \frac{\rho^2}{\Delta}\, dr^2 \\ & - \rho^2 d\theta^2 - \left((r^2 + a^2)\sin^2\theta + \frac{2mra^2\sin^4\theta}{\rho^2}\right) d\phi^2, \end{aligned}} \tag{4.70}$$

where

$$\Delta \equiv r^2 + a^2 - 2mr \tag{4.71}$$

and

$$\rho^2 \equiv r^2 + a^2\cos^2\theta. \tag{4.72}$$

It can be seen that if $a = 0$, the solution reduces to that obtained by Schwarzschild, which suggests that $m = GM/c^2$, where M is the mass of the object. This claim can be further justified by comparing the 00-component of the weak-field approximation with the Newtonian potential, as in the Schwarzschild case. Arguments exist to justify the claim that the other constant a, which has the dimensions of length, is related to the object's angular momentum per unit mass. One of these is based on a comparison of the Kerr solution with the approximate solution obtained by Lense and Thirring for a rotating sphere of constant density in the weak-field limit. The conclusion is that

$$a = J/Mc, \tag{4.73}$$

where J is the angular momentum of the rotating source.

However, as regards actually measuring things, it should be borne in mind that we can only *infer* the actual mass of a body such as the Sun or the Earth from planetary or satellite data; and we cannot directly *measure* the angular momentum of such bodies at all, because we are always hampered by a lack of data concerning interior velocity profiles.

---

[23]The derivation of this solution uses techniques that are beyond the scope of this introductory text. See Kerr, 1963.

[24]Anyone considering checking that the solution satisfies the empty-spacetime field equations is warned that it requires an extremely long calculation. The symbolic computing system Maple coupled with the package GRTensorII makes short work of the job. For details visit `http://grtensor.phy.queensu.ca/`.

[25]See Boyer and Lindquist, 1967.

The value astronomers give for the Sun's mass is 2×10^{30} kg, and the angular momentum of the Sun is thought to be about $1.6 \times 10^{41}\,\mathrm{kg\,m^2\,s^{-1}}$. So for the Sun $a \approx 267$ m—considerably smaller than the Sun's radius. Estimates for J for the Earth indicate a rough value for a of about 1 m.

From the equation giving the line element we get

$$[g_{\mu\nu}] = \begin{bmatrix} \left(1 - \dfrac{2mr}{\rho^2}\right) c^2 & 0 & 0 & \dfrac{2mcra\sin^2\theta}{\rho^2} \\ 0 & -\dfrac{\rho^2}{\Delta} & 0 & 0 \\ 0 & 0 & -\rho^2 & 0 \\ \dfrac{2mcra\sin^2\theta}{\rho^2} & 0 & 0 & -\left((r^2+a^2)\sin^2\theta + \dfrac{2mra^2\sin^4\theta}{\rho^2}\right) \end{bmatrix} \tag{4.74}$$

for the components of the metric tensor, from which we can deduce that the contravariant components are given by (see Exercise 3)

$$[g^{\mu\nu}] = \begin{bmatrix} \dfrac{1}{c^2\Delta}\left(r^2 + a^2 + \dfrac{2mra^2\sin^2\theta}{\rho^2}\right) & 0 & 0 & \dfrac{2mra}{c\rho^2\Delta} \\ 0 & -\dfrac{\Delta}{\rho^2} & 0 & 0 \\ 0 & 0 & -\dfrac{1}{\rho^2} & 0 \\ \dfrac{2mra}{c\rho^2\Delta} & 0 & 0 & -\dfrac{(1 - 2mr/\rho^2)}{\Delta\sin^2\theta} \end{bmatrix}. \tag{4.75}$$

Symmetries of the Kerr solution. The components of the metric tensor do not depend on t and ϕ, which means that the Kerr solution is both *stationary* and *axially symmetric*. These symmetries can also be expressed by saying that the metric tensor is invariant under either of the coordinate transformations:

(a) $t' = t + \text{const}$, r, θ, ϕ unchanged;
(b) $\phi' = \phi + \text{const}$, t, r, θ unchanged.

It is also invariant under the transformation:

(c) $t' = -t$, $\phi' = -\phi$, r, θ unchanged,

but not under either of the following

(d) $t' = -t$, r, θ, ϕ unchanged;
(e) $\phi' = -\phi$, t, r, θ unchanged.

in which only one of t and ϕ change sign.

This kind of behavior is characteristic of a spinning object. If we change the direction of time, or measure the angle of rotation in the opposite direction, then we expect some quantities to change sign. However, if we do both, then (for a steadily rotating object) things should remain the same, since running events backwards in time gives a rotation in the opposite direction.

We can use the fact that the metric tensor is independent of ϕ to establish the phenomenon generally known as the *dragging of inertial frames*, discussed below.

Dragging of inertial frames. In the language of Section 2.1, ϕ is a cyclic coordinate, so for a test particle following a timelike geodesic

$$\frac{\partial L}{\partial \dot{\phi}} = \text{const},$$

where

$$L = \left(1 - \frac{2mr}{\rho^2}\right) c^2 \dot{t}^2 + \frac{4mcra\sin^2\theta}{\rho^2}\dot{t}\dot{\phi} - \frac{\rho^2}{\Delta}\dot{r}^2 - \rho^2\dot{\theta}^2$$
$$- \left((r^2 + a^2)\sin^2\theta + \frac{2mra^2\sin^4\theta}{\rho^2}\right)\dot{\phi}^2,$$

and dots denote differentiation with respect to proper time τ. This gives

$$\frac{4mcra\sin^2\theta}{\rho^2}\dot{t} - 2\left((r^2 + a^2)\sin^2\theta + \frac{2mra^2\sin^4\theta}{\rho^2}\right)\dot{\phi} = \text{const}, \qquad (4.76)$$

which is the generalization of equation (4.23) of the Schwarzschild solution. Putting $\theta = \pi/2$ and $a = 0$ recovers equation (4.23).

Suppose now we have the case of a particle moving such that the constant in equation (4.76) is zero. For such a particle $\dot{\phi} \to 0$ as $r \to \infty$, so in the infinite distance it has no azimuthal velocity, but as it approaches the rotating object (i.e., for finite r) we see that

$$\boxed{\frac{d\phi}{dt} = \frac{\dot{\phi}}{\dot{t}} = \frac{2mcra}{(r^2 + a^2)\rho^2 + 2mra^2\sin^2\theta}.} \qquad (4.77)$$

In the Schwarzschild case, we would have $d\phi/dt = 0$, but here we have the non-zero result above. Thus a particle in free-fall gets swept along sideways by the rotating object. We say that the inertial frames are being *dragged around* by the rotating object. Mach would have been pleased by this discovery.[26]

Although the Kerr solution describes spacetime exterior to the rotating source, rather than interior to it, we now see that general relativity does indeed predict an effect not unlike the dragging of air close to a spinning ball.

[26] See Mach, 1893; also Weinberg, 1972, p.16.

In April 2004, after years of delay, four perfectly spherical gyroscope balls were finally sent into polar (Earth) orbit to begin measuring both the geodesic effect (see the remarks at the end of Section 4.7) as well as the much smaller Kerr frame-dragging.[27] (See Fig. 4.15.)

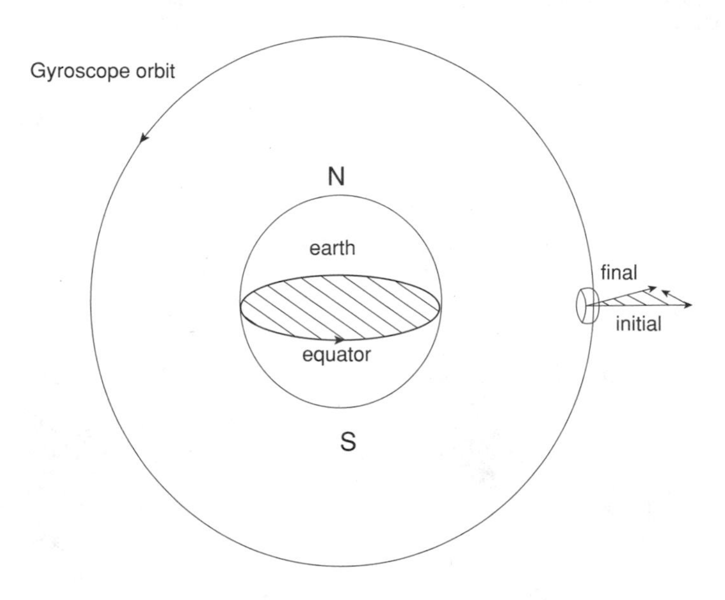

Fig. 4.15. Deviation, due to frame-dragging only, of a gyroscope orbiting the earth.

Some idea of how minuscule the Kerr effect is for motion near the surface of the Earth can be got from the following observations. Outside the Earth, $m \ll r$, $a \ll r$, and $t \approx \tau$, so equation (4.77) gives us the approximate equation

$$\frac{d\phi}{d\tau} = \frac{2mca}{r^3}$$

for estimating the rate of drag. Evaluating this using typical values of $2m = 8.89 \times 10^{-3}$ m, $c = 3 \times 10^8$ m s^{-1}, $a = 1$ m, and (for a satellite 400 miles above the Earth) $r = 7.04 \times 10^6$ m gives a value of 7.64×10^{-15} rad s^{-1} for $d\phi/d\tau$. This is equivalent to 2.41×10^{-7} radians, or 49.7 milliseconds of arc per year.

Event horizons and stationary-limit surfaces. In our discussion of black holes in Section 4.8, we noted two properties of the spherical surface given by $r = 2m$. One was that as the radial coordinate r_E of an emitter approached $2m$, the redshift of the radiation became infinitely large; in short, the surface is a *surface of infinite redshift*. The other was the impossibility of

[27] A twenty-five-page description of the 2004 Stanford Gravity Probe B experiment may be found at `http://einstein.stanford.edu`.

an external observer seeing events that occur inside the surface $r = 2m$; that is, the surface is an *event horizon*. We shall see that for the Kerr solution (with $a \neq 0$) things are a little different, in that there can be two event horizons and two surfaces of infinite redshift. Moreover, the event horizons do not coincide with the surfaces of infinite redshift.

What characterizes an event horizon is that photons originating on or inside it will never reach an outside observer, while photons originating outside it will eventually do so. For photons, we have (on omitting metric components that are zero)

$$0 = g_{00}\, dt^2 + 2g_{03}\, dt\, d\phi + g_{11}\, dr^2 + g_{22}\, d\theta^2 + g_{33}\, d\phi^2, \tag{4.78}$$

from which we get

$$\left(\frac{dr}{dt}\right)^2 = \frac{\Delta}{\rho^2}\left[g_{00} + g_{22}\left(\frac{d\theta}{dt}\right)^2 + g_{33}\left(\frac{d\phi}{dt}\right)^2 + 2g_{03}\frac{d\phi}{dt}\right]. \tag{4.79}$$

The expression on the right will approach zero as $\Delta \to 0$, no matter what the value of the square-bracketed term is. That is, the radial coordinate velocity dr/dt will tend towards zero the closer a photon gets to the surface given by $\Delta = 0$, no matter how it is moving. Hence we conclude that in the Kerr solution event horizons are given by $\Delta = 0$. This yields the following quadratic equation in r:

$$r^2 - 2mr + a^2 = 0. \tag{4.80}$$

Provided $m^2 > a^2$, this has the two solutions

$$r = m \pm \sqrt{m^2 - a^2}, \quad \text{or} \quad r = m(1 \pm \sqrt{1 - a^2/m^2})\,, \tag{4.81}$$

giving *two* event horizons $H_\pm$. We see that as $a \to 0$, the surface H_+ tends towards the event horizon $r = 2m$ of the Schwarzschild solution, while the surface H_- shrinks towards the point given by $r = 0$.

An argument like that in Section 4.3 gives

$$\frac{\lambda_R}{\lambda_E} = \frac{1 - (2mr/\rho^2)_R}{1 - (2mr/\rho^2)_E} \tag{4.82}$$

for the wavelength generalization of equation (4.14), showing that the redshift approaches infinity as emitters approach the surfaces given by

$$1 - (2mr/\rho^2) = 0, \quad \text{i.e.,} \quad r^2 - 2mr + a^2\cos^2\theta = 0.$$

Provided $m^2 > a^2$ we can again solve for r to get

$$r = m \pm \sqrt{m^2 - a^2\cos^2\theta}, \tag{4.83}$$

and in this way we get two surfaces $S_\pm$ of infinite redshift, different from the event horizons. We see that as $a \to 0$, the surface S_+ tends towards the infinite-redshift surface $r = 2m$ of the Schwarzschild solution, while the surface S_- shrinks towards the point given by $r = 0$.

Consider now a photon which at some instant is traveling in the azimuthal direction, so at that instant $dr/dt = d\theta/dt = 0$. Equation (4.79) then gives

$$0 = \frac{\Delta}{\rho^2}\left[g_{00} + g_{33}\left(\frac{d\phi}{dt}\right)^2 + 2g_{03}\frac{d\phi}{dt}\right],$$

so, on solving for $d\phi/dt$, we get

$$d\phi/dt = \left(-g_{03} \pm \sqrt{g_{03}^2 - g_{33}g_{00}}\right)/g_{33}. \tag{4.84}$$

So if the photon is situated where $g_{00} = 0$, then $d\phi/dt$ is either $-2g_{03}/g_{33}$, or zero. The zero value corresponds to the case where the photon is traveling against the spin and experiences such a large dragging effect that it appears to be unable to move azimuthally, while the non-zero value corresponds to traveling with the spin. Since $g_{00} = 0$ yields $r^2 - 2mr + a^2\cos^2\theta = 0$, this gives an alternate characterization of the surfaces $S_\pm$ as *stationary-limit surfaces.*

To get some idea of the relationship between the four surfaces $H_\pm$ and $S_\pm$ we shall make use of the Cartesian-like coordinates (x, y, z) used by Kerr[28] to obtain his solution. In terms of the coordinates (r, θ, ϕ), these are defined by

$$\begin{aligned} x &\equiv (r^2 + a^2)^{1/2}\sin\theta\cos(\phi + \alpha(r)), \\ y &\equiv (r^2 + a^2)^{1/2}\sin\theta\sin(\phi + \alpha(r)), \\ z &\equiv r\cos\theta, \end{aligned} \tag{4.85}$$

where

$$\alpha(r) = a\int_\infty^r \frac{dr}{\Delta} + \arctan\left(\frac{a}{r}\right).$$

When plotted in Euclidean space (using (x, y, z) as Cartesian coordinates) the surfaces $r =$ const are confocal ellipsoids, the surfaces $\theta =$ const are hyperboloids, and the surfaces $\phi =$ const are best described as distorted half-planes. As $a \to 0$, they become respectively the spheres, cones, and half-planes of spherical coordinates, as described in Example 1.1.1. For $r = 0$, the ellipsoid degenerates to the disc given by

$$x^2 + y^2 \le a^2, \qquad z = 0.$$

It can be shown (essentially by looking to see where the invariant quantity $R_{\mu\nu\sigma\tau}R^{\mu\nu\sigma\tau}$ becomes infinite) that the edge of this disc is a singularity; as $a \to 0$, this ring singularity shrinks to the point singularity given by $r = 0$ in the Schwarzschild solution.

[28] Usually referred to as Kerr–Schild coordinates.

Figure 4.16 serves to illustrate the relationship between the four surfaces $H_\pm$ and $S_\pm$, and the ring singularity. The curves shown are cross-sections through surfaces and the black dots represent the ring singularity; the full three-dimensional picture is got by revolving these about the vertical axis. As $a \to 0$, H_+ and S_+ both tend towards the event horizon $r = 2m$ of the

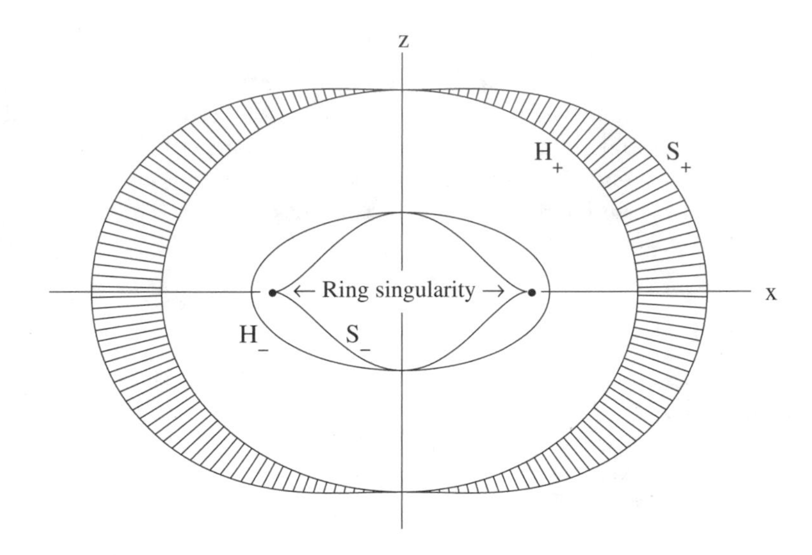

Fig. 4.16. Cross-sections of the surfaces $H_\pm$ and $S_\pm$, and the ring singularity of the Kerr solution. The shaded region represents the ergosphere.

Schwarzschild solution, while H_-, S_-, and the ring singularity shrink to the point given by $r = 0$. Because the surfaces H_- and S_- are inside the event horizon H_+ they are not visible to an external observer. However, they are of some theoretical interest and are discussed in more advanced texts.[29]

Note that it would be possible for an external observer to see a particle crossing (inwards or outwards) the stationary-limit surface S_+. It is also possible to imagine a particle traveling inwards and, after crossing S_+, decaying into two particles, one of which falls into the rotating black hole, while the other one escapes. More advanced texts show that the escaping particle can emerge with more energy than the combination had before entering—energy that has come from the angular momentum of the rotating object. Because of this possibility, the region between the sphere H_+ and the ellipsoid S_+ is

[29]See, for example, Misner, Thorne and Wheeler (1973), or Ohanian and Ruffini (1994).

referred to as the *ergoregion*, or *ergosphere* (from the Greek $\epsilon\rho\gamma o\nu$ meaning work, or energy).

The existence of the four surfaces $H_\pm$ and $S_\pm$ depends on having $a^2 < m^2$. If $a^2 > m^2$, then these do not exist and the ring singularity is exposed to view—it is a *naked singularity*.[30] The case $a^2 = m^2$ is that of *maximal rotation*, and its implications are left as an exercise. (See Exercise 6.)

We should end this section by noting that our discussion of event horizons and stationary-limit surfaces is somewhat unsatisfactory. The reason is that the Boyer–Lindquist coordinate system is inadequate for a proper discussion of these surfaces. It is adequate for discussing regions of spacetime away from the surfaces, but as the surfaces are approached, at least one of the metric tensor components tends to 0 or ∞. For this reason, any results should be regarded as having been obtained by a limiting approach to the surfaces. For a proper discussion one should work in a coordinate system analogous to the Eddington–Finkelstein coordinates used to discuss Schwarzschild black holes in Section 4.8.

Exercises 4.10

1. Show that with $a = 0$ the Kerr line element reduces to the Schwarzschild line element.

2. Check that the expression J/Mc, as used in equation (4.73), has the dimensions of length.

3. Verify that the contravariant components $g^{\mu\nu}$ of the Kerr metric tensor are as given by equation (4.75).

4. Check the claims made for the coordinate transformations (a)–(e) used to exhibit the symmetries of the Kerr solution.

5. The transformation from the Boyer–Lindquist coordinates (r, θ, ϕ) to the Kerr–Schild coordinates (x, y, z) is often expressed as

$$x + iy = (r + ia)\sin\theta\, \exp i(\phi + \textstyle\int_\infty^r (a/\Delta)\, dr),$$
$$z = r\cos\theta.$$

   Show that these lead to the equations (4.85).

6. In a physically realistic situation, m is positive, but a can be positive, negative, or zero. What happens to the surfaces $H_\pm$ and $S_\pm$ as $|a| \to m$?

[30] See, for example, Shapiro and Teukolsky, 1983, p. 358.

Problems 4

1. We have referred to the surface in space given by $\theta = \pi/2$ as the equatorial plane. Show that it is not flat.

2. A free particle of rest mass μ just misses an observer fixed in space at $r = r_0$. Show that he assigns to it an energy E given by

$$E = \mu c^2 (1 - 2m/r_0)^{-1/2} k,$$

where k is the integration constant of equation (4.22). (Use Exercise 3.1.2.)

3. Deduce from the previous problem that "the energy at infinity" of a particle is $\mu c^2 k$, and hence that a particle can escape to infinity only if $k \geq 1$.

4. Show that the proper time for a photon to complete one revolution at $r = 3m$, as measured by an observer stationed at $r = 3m$, is $6\pi m/c$. What orbital period does a very distant observer assign to the photon?

5. Show, by perturbing the geodesic in the equatorial plane, that the circular orbit of a photon at $r = 3m$ is unstable.

6. By considering the conservation of energy and angular momentum, show that the path of a particle moving in the equatorial plane under a Newtonian gravitational force due to a spherical object of mass M situated at the origin is given by

$$(du/d\phi)^2 + u^2 = E + 2GMu/h^2,$$

where $u \equiv 1/r$, h is the angular momentum per unit mass, and E is a constant.

7. Find the Schwarzschild radius of a spherical object with the same mass as that of the Earth.
(Take $M_\oplus = 6 \times 10^{24}\,\mathrm{kg}$, $G = 6.67 \times 10^{-11}\,\mathrm{N\,m^2\,kg^{-2}}$, $c = 3 \times 10^8\,\mathrm{m\,s^{-1}}$.)

8. Suppose we have two spherical objects that are just black holes; that is, for each r_B is its Schwarzschild radius. If one has the same mass as the Earth, and the other the same mass as the Sun, which has the greater density?
(Take $M_\oplus = 6 \times 10^{24}\,\mathrm{kg}$, $M_\odot = 2 \times 10^{30}\,\mathrm{kg}$.)

9. What form do equations (4.5) and (4.6) take in isotropic coordinates?

10. (a) It is stated in some sources that for a gyroscope in a circular polar orbit, 400 miles above the surface the Earth, the annual geodesic effect amounts to about 6600 milliseconds of arc. Verify this number.

(b) The Kerr effect is much smaller, being about 49.7 milliseconds of arc per year. Explain why a polar orbit is used to detect the Kerr effect, whereas any circular orbit could be used for the geodesic effect.

(c) The reference star (HR 8703) for the Gravity Probe B experiment lies in Pegasus, which is a constellation close to the celestial equator. For the gyroscope in a polar orbit, but pointing instead to our North Star, what would be the annual deviations of the geodesic and Kerr effects respectively?

Take $m = 4.445 \times 10^{-3}$ m, $c = 3 \times 10^{8}\,\mathrm{m\,s^{-1}}$, and the radius of the Earth to be 6.37×10^{6} m.

5

Gravitational radiation

5.0 Introduction

If we ask what characterizes radiation, our answers might include the transmission of energy and information through space, or the existence of a wave equation which some quantity satisfies. These aspects are, of course, related, in that there is a characteristic speed of transmission which is determined by the wave equation. In Newtonian gravitational theory energy (and information) is transmitted via the gravitational field which is determined by the gravitational potential V. In empty space V satisfies $\nabla^2 V = 0$, which is not a wave equation, but might be regarded as the limit of a wave equation in which the characteristic speed of transmission tends to infinity. Put another way, gravitational effects are, according to Newton's theory, transmitted instantaneously, which is thoroughly unsatisfactory from the relativistic point of view. Moreover, with an infinite speed of transmission it is impossible to associate a wavelength with a given frequency of oscillation.

Einstein's theory, being a relativistic theory, does not suffer from these defects, and as we shall see, it yields a wave equation for the propagation of gravitational disturbances with a characteristic speed equal to c. A discussion of gravitational radiation using the exact field equations is virtually impossible, because of their extreme nonlinearity (although considerable progress has been made in this direction over the last 20 years), and we shall therefore resort to a *linearization* of the equations appropriate for a weak gravitational field. This leads to the emergence of a wave equation, and allows us to compare gravitational with electromagnetic radiation.

Electromagnetic radiation is generated by accelerating charges, and by analogy we expect accelerating masses to produce gravitational radiation. By the same analogy, we might expect gravitational radiation to be predominantly dipole, but this is not the case. The *mass dipole moment* of a system of particles is, by definition, the 3-vector

$$\mathbf{d} = \sum_{\text{all particles}} m\mathbf{x},$$

where $\mathbf{x}$ is the position vector of a particle of mass m. Hence $\dot{\mathbf{d}}$ is the total momentum of the system, so $\ddot{\mathbf{d}} = \mathbf{0}$ by virtue of conservation of momentum, and it is because of this that we get no dipole radiation. We shall see in Section 5.3 that it is the second time derivative of the *second moment of the mass distribution* of the source that produces the radiation, showing that it is predominantly *quadrupole*.[1] This second moment is the tensor $\mathbf{I}$ with components defined by

$$I^{ij} = \sum_{\text{all particles}} m x^i x^j,$$

which for a continuous distribution takes the form of a volume integral:

$$I^{ij} = \int \rho x^i x^j \, dV.$$

We make use of both these forms in our discussion of generation of radiation in Section 5.3.

5.1 What wiggles?

As explained in the previous section, our approach to gravitational radiation is via a linearization of Einstein's theory appropriate for a weak field. This means that over extensive regions of spacetime there exist nearly Cartesian coordinate systems in which

$$g_{\mu\nu} = \eta_{\mu\nu} + h_{\mu\nu}, \tag{5.1}$$

where the $h_{\mu\nu}$ are small compared with unity, and such coordinate systems will be used throughout this chapter. The rules of the "linearization game" are as follows:

(a) $h_{\mu\nu}$ together with its first derivatives $h_{\mu\nu,\rho}$ and higher derivatives are small, and all products of these are ignored;
(b) suffixes are raised and lowered using $\eta^{\mu\nu}$ and $\eta_{\mu\nu}$, rather than $g^{\mu\nu}$ and $g_{\mu\nu}$.

The situation is like that of Section 3.6, but *without* the quasi-static condition. As a consequence of (a) and (b), all quantities having the kernel letter h are small, and products of them are ignored. The normal symbol for equality will be used to indicate equality up to first order in small quantities, as well as exact equality.

[1] In the classical radiation field associated with a quantum-mechanical particle of integral spin s, the $2s$-pole radiation predominates. Hence gravitons (quadrupole) should have spin 2, just as photons (dipole) have spin 1.

With these preliminaries explained, we have (see Exercise 2.7.1) $g^{\mu\nu} = \eta^{\mu\nu} - h^{\mu\nu}$, and

$$\Gamma^{\mu}_{\nu\sigma} = \tfrac{1}{2}\eta^{\mu\beta}(h_{\sigma\beta,\nu} + h_{\nu\beta,\sigma} - h_{\nu\sigma,\beta}) = \tfrac{1}{2}(h^{\mu}_{\sigma,\nu} + h^{\mu}_{\nu,\sigma} - h_{\nu\sigma}{}^{,\mu}), \tag{5.2}$$

on putting $\eta^{\mu\beta}h_{\nu\sigma,\beta} = h_{\nu\sigma}{}^{,\mu}$. So the Ricci tensor is

$$R_{\mu\nu} = \Gamma^{\alpha}_{\mu\alpha,\nu} - \Gamma^{\alpha}_{\mu\nu,\alpha} = \tfrac{1}{2}(h_{,\mu\nu} - h^{\alpha}_{\nu,\mu\alpha} - h^{\alpha}_{\mu,\nu\alpha} + h_{\mu\nu,\alpha}{}^{\alpha}), \tag{5.3}$$

where $h \equiv h^{\mu}_{\mu} = \eta^{\mu\nu}h_{\mu\nu}$, and the curvature scalar is

$$R \equiv g^{\mu\nu}R_{\mu\nu} = \eta^{\mu\nu}R_{\mu\nu} = h_{,\alpha}{}^{\alpha} - h^{\alpha\beta}{}_{,\alpha\beta}, \tag{5.4}$$

on relabeling suffixes. The covariant form of the field equations (3.38) then yields

$$h_{,\mu\nu} - h^{\alpha}_{\nu,\mu\alpha} - h^{\alpha}_{\mu,\nu\alpha} + h_{\mu\nu,\alpha}{}^{\alpha} - \eta_{\mu\nu}(h_{,\alpha}{}^{\alpha} - h^{\alpha\beta}{}_{,\alpha\beta}) = 2\kappa T_{\mu\nu},$$

and this simplifies to

$$\bar{h}_{\mu\nu,\alpha}{}^{\alpha} + (\eta_{\mu\nu}\bar{h}^{\alpha\beta}{}_{,\alpha\beta} - \bar{h}^{\alpha}_{\nu,\mu\alpha} - \bar{h}^{\alpha}_{\mu,\nu\alpha}) = 2\kappa T_{\mu\nu}, \tag{5.5}$$

on putting

$$\boxed{\bar{h}_{\mu\nu} \equiv h_{\mu\nu} - \tfrac{1}{2}h\eta_{\mu\nu}.} \tag{5.6}$$

A further simplification may be effected by means of a gauge transformation, a concept which we now explain.

A *gauge transformation* is a small change of coordinates defined by

$$x^{\mu'} \equiv x^{\mu} + \xi^{\mu}(x^{\alpha}), \tag{5.7}$$

where the ξ^{μ} are of the same order of smallness as the $h_{\mu\nu}$. Such a small change of coordinates takes a nearly Cartesian coordinate system into one of the same kind. The matrix element $X^{\mu'}_{\nu} \equiv \partial x^{\mu'}/\partial x^{\nu}$ is given by

$$X^{\mu'}_{\nu} = \delta^{\mu}_{\nu} + \xi^{\mu}{}_{,\nu}, \tag{5.8}$$

and a straightforward calculation (see Exercise 5.1.3) shows that under a gauge transformation

$$h^{\mu'\nu'} = h^{\mu\nu} - \xi^{\mu,\nu} - \xi^{\nu,\mu}, \tag{5.9}$$

$$h' = h - 2\xi^{\mu}{}_{,\mu}, \tag{5.10}$$

$$\bar{h}^{\mu'\nu'} = \bar{h}^{\mu\nu} - \xi^{\mu,\nu} - \xi^{\nu,\mu} + \eta^{\mu\nu}\xi^{\alpha}{}_{,\alpha}. \tag{5.11}$$

The inverse matrix element $X^{\mu}_{\nu'} \equiv \partial x^{\mu}/\partial x^{\nu'}$ is given by

$$X^{\mu}_{\nu'} = \delta^{\mu}_{\nu} - \xi^{\mu}{}_{,\nu}$$

(see Exercise 5.1.3), so that

$$\bar{h}^{\mu'\alpha'}{}_{,\alpha'} = \bar{h}^{\mu'\alpha'}{}_{,\beta} X^{\beta}_{\alpha'} = \bar{h}^{\mu'\alpha'}{}_{,\beta} \delta^{\beta}_{\alpha} = \bar{h}^{\mu'\alpha'}{}_{,\alpha} = \bar{h}^{\mu\alpha}{}_{,\alpha} - \xi^{\mu}{}_{,\alpha}{}^{\alpha}, \tag{5.12}$$

on using equation (5.11) and simplifying.

If therefore we choose ξ^μ to be a solution of

$$\xi^{\mu}{}_{,\alpha}{}^{\alpha} = \bar{h}^{\mu\alpha}{}_{,\alpha}, \tag{5.13}$$

then we have $\bar{h}^{\mu'\alpha'}{}_{,\alpha'} = 0$. In the new coordinate system, each term in the bracketed expression on the left of equation (5.5) is separately zero, and, on dropping primes, the equation reduces to

$$\bar{h}_{\mu\nu,\alpha}{}^{\alpha} = 2\kappa T_{\mu\nu}. \tag{5.14}$$

This simplified equation is valid whenever $\bar{h}^{\mu\nu}$ satisfies the *gauge condition*

$$\bar{h}^{\mu\alpha}{}_{,\alpha} = 0, \tag{5.15}$$

and the above considerations show that we can always arrange for this to be satisfied.

This simplification is an exact parallel of that introduced into electromagnetism by adopting the Lorentz gauge condition (see Sec. A.8). The quantities $\bar{h}^{\mu\nu}$ correspond to the 4-potential A^μ, and the gauge condition (5.15) corresponds to the Lorentz gauge condition $A^{\mu}{}_{,\mu} = 0$ (see Exercise A.8.1). A gauge transformation $A_\mu \to A_\mu - \psi_{,\mu}$ will preserve the Lorentz gauge condition if and only if $\psi_{,\mu}{}^{\mu} = 0$. Correspondingly, as equation (5.12) shows, a gauge transformation (5.7) will preserve the gauge condition (5.15) if and only if

$$\xi^{\mu}{}_{,\alpha}{}^{\alpha} = 0. \tag{5.16}$$

Let us introduce[2] the *d'Alembertian* $\Box^2$ defined by

$$\Box^2 \equiv -\eta^{\alpha\beta}\partial_\alpha\partial_\beta, \tag{5.17}$$

so that

$$\Box^2 = \partial^2/\partial x^2 + \partial^2/\partial y^2 + \partial^2/\partial z^2 - c^{-2}\partial^2/\partial t^2 = \nabla^2 - c^{-2}\partial^2/\partial t^2, \tag{5.18}$$

on putting $x^0 \equiv ct$, $x^1 \equiv x$, $x^2 \equiv y$, $x^3 \equiv z$. Then for any quantity f,

$$f_{,\alpha}{}^{\alpha} = \eta^{\alpha\beta} f_{,\alpha\beta} = -\Box^2 f,$$

and we see that the results above may be summarized as follows.

The quantities $\bar{h}^{\mu\nu} \equiv h^{\mu\nu} - \frac{1}{2}h\eta^{\mu\nu}$ satisfy

$$\boxed{\Box^2 \bar{h}^{\mu\nu} = -2\kappa T^{\mu\nu},} \tag{5.19}$$

[2]See footnote 9 in Sec. A.8.

provided the gauge condition

$$\bar{h}^{\mu\nu}{}_{,\nu} = 0 \tag{5.20}$$

holds. The remaining gauge freedom $x^\mu \to x^\mu + \xi^\mu$ preserves the gauge condition provided ξ^μ satisfies

$$\boxed{\Box^2 \xi^\mu = 0.} \tag{5.21}$$

Inasmuch as equation (5.19) is a wave equation with source term equal to $-2\kappa T^{\mu\nu} \equiv (16\pi G/c^4)T^{\mu\nu}$, the answer to the question posed in the section heading is $\bar{h}^{\mu\nu}$, a quantity related to $h_{\mu\nu}$, which represents a perturbation of the metric tensor $g_{\mu\nu}$ away from the flat metric tensor $\eta_{\mu\nu}$. In empty spacetime equation (5.19) reduces to $\Box^2 \bar{h}^{\mu\nu} = 0$, and we see that gravitational radiation propagates through empty spacetime with the speed of light.

Exercises 5.1

1. Check expressions (5.3), (5.4), and (5.5) given for the Ricci tensor, the curvature scalar, and the field equations.

2. If $\bar{h}$ is defined by $\bar{h} \equiv \bar{h}^\mu_\mu$, show that $\bar{h} = -h$, and hence that $h_{\mu\nu} = \bar{h}_{\mu\nu} - \frac{1}{2}\bar{h}\eta_{\mu\nu}$.

3. Use the equality $g^{\mu\nu} = \eta^{\mu\nu} - h^{\mu\nu}$ to check equation (5.9). Deduce equation (5.10), where $h' \equiv \eta_{\mu\nu} h^{\mu'\nu'}$, and hence verify equation (5.11). Show also that $X^\mu_{\nu'} = \delta^\mu_\nu - \xi^\mu{}_{,\nu}$.

4. Equations (5.19) and (5.20) together imply that $T^{\mu\nu}{}_{,\nu} = 0$. Is this consistent with $T^{\mu\nu}{}_{;\nu} = 0$?

5.2 Two polarizations

The simplest sort of solution to the wave equation $\Box^2 \bar{h}^{\mu\nu} = 0$ of empty spacetime is that representing a plane wave, given by

$$\bar{h}^{\mu\nu} = \Re\left[A^{\mu\nu} \exp(ik_\alpha x^\alpha)\right], \tag{5.22}$$

where $[A^{\mu\nu}]$ is the *amplitude matrix* having constant entries, $k^\mu \equiv \eta^{\mu\alpha} k_\alpha$ is the *wave* 4*-vector* in the direction of propagation, and $\Re$ denotes that we take the real part of the bracketed expression following it. It follows from $\Box^2 \bar{h}^{\mu\nu} = 0$ that k^μ is null, and from the gauge condition (5.20) that

$$A^{\mu\nu} k_\nu = 0. \tag{5.23}$$

Since $\bar{h}^{\mu\nu} = \bar{h}^{\nu\mu}$, we see that the amplitude matrix has ten different (complex) entries, but the condition (5.23) gives four conditions on these, cutting their number down to six (see Exercise 5.2.1). However, we still have the gauge freedom $x^\mu \to x^\mu + \xi^\mu$ (subject to the condition (5.21)), and as we shall see, this may be used to reduce the number still further, so that ultimately there are just *two* entries in the amplitude matrix which may be independently specified. This results in two possible polarizations for plane gravitational waves.

To fix our ideas, let us consider a plane wave propagating in the x^3 direction, so that

$$k^\mu = (k, 0, 0, k) \quad \text{and} \quad k_\mu = (k, 0, 0, -k), \tag{5.24}$$

where $k > 0$. Thus $k = \omega/c$, where ω is the angular frequency. Equation (5.23) gives $A^{\mu 0} = A^{\mu 3}$, which implies that all the $A^{\mu\nu}$ may be expressed in terms of A^{00}, A^{01}, A^{02}, A^{11}, A^{12}, and A^{22}:

$$[A^{\mu\nu}] = \begin{bmatrix} A^{00} & A^{01} & A^{02} & A^{00} \\ A^{01} & A^{11} & A^{12} & A^{01} \\ A^{02} & A^{12} & A^{22} & A^{02} \\ A^{00} & A^{01} & A^{02} & A^{00} \end{bmatrix}. \tag{5.25}$$

Consider now a gauge transformation generated by

$$\xi^\mu = -\Re\left[i\varepsilon^\mu \exp(ik_\alpha x^\alpha)\right],$$

where the ε^μ are constants. This satisfies the condition (5.21), as required, and has

$$\xi^\mu_{,\nu} = \Re\left[\varepsilon^\mu k_\nu \exp(ik_\alpha x^\alpha)\right]. \tag{5.26}$$

In the new gauge the amplitude matrix is defined by

$$\bar{h}^{\mu'\nu'} = \Re\left[A^{\mu'\nu'} \exp(ik_{\alpha'} x^{\alpha'})\right],$$

and since $\exp(ik_{\alpha'} x^{\alpha'})$ differs from $\exp(ik_\alpha x^\alpha)$ by only a first-order quantity, substitution in equation (5.11) and using equation (5.26) gives

$$A^{\mu'\nu'} = A^{\mu\nu} - \varepsilon^\mu k^\nu - k^\mu \varepsilon^\nu + \eta^{\mu\nu}(\varepsilon^\alpha k_\alpha).$$

If we feed in k^μ from equation (5.24) and $A^{\mu\nu}$ from equation (5.25), then we obtain

$$\begin{aligned} A^{0'0'} &= A^{00} - k(\varepsilon^0 + \varepsilon^3), & A^{1'1'} &= A^{11} - k(\varepsilon^0 - \varepsilon^3), \\ A^{0'1'} &= A^{01} - k\varepsilon^1, & A^{1'2'} &= A^{12}, \\ A^{0'2'} &= A^{02} - k\varepsilon^2, & A^{2'2'} &= A^{22} - k(\varepsilon^0 - \varepsilon^3). \end{aligned} \tag{5.27}$$

So conveniently choosing our constants ε^μ to be

$$\begin{aligned} \varepsilon^0 &= (2A^{00} + A^{11} + A^{22})/4k, & \varepsilon^1 &= A^{01}/k, \\ \varepsilon^2 &= A^{02}/k, & \varepsilon^3 &= (2A^{00} - A^{11} - A^{22})/4k, \end{aligned}$$

we obtain $A^{0'0'} = A^{0'1'} = A^{0'2'} = 0$ and $A^{1'1'} = -A^{2'2'}$.

On dropping the primes, we see that in the new gauge the amplitude matrix has just four entries, A^{11}, A^{12}, A^{21}, A^{22}, and of these only two may be independently specified, because $A^{11} = -A^{22}$, and $A^{12} = A^{21}$. This new gauge, *which is determined by the wave itself*, is known as the *transverse traceless gauge*, or *TT gauge* for short. In this gauge $\bar{h} \equiv \bar{h}^\mu_\mu = 0$ (because $A^{00} = A^{33} = 0$ and $A^{11} = -A^{22}$), and it follows that $h = 0$ so there is no difference between $h_{\mu\nu}$ and $\bar{h}_{\mu\nu}$ (see Exercise 5.1.2). It is because $h = \bar{h} = 0$ that the gauge is called traceless, and it is because $h_{0\mu} = \bar{h}_{0\mu} = 0$ that it is called transverse. We shall work in the TT gauge for the remainder of this section.

If we introduce two *linear polarization matrices* $[e_1^{\mu\nu}]$ and $[e_2^{\mu\nu}]$, defined by

$$[e_1^{\mu\nu}] = \begin{bmatrix} 0 & 0 & 0 & 0 \\ 0 & 1 & 0 & 0 \\ 0 & 0 & -1 & 0 \\ 0 & 0 & 0 & 0 \end{bmatrix}, \quad [e_2^{\mu\nu}] = \begin{bmatrix} 0 & 0 & 0 & 0 \\ 0 & 0 & 1 & 0 \\ 0 & 1 & 0 & 0 \\ 0 & 0 & 0 & 0 \end{bmatrix}, \tag{5.28}$$

we see that the general amplitude matrix is a linear combination of them:

$$A^{\mu\nu} = \alpha e_1^{\mu\nu} + \beta e_2^{\mu\nu}, \tag{5.29}$$

where α and β are (complex) constants.

The significance of these matrices may be appreciated by reviewing the analogous situation in electromagnetic radiation, where the plane-wave solution to $\Box^2 A^\mu = 0$ is

$$A^\mu = \Re\left[B^\mu \exp(ik_\alpha x^\alpha)\right], \quad B^\mu = \text{constant}.$$

The Lorentz gauge condition $A^\mu{}_{,\mu} = 0$ implies that $B^\mu k_\mu = 0$, which reduces the number of independent components of the *amplitude vector* B^μ to three. If, as before, we consider a wave propagating in the x^3 direction, so that $k^\mu = (k, 0, 0, k)$, then $B^\mu k_\mu = 0$ implies that $B^0 = B^3$, so

$$B^\mu = (B^0, B^1, B^2, B^0),$$

which is analogous to equation (5.25). Changing the gauge by putting $A'_\mu = A_\mu - \psi_{,\mu}$, where

$$\psi = -\Re\left[i\varepsilon \exp(ik_\alpha x^\alpha)\right],$$

preserves the Lorentz condition (because $\Box^2\psi = 0$), and transforms B^μ to

$$(B')^\mu = B^\mu - \varepsilon k^\mu.$$

So

$$(B')^0 = B^0 - \varepsilon k, \quad (B')^1 = B^1, \quad (B')^2 = B^2,$$

which are analogous to equations (5.27). If therefore we choose $\varepsilon = B^0/k$, then, on dropping primes, we have $B^0 = 0$, and in the new gauge the amplitude vector has just two components (B^1 and B^2) which may be independently specified. This leads to two *linear polarization vectors*:

$$e_1^\mu = (0,1,0,0) \quad \text{and} \quad e_2^\mu = (0,0,1,0), \tag{5.30}$$

and the general amplitude vector is a linear combination of these:

$$B^\mu = \alpha e_1^\mu + \beta e_2^\mu.$$

If $B^\mu = \alpha e_1^\mu$, then the force on a free test charge is in the x^1 direction with a magnitude that varies sinusoidally as the wave passes, causing it to oscillate in the x^1 direction, whereas if $B^\mu = \beta e_2^\mu$ the oscillations take place in the x^2 direction. (These facts may be derived by using equation (A.59).) The particular combinations $B^\mu = \alpha(e_1^\mu \pm i e_2^\mu)$ give *circularly polarized waves* in which the mutually orthogonal oscillations combine, so that the test charge moves in a circle. The polarization matrices (5.28) have a similar effect on free test particles as the gravitational wave passes, as we shall now show.

A short calculation (see Exercise 5.2.2) shows that in the TT gauge $\Gamma^\mu_{00} = 0$, and this implies that the geodesic equation (2.71) is satisfied by $\dot{x}^\mu \equiv dx^\mu/d\tau = c\delta^\mu_0$. (This gives $g_{\mu\nu}\dot{x}^\mu\dot{x}^\nu = c^2$, as required, since in the TT gauge $h_{00} = 0$, so $g_{00} = 1$.) Hence curves having constant spatial coordinates are timelike geodesics, and may be taken as the world lines of a cloud of test particles. It follows that a small spacelike vector $\xi^\mu = (0, \xi^1, \xi^2, \xi^3)$ which gives the spatial separation between two nearby particles of the cloud is constant (see also Problem 5.1). However, this does *not* mean that their spatial separation d is constant, for d is given by

$$d^2 = \tilde{g}_{ij}\xi^i\xi^j,$$

where

$$\tilde{g}_{ij} \equiv -g_{ij} = \delta_{ij} - h_{ij},$$

and the h_{ij} are not constant. If we put

$$\zeta^i \equiv \xi^i + \tfrac{1}{2}h^i_k\xi^k, \tag{5.31}$$

then (to first order in $h_{\mu\nu}$)

$$\delta_{ij}\zeta^i\zeta^j = (\delta_{ij} - h_{ij})\xi^i\xi^j = d^2, \tag{5.32}$$

as a short calculation shows (see Exercise 5.2.3). So ζ^i may be regarded as a faithful position vector giving *correct* spatial separations when contracted with the Euclidean metric tensor δ_{ij}.

Note that in the TT gauge $h^3_i = 0$, so equation (5.31) gives $\zeta^3 = \xi^3 =$ constant. Hence if the test particle separation lies in the direction of propagation of the wave, then it is unaffected by the passage of the wave, showing that a gravitational wave is *transverse*.

Let us now select a particular test particle as a reference particle, and refer the motion of others to it by means of ζ^i, using equation (5.31). If $A^{\mu\nu} = \alpha e_1^{\mu\nu}$, with α real and positive for convenience, and $\xi^i = (\xi^1, \xi^2, 0)$, then this equation gives

$$\zeta^i = (\xi^1, \xi^2, 0) - \tfrac{1}{2}\alpha \cos k(x^0 - x^3)(\xi^1, -\xi^2, 0). \tag{5.33}$$

So if we consider those particles which, when $\cos k(x^0 - x^3) = 0$, form a circle

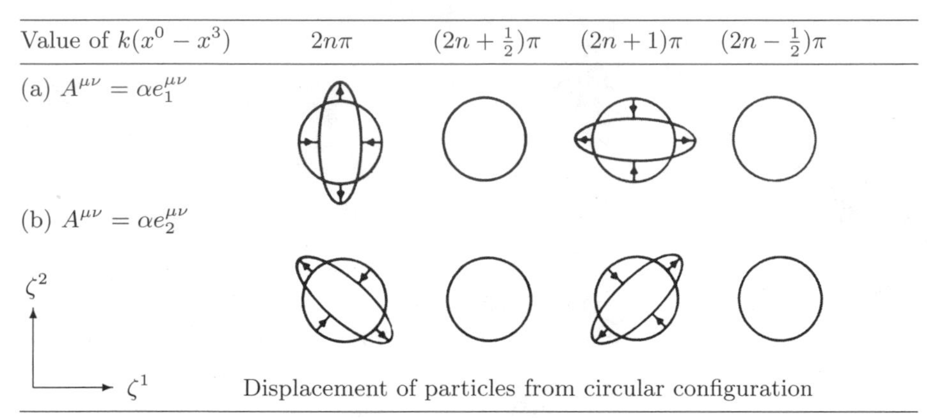

Table 5.1. Effect of a plane wave on a transverse ring of test particles.

with the reference particle as center, this circle lying in a plane perpendicular to the direction of propagation, then as the wave passes, these particles remain coplanar, and at other times their positions are as shown in row (a) of Table 5.1. All this follows from equation (5.33).

If, however, $A^{\mu\nu} = \alpha e_2^{\mu\nu}$, again with α real and positive for convenience, then

$$\zeta^i = (\xi^1, \xi^2, 0) - \tfrac{1}{2}\alpha \cos k(x^0 - x^3)(\xi^2, \xi^1, 0), \tag{5.34}$$

resulting in a sequence of diagrams as shown in row (b) of Table 5.1, which may be obtained from those in row (a) by a 45° rotation.[3]

In this way we see how the two polarizations of a plane gravitational wave affect the relative displacements of test particles. As in electromagnetic radiation, we may also have circularly polarized waves in which $A^{\mu\nu} = \alpha(e_1^{\mu\nu} \pm ie_2^{\mu\nu})$ (see Problem 5.2).

[3]This expressive means of showing the effect of the polarization mode on a cloud of test particles is borrowed from Misner, Thorne, and Wheeler. See Misner, Thorne, and Wheeler, 1973, §35.6.

Exercises 5.2

1. Show that equation (5.23) implies that each $A^{0\mu}$ may be expressed in terms of the A^{ij} for any (null) k^μ.

2. Show that in the TT gauge associated with any plane wave, $\Gamma^\mu_{00} = 0$ and $\Gamma^\mu_{0\nu} = \frac{1}{2}h^\mu_{\nu,0}$.

3. Verify equation (5.32). (Recall that $\eta_{ij} = -\delta_{ij}$.)

4. Check equations (5.33) and (5.34).

5. In constructing Table 5.1 we took α to be real and positive. What is the effect of having $\alpha = |\alpha|\, e^{i\theta}$, $\theta \neq 0$?

5.3 Simple generation and detection

Equation (5.19) gives the relation between the gravitational radiation, represented by $\bar{h}^{\mu\nu}$, and its source, represented by $T^{\mu\nu}$. The solution of this equation is well-known from electromagnetism, and may be expressed as a retarded integral:

$$\boxed{\bar{h}^{\mu\nu}(x^0, \mathbf{x}) = \frac{\kappa}{2\pi} \int \frac{T^{\mu\nu}(x^0 - |\mathbf{x} - \mathbf{x}'|, \mathbf{x}')}{|\mathbf{x} - \mathbf{x}'|} dV'.} \tag{5.35}$$

Here $\mathbf{x}$ represents the spatial coordinates of the *field point* at which $\bar{h}^{\mu\nu}$ is determined, $\mathbf{x}'$ represents those of a point of the source, and $|\mathbf{x} - \mathbf{x}'|$ is the spatial distance between them. The volume integral is taken over the region of spacetime occupied by the points of the source at the retarded times $x^0 - |\mathbf{x} - \mathbf{x}'|$. This region of spacetime is the intersection of the past half of the null cone at the field point with the world tube of the source.[4]

Suppose now that the source is some sort of matter distribution localized near the origin O, and that the source particles have speeds which are small compared with c. If we take our field point at a distance r from O that is large compared to the maximum displacements of the source particles from O, then equation (5.35) may be approximated by[5]

$$\bar{h}^{\mu\nu}(ct, \mathbf{x}) = -\frac{4G}{c^4 r} \int T^{\mu\nu}(ct - r, \mathbf{x}')\, dV', \tag{5.36}$$

[4]See, for example, Landau and Lifshitz, 1980, §§62, 63.

[5]See, for example, Landau and Lifshitz, 1980, §§66, 67. Note that the assumption of small speeds is equivalent to the dimensions of the source being small compared with the wavelength. If it were not made, then equation (5.41) would contain extra terms indicating radiation from moments higher than the quadrupole, and for this reason the assumption is sometimes referred to as the *quadrupole assumption*.

on putting $\kappa = -8\pi G/c^4$ and $x^0 = ct$. This approximation is appropriate for looking at the gravitational radiation in the *far zone* or *wave zone*, and by comparison with electromagnetic theory, we expect that (over not too large regions of space) it looks like a plane wave, in which case the radiative part of $\bar{h}^{\mu\nu}$ is completely determined by its spatial part $\bar{h}^{ij}$, as Exercise 5.2.1 shows. It follows that we need only consider $\int T^{ij} dV$ (at the retarded time), a neat expression for which may be obtained as follows.

The stress tensor of the source satisfies the conservation equation $T^{\mu\nu}{}_{,\nu} = 0$ (see Exercise 5.1.4). That is,

$$T^{00}{}_{,0} + T^{0k}{}_{,k} = 0, \tag{5.37}$$

$$T^{i0}{}_{,0} + T^{ik}{}_{,k} = 0. \tag{5.38}$$

Consider the integral identity

$$\int (T^{ik} x^j)_{,k} dV = \int T^{ik}{}_{,k} x^j dV + \int T^{ij} dV,$$

where the integrals are taken over a region of space enclosing the source, so that $T^{\mu\nu} = 0$ on the boundary of the region. The integral on the left is zero, as may be seen by converting it to a surface integral over the boundary using Gauss's theorem. Hence on using equation (5.38),

$$\int T^{ij} dV = -\int T^{ik}{}_{,k} x^j dV = \int T^{i0}{}_{,0} x^j dV = \frac{1}{c}\frac{d}{dt}\int T^{i0} x^j dV.$$

Interchanging i and j and adding gives

$$\int T^{ij} dV = \frac{1}{2c}\frac{d}{dt}\int \left(T^{i0} x^j + T^{j0} x^i\right) dV. \tag{5.39}$$

But

$$\int (T^{0k} x^i x^j)_{,k} dV = \int T^{0k}{}_{,k} x^i x^j dV + \int \left(T^{0i} x^j + T^{0j} x^i\right) dV,$$

where again the integral on the left vanishes by Gauss's theorem. Hence on using equation (5.37) we have

$$\int \left(T^{0i} x^j + T^{0j} x^i\right) dV = \frac{1}{c}\frac{d}{dt}\int T^{00} x^i x^j dV. \tag{5.40}$$

Combining equations (5.39) and (5.40) gives

$$\int T^{ij} dV = \frac{1}{2c^2}\frac{d^2}{dt^2}\int T^{00} x^i x^j dV.$$

For slowly moving source particles $T^{00} \approx \rho c^2$, where ρ is the proper density, and equation (5.36) yields the approximate expression

$$\boxed{\bar{h}^{ij}(ct, \mathbf{x}) = -\frac{2G}{c^4 r}\frac{d^2}{dt^2}\left[\int \rho x^i x^j dV\right]_{\text{ret}},} \tag{5.41}$$

the notation indicating that the integral is evaluated at the retarded time $t - r/c$. The integrand is recognizable as the second moment of the mass distribution, and indicates the essentially quadrupole nature of gravitational radiation (compare remarks at the end of Sec. 5.0).

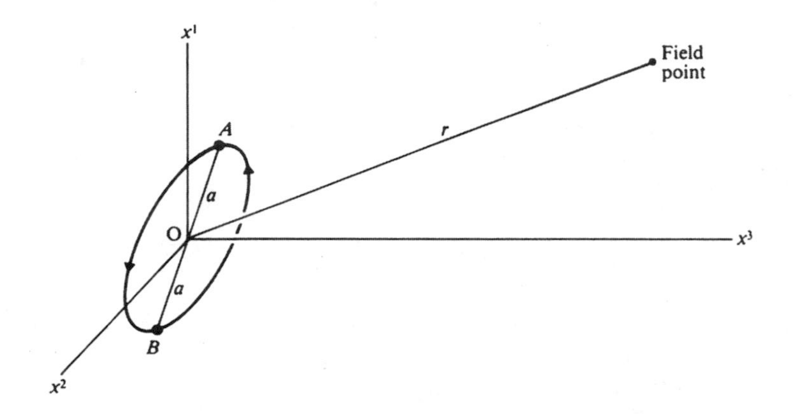

Fig. 5.1. A rotating dumbbell as a source of gravitational radiation.

As an illustration of the above ideas, let us take as a source a dumbbell consisting of two particles A and B of equal mass M connected by a light rod of length $2a$, the dumbbell rotating about the x^3 axis in the positive sense with angular constant speed ω, so that the particles remain in the plane $x^3 = 0$ with the midpoint of the rod at O (see Fig. 5.1). The positions of A and B at time t may be taken to be

$$x^j = \pm(a\cos\omega t, a\sin\omega t, 0),$$

and equation (5.41) gives (on replacing the integral by a sum over the two source particles)

$$\begin{aligned}
\left[\bar{h}^{ij}(ct, \mathbf{x})\right] &= -\frac{4GMa^2}{c^4 r}\frac{d^2}{dt^2}\begin{bmatrix} \cos^2\omega t & \cos\omega t\sin\omega t & 0 \\ \cos\omega t\sin\omega t & \sin^2\omega t & 0 \\ 0 & 0 & 0 \end{bmatrix}_{\text{ret}} \\
&= \frac{8GMa^2\omega^2}{c^4 r}\begin{bmatrix} \cos 2\omega(t - r/c) & \sin 2\omega(t - r/c) & 0 \\ \sin 2\omega(t - r/c) & -\cos 2\omega(t - r/c) & 0 \\ 0 & 0 & 0 \end{bmatrix}.
\end{aligned} \tag{5.42}$$

This clearly represents a gravitational wave of angular frequency 2ω.

For field points near to a point on the x^3-axis where $r \approx x^3$, we have

$$[\bar{h}^{ij}] \approx \frac{8GMa^2\omega^2}{c^4 r} \Re \left[(e_1^{ij} - ie_2^{ij}) \exp \frac{2i\omega}{c}(x^0 - x^3) \right], \tag{5.43}$$

and to an observer on the x^3-axis the wave looks like a circularly polarized plane wave (see Sec. 5.2). Note that this plane-wave approximation automatically has $\bar{h}^{ij}$ in its TT gauge. This does not happen for the plane-wave approximation which an observer at a field point in the plane $x^3 = 0$ holds to be valid, and a transformation to its TT gauge is needed to find its polarization mode (see Exercise 5.3.1).

Equation (5.42) generalizes to give the $\bar{h}^{ij}$ produced by a straight bar, having its center of mass at O and rotating in the plane $x^3 = 0$ with angular speed ω, simply by replacing $2Ma^2$ by the moment of inertia I of the bar about the x^3-axis. It may be shown by methods beyond the scope of this book that the rate at which such an object loses energy by radiation is given by

$$\frac{dE}{dt} = \frac{32GI^2\omega^6}{5c^5}. \tag{5.44}$$

This expression is obtained by looking at the energy–momentum carried by the gravitational field itself, which is quadratic in $h_{\mu\nu}$ and its derivatives, and consequently neglected in the linearized theory. As a consequence of the energy loss, ω must decrease, but in the linearized theory it remains constant.

If we feed into expression (5.44) typical laboratory values for I and ω, we find that the power of a laboratory generator is so small that we must look to astrophysical phenomena as possible sources for observable gravitational radiation (see Exercise 5.3.2). These include continuous generators, such as binary stars and pulsating neutron stars, as well as impulsive generators, such as colliding black holes, which would be expected to give off bursts of radiation.

Let us now consider detection. Our discussion of polarization in Section 5.2 shows that the effect of a gravitational wave on a cloud of free test particles is a variation in their separations; it is as if a varying tidal force were acting on the cloud. If the test particles were not free, but constrained to be the constituent particles of an elastic body, then this tidal force would give rise to vibrations in the body, and here we have the rudiments of a gravitational-wave detector. If the incident radiation were a plane wave of a given frequency, then the responsiveness of such a detector would be enhanced if its fundamental frequency of vibration were to coincide with that of the wave.

Because of the extreme low power of laboratory generators and the extreme distance of astrophysical generators, detection of the predicted radiation was for a long time thought impossible. However, over the past two decades, considerable ingenuity and effort have gone into the design and construction of detectors, including mechanical ones based on the principle outlined above. Observation of the extraordinary binary neutron star PSR 1913+16 indicates that it is losing energy at a rate attributable to gravitational radiation. Perhaps one day we shall see gravitational astronomy takes its place alongside optical and radio astronomy.

Exercises 5.3

1. Obtain the plane-wave approximation to the radiation from a rotating dumbbell, held to be valid by an observer at a field point in the plane of rotation, in its TT gauge, and hence show that it is linearly polarized. (To relate matters to the theory of Sec. 5.2, you will find it convenient to put the field point on the x^3 axis, and to let the dumbbell rotate about one of the other coordinate axes.)

2. Show that the power generated by:
   (a) a steel bar of mass 2×10^5 kg (about 200 tons) and length 10 m, rotating at an angular speed of 50 rad s^{-1} about an axis through its center of mass which is perpendicular to its length, is about 7.6×10^{-30} W;
   (b) a binary star, with equal components each of one solar mass, describing circular orbits with a period of one month, is about 5.8×10^{16} W. (Take $G = 6.67 \times 10^{-11}\,\mathrm{N\,m^2\,kg^{-2}}$, $c = 3 \times 10^8\,\mathrm{m\,s^{-1}}$, $M_\odot = 2 \times 10^{30}$ kg.)

Problems 5

1. Show that if in the TT gauge associated with a plane wave $\dot{x}^\mu = c\delta^\mu_0$, then the equation of geodesic deviation (3.41) has ξ^μ = constant as a solution.

2. The separation vector ξ^i of two test particles reacting to a circularly polarized wave propagating in the x^3 direction takes the form $\xi^i = (\xi^1, \xi^2, 0) =$ constant (see Sec. 5.2 for details). Show that one of the particles moves in a circle with respect to the other.

3. Use equation (5.36) to find $\bar{h}^{\mu\nu}$ in the far zone, due to a single particle of mass M situated at the origin. Obtain the corresponding line element $c^2 d\tau^2 = g_{\mu\nu} dx^\mu dx^\nu$, and compare it with the approximation of the Schwarzschild line element in isotropic coordinates (equation (4.67)) valid for large ρ.

4. Four particles of equal mass are situated at the ends of the arms of a light rigid cross having arms of equal length, and the whole configuration rotates freely about an axis through its center of mass perpendicular to its plane. Show that in the far zone there is no quadrupole radiation. (Use equations (5.41) with the integral replaced by a sum over the four source particles.)

6

Elements of cosmology

6.0 Introduction

The fundamental force keeping solar systems, binary stars, and galaxies together is the force of gravity (as opposed to electric, magnetic, and nuclear forces), and it is not unreasonable to suppose that the force governing the large-scale motions of the entire universe is primarily gravitational. If there is some other force governing these motions, there has to date been no evidence for it, neither in the solar system, nor in the observable galaxies. By the universe we mean all detectable components in the sky: stars, galaxies, constellations, pulsars, quasars, as well as such things as cosmic rays and background radiation. If this directly observable universe is part of a much grander system of universe-within-universes (C.V.I. Charlier's hypothesis[1]) then there is little we can say.

General relativity is a satisfactory theory of gravitation, correctly predicting the motions of particles and photons in curved spacetime, but in order to apply it to the universe we must make some simplifying assumptions. We shall grossly idealize the universe, and model it by a simple macroscopic fluid, devoid of shear-viscous, bulk-viscous, and heat-conductive properties. Its stress tensor $T_{\mu\nu}$ is then that of a perfect fluid, so

$$T_{\mu\nu} = (\rho + p)u_\mu u_\nu - p g_{\mu\nu}, \tag{6.1}$$

where ρ is its proper density, p is its pressure, u_μ is the (covariant) world velocity of the fluid particles (stars, etc.) and for convenience we have adopted units in which $c = 1$.

Any results we obtain from general relativity should agree with observation. In astronomy, it is never easy to give numbers exactly, but the major items of data that we possess for the universe include the following observed properties:

[1]See Charlier, 1922.

(i) *Homogeneity.* The number of galaxies per unit volume, and therefore the density ρ, appear to be uniform throughout large regions of space. A ball-park figure for the visible mass associated with galaxies is $\rho = 2 \times 10^{-28}\,\mathrm{kg\,m^{-3}}$.
(ii) *Isotropy.* The number of galaxies per unit solid angle appears to be the same in all directions.
(iii) *Redshift.* There is a redshift $z \equiv \Delta\lambda/\lambda_0$ for the wavelength of light emitted by galaxies, and z increases with distance.
(iv) *Olbers' paradox data.* The night sky is not as bright as day. The universe cannot therefore be (spatially) infinite if it is also homogeneous, unless there is a mechanism beyond the inverse-square law for weakening the energy from distant stars and galaxies.

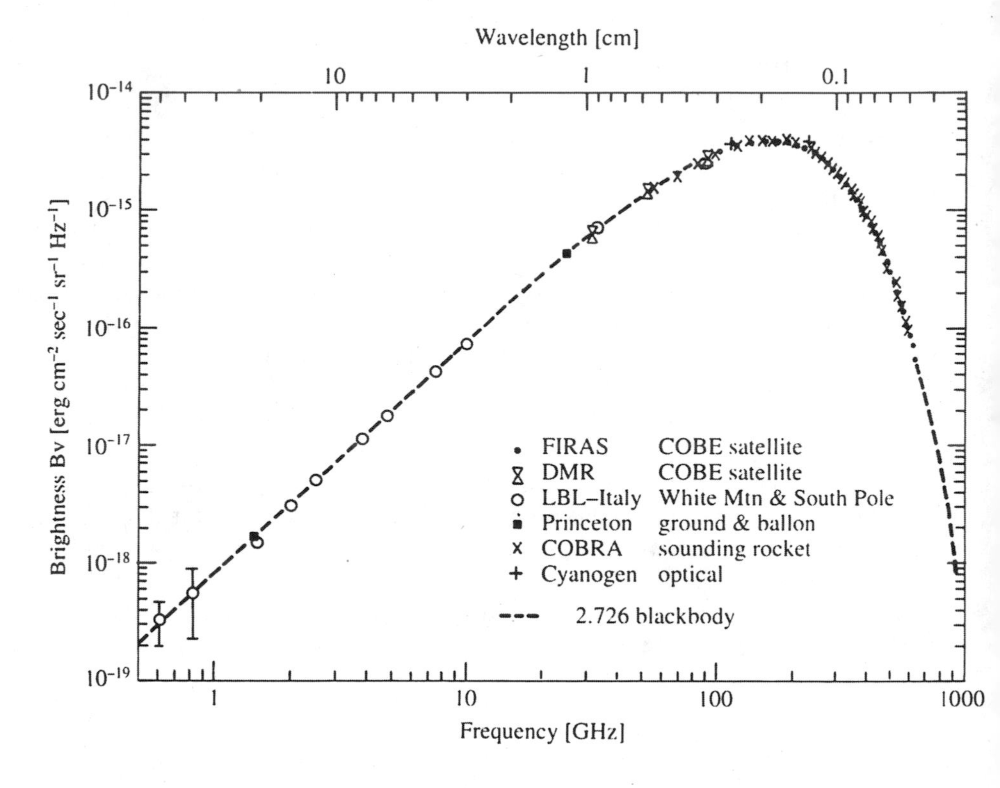

Fig. 6.1. Measurements of the spectrum of the cosmic microwave background radiation. Results since 1980 are plotted, except for cyanogen (at 114 and 227 GHz) where only results since 1989 are plotted. No error bars are plotted for the individual COBRA data points; otherwise, where error bars are not visible, they are smaller than the data points. The COBE FIRAS spectrum deviates from a 2.726 K blackbody by < 0.03% of the peak brightness. By permission of M. Bensadoun.

(v) *Background microwave radiation.* Isotropic radiation, apparently corresponding to blackbody radiation of about 2.7 K, discovered by Penzias and Wilson in 1965.[2] (See Fig. 6.1.)
(vi) *Ages of meteorites.* Radioactive dating gives the age of solar-system meteorites, and rocks from the Moon and Earth, as at least 4.5×10^9 yrs.[3]
(vii) *Temperature and luminosity of white dwarfs.* Studies of the temperature and luminosity of white dwarf stars indicate ages between 8×10^9 yrs and 12×10^9 yrs.[4]

In subsequent sections, we shall see how some aspects of the above are incorporated in our treatment of cosmology, but let us first say a little about redshift. If the observed redshift is interpreted as due to a velocity of recession, then the observations may be incorporated in a simple law that states that the speed of recession is proportional to distance. The constant of proportionality is known as *Hubble's constant.* The observations which determine this "constant" were made over the last few decades, a relatively short period. We shall see as the theory develops that it is in fact a function $H(t)$ of time. Estimates for its present-day value are subject to ongoing revision, ranging between 50 and 90 km s^{-1} Mpc^{-1}, which simplifies to the range 19.6×10^9 to 10.8×10^9 yrs for $1/H$.[5]

Exercise 6.0

1. The redshift z is defined by $z \equiv \Delta\lambda/\lambda_0$, where λ_0 is the proper wavelength, and $\Delta\lambda$ is the difference between the observed wavelength and the proper wavelength. If z is the Doppler shift due to a speed of recession v, show that on the basis of special relativity $z \approx v/c$, for v small compared with the speed of light c.

6.1 Robertson–Walker line element

It is outside our syllabus to rederive the independent work of Friedmann, Robertson, and Walker,[6] and others, on metrics, maximally symmetric subspaces, and descriptions of spacetimes that comply with the *cosmological principle.* (This is the hypothesis that the universe is spatially homogeneous and isotropic.) We take on faith the famous Robertson–Walker line element, adding

[2]See Penzias and Wilson, 1965.
[3]See Ohanian and Ruffini, 1994, p. 532.
[4]See Winget *et al.*, 1987.
[5]Note that when people say that the age of the universe is thus about 15×10^9 years, they are taking the inverse of the present-day value of H. However, we shall see that in the most common Friedmann model the age is given by $2/3H$, which corresponds to an age-range of 7.2×10^9 to 13.1×10^9 years.
[6]See Friedmann, 1922; Robertson, 1935 and 1936; and Walker, 1936.

only some minor intuitive ideas, and this is the starting point for our discussions.

With a timelike coordinate t, and spatial coordinates r, θ, ϕ, this line element is

$$\boxed{d\tau^2 = dt^2 - (R(t))^2 \left((1 - kr^2)^{-1} dr^2 + r^2 d\theta^2 + r^2 \sin^2\theta \, d\phi^2\right),} \tag{6.2}$$

where $R(t)$ is a dimensionless scale factor depending only on the time t, and k is either 0, 1, or -1, and is related to the spatial curvature. (Again we have taken $c = 1$ for convenience.) The spatial geometry is determined by the line element

$$ds^2 = (1 - kr^2)^{-1} dr^2 + r^2 d\theta^2 + r^2 \sin^2\theta \, d\phi^2. \tag{6.3}$$

A three-dimensional manifold with such a line element is clearly flat if $k = 0$, but for $k = \pm 1$ it is curved (see Problem 6.1). For $k = 1$ it is a space of constant positive curvature, the three-dimensional counterpart of a sphere, and the space is *closed* in the sense that it has finite volume. For $k = -1$ it is a space of constant negative curvature, and is *open* in the sense that its volume is infinite. To justify these remarks would involve us in a long digression into differential geometry.[7] The scale factor $R(t)$ simply "blows up" these spaces in a uniform manner, so that they expand or contract as dR/dt is positive or negative.

The scale factor $R(t)$ operates on the whole spatial part, regardless of direction, and depends on a *commonly measured time* t, which we rationalize as follows. Allow each galaxy to carry its own clock measuring its own proper time τ. These clocks may (ideally) have been synchronized when $R(t) = 0$, that is, at the beginning of the expansion. Because the universe is homogeneous and isotropic, there is no reason for clocks in different places to differ in the measurements of their own proper times. Furthermore, if we tie the coordinate system (t, r, θ, ϕ) to the galaxies, so that their world lines are given by $(r, \theta, \phi) =$ constant, then we have a *co-moving coordinate system* and the time t is nothing more than the proper time τ. This commonly measured time is often referred to as *cosmic time.*

To gain an intuitive idea of the significance of the Robertson–Walker line element it is useful to imagine a balloon with spots on it to represent the galaxies, the balloon expanding (or contracting) with time. The distance between spots would depend only on a time-varying scale factor $R(t)$, and each spot could be made to possess the same clock time t. The spatial origin of such a co-moving coordinate system might lie on *any one* of the spots.

The line element (6.2) is our trial solution for cosmological models, and our next task is to feed it into the field equations (3.39) using the form (6.1) for $T_{\mu\nu}$. As we shall see, this yields relations involving R, k, ρ, and p, and

[7] See, for example, Misner, Thorne, and Wheeler, 1973, §27.6, in particular Box 27.2.

gives a variety of models for comparison with the observed universe. Because of the assumed homogeneity, ρ and p are functions of t alone.

6.2 Field equations

If we label our coordinates according to $t \equiv x^0$, $r \equiv x^1$, $\theta \equiv x^2$, $\phi \equiv x^3$, then the nonzero connection coefficients are

$$\begin{aligned}
&\Gamma^0_{11} = R\dot{R}/(1-kr^2), &&\Gamma^0_{22} = R\dot{R}r^2, &&\Gamma^0_{33} = R\dot{R}r^2\sin^2\theta,\\
&\Gamma^1_{01} = \dot{R}/R, &&\Gamma^1_{11} = kr/(1-kr^2), &&\Gamma^1_{22} = -r(1-kr^2),\\
&\Gamma^1_{33} = -r(1-kr^2)\sin^2\theta, && && \\
&\Gamma^2_{02} = \dot{R}/R, &&\Gamma^2_{12} = 1/r, &&\Gamma^2_{33} = -\sin\theta\cos\theta,\\
&\Gamma^3_{03} = \dot{R}/R, &&\Gamma^3_{13} = 1/r, &&\Gamma^3_{23} = \cot\theta.
\end{aligned} \tag{6.4}$$

These were obtained in Example 2.1.2, but we have changed the notation so that derivatives with respect to t are now denoted by dots. Feeding the connection coefficients into

$$R_{\mu\nu} \equiv \Gamma^\sigma_{\mu\sigma,\nu} - \Gamma^\sigma_{\mu\nu,\sigma} + \Gamma^\rho_{\mu\sigma}\Gamma^\sigma_{\rho\nu} - \Gamma^\rho_{\mu\nu}\Gamma^\sigma_{\rho\sigma}$$

(and remembering that $\Gamma^\mu_{\nu\sigma} = \Gamma^\mu_{\sigma\nu}$) gives

$$\begin{aligned}
R_{00} &= 3\ddot{R}/R,\\
R_{11} &= -(R\ddot{R} + 2\dot{R}^2 + 2k)/(1-kr^2),\\
R_{22} &= -(R\ddot{R} + 2\dot{R}^2 + 2k)r^2,\\
R_{33} &= -(R\ddot{R} + 2\dot{R}^2 + 2k)r^2\sin^2\theta.\\
R_{\mu\nu} &= 0, \quad \mu \neq \nu
\end{aligned} \tag{6.5}$$

With $c = 1$, $u^\mu u_\mu = 1$, so

$$T = T^\mu_\mu = (\rho + p) - 4p = \rho - 3p.$$

In our co-moving coordinate system, $u^\mu = \delta^\mu_0$, so

$$u_\mu = g_{\mu\nu}\delta^\nu_0 = g_{\mu 0} = \delta^0_\mu.$$

Hence

$$T_{\mu\nu} = (\rho + p)\delta^0_\mu\delta^0_\nu - pg_{\mu\nu}$$

and

$$\begin{aligned}
T_{\mu\nu} - \tfrac{1}{2}Tg_{\mu\nu} &= (\rho + p)\delta^0_\mu\delta^0_\nu - pg_{\mu\nu} - \tfrac{1}{2}(\rho - 3p)g_{\mu\nu}\\
&= (\rho + p)\delta^0_\mu\delta^0_\nu - \tfrac{1}{2}(\rho - p)g_{\mu\nu}.
\end{aligned}$$

Extracting $g_{\mu\nu}$ from the line element (6.2), we see that

$$\begin{aligned} T_{00} - \tfrac{1}{2}Tg_{00} &= \tfrac{1}{2}(\rho + 3p), \\ T_{11} - \tfrac{1}{2}Tg_{11} &= \tfrac{1}{2}(\rho - p)R^2/(1 - kr^2), \\ T_{22} - \tfrac{1}{2}Tg_{22} &= \tfrac{1}{2}(\rho - p)R^2r^2, \\ T_{33} - \tfrac{1}{2}Tg_{33} &= \tfrac{1}{2}(\rho - p)R^2r^2\sin^2\theta, \\ T_{\mu\nu} - \tfrac{1}{2}Tg_{\mu\nu} &= 0, \quad \mu \neq \nu. \end{aligned}$$

So the field equations in the form (3.39) yield just two equations:

$$3\ddot{R}/R = \tfrac{1}{2}\kappa(\rho + 3p), \tag{6.6}$$

$$R\ddot{R} + 2\dot{R}^2 + 2k = -\tfrac{1}{2}\kappa(\rho - p), \tag{6.7}$$

where (with $c = 1$) $\kappa = -8\pi G$. The fact that the three (nontrivial) spatial equations are equivalent is essentially due to the homogeneity and isotropy of the Robertson–Walker line element.

Eliminating $\ddot{R}$ from equations (6.6) and (6.7) gives

$$\boxed{\dot{R}^2 + k = (8\pi G/3)\rho R^2.} \tag{6.8}$$

We shall refer to this equation as the *Friedmann equation*. Note that the pressure has completely canceled out of this equation.

We know from Section 3.1 that $T^{\mu\nu}{}_{;\mu} = 0$ yields the continuity equation and the equations of motion of the fluid particles. With $c = 1$ these become (when adapted to curved spacetime)

$$(\rho u^\mu)_{;\mu} + pu^\mu{}_{;\mu} = 0, \tag{6.9}$$

$$(\rho + p)u^\nu{}_{;\mu}u^\mu = (g^{\mu\nu} - u^\mu u^\nu)p_{,\mu}. \tag{6.10}$$

The continuity equation (6.9) may be written as

$$\rho_{,\mu}u^\mu + (\rho + p)(u^\mu{}_{,\mu} + \Gamma^\mu_{\nu\mu}u^\nu) = 0,$$

and with $u^\mu = \delta^\mu_0$ this reduces to

$$\boxed{\dot{\rho} + (\rho + p)(3\dot{R}/R) = 0,} \tag{6.11}$$

which does contain the pressure. As for the equation of motion (6.10), both sides turn out to be identically zero, and it is automatically satisfied. This means that the fluid particles (galaxies) follow geodesics, which was to be expected, since with p a function of t alone, there is no pressure gradient (i.e., no 3-gradient ∇p) to push them off geodesics.

We make use of equations (6.8) and (6.11) in the next section where we discuss the standard Friedmann models of the universe.

Exercises 6.2

1. Check the Ricci tensor components given by equation (6.5).
2. Show that equation (6.11) may also be derived by eliminating $\ddot{R}$ from equation (6.6) and the derivative of equation (6.8) with respect to t.
3. Verify that equation (6.10) is automatically satisfied.

6.3 The Friedmann models

Observational evidence to date suggests that the universe is matter-dominated, and that the pressure is negligible when compared with the density. The standard Friedmann models arise from setting $p = 0$, and our discussion will be confined to these models only.

With $p = 0$, we see that

$$\rho R^3 = \text{constant} \tag{6.12}$$

is an integral of the continuity equation (6.11). As we shall see, this leads to three possible models, each of which has $R(t) = 0$ at some point in time, and it is natural to take this point as the origin of t, so that $R(0) = 0$, and t is then the age of the universe (compare remarks in Sec. 6.1). Let us use a subscript zero to denote present-day values of quantities, so that t_0 is the present age of the universe, and $R_0 \equiv R(t_0)$ and $\rho_0 \equiv \rho(t_0)$ are the present-day values of R and ρ. We may then write equation (6.12) as

$$\rho R^3 = \rho_0 R_0^3. \tag{6.13}$$

The Friedmann equation (6.8) then becomes

$$\dot{R}^2 + k = A^2/R, \tag{6.14}$$

where $A^2 \equiv 8\pi G\rho_0 R_0^3/3$ $(A > 0)$. *Hubble's "constant"* $H(t)$ is defined by

$$\boxed{H(t) \equiv \dot{R}(t)/R(t),} \tag{6.15}$$

and we denote its present-day value by $H_0 \equiv H(t_0)$. Equation (6.8) gives

$$\frac{k}{R_0^2} = \frac{8\pi G\rho_0}{3} - H_0^2 = \frac{8\pi G}{3}\left(\rho_0 - \frac{3H_0^2}{8\pi G}\right).$$

Hence $k > 0$, $k = 0$, or $k < 0$ as $\rho_0 > \rho_c$, $\rho_0 = \rho_c$, or $\rho_0 < \rho_c$ respectively, where ρ_c is a *critical density* given by

$$\boxed{\rho_c \equiv 3H_0^2/8\pi G.} \tag{6.16}$$

The *deceleration parameter* q_0 is defined to be the present-day value of $-R\ddot{R}/\dot{R}^2$. Using equations (6.6) (with $p = 0$) and (6.7) gives

$$q_0 = 4\pi G\rho_0/3H_0^2 = \rho_0/2\rho_c. \tag{6.17}$$

The three Friedmann models arise from integrating equation (6.14) for the three possible values of k: $k = 0, \pm 1$.

(i) *Flat model.* $k = 0$; hence $\rho_0 = \rho_c$, $q_0 = \frac{1}{2}$.
Equation (6.14) gives

$$dR/dt = A/R^{1/2},$$

and integrating gives

$$R(t) = (3A/2)^{2/3}t^{2/3}. \tag{6.18}$$

This model is also known as the Einstein–de Sitter model, for reasons mentioned in Section 6.6. Its graph is plotted in Figure 6.2. Note that $\dot{R} \to 0$ as $t \to \infty$.

(ii) *Closed model.* $k = 1$; hence $\rho_0 > \rho_c$, $q_0 > \frac{1}{2}$.
Equation (6.14) gives

$$\frac{dR}{dt} = \left(\frac{A^2 - R}{R}\right)^{1/2},$$

so

$$t = \int_0^R \left(\frac{R}{A^2 - R}\right)^{1/2} dR.$$

Putting $R \equiv A^2 \sin^2(\psi/2)$ gives

$$t = A^2 \int_0^\psi \sin^2(\psi/2)\, d\psi = \tfrac{1}{2}A^2 \int_0^\psi (1 - \cos\psi)\, d\psi = \tfrac{1}{2}A^2(\psi - \sin\psi).$$

So

$$R = \tfrac{1}{2}A^2(1 - \cos\psi), \quad t = \tfrac{1}{2}A^2(\psi - \sin\psi), \tag{6.19}$$

and these two equations give $R(t)$ via the parameter ψ. The graph of $R(t)$ is a cycloid, and is shown in Figure 6.2.

(iii) *Open model.* $k = -1$; hence $\rho_0 < \rho_c$, $q_0 < \frac{1}{2}$.
Equation (6.14) gives

$$\frac{dR}{dt} = \left(\frac{A^2 + R}{R}\right)^{1/2},$$

so

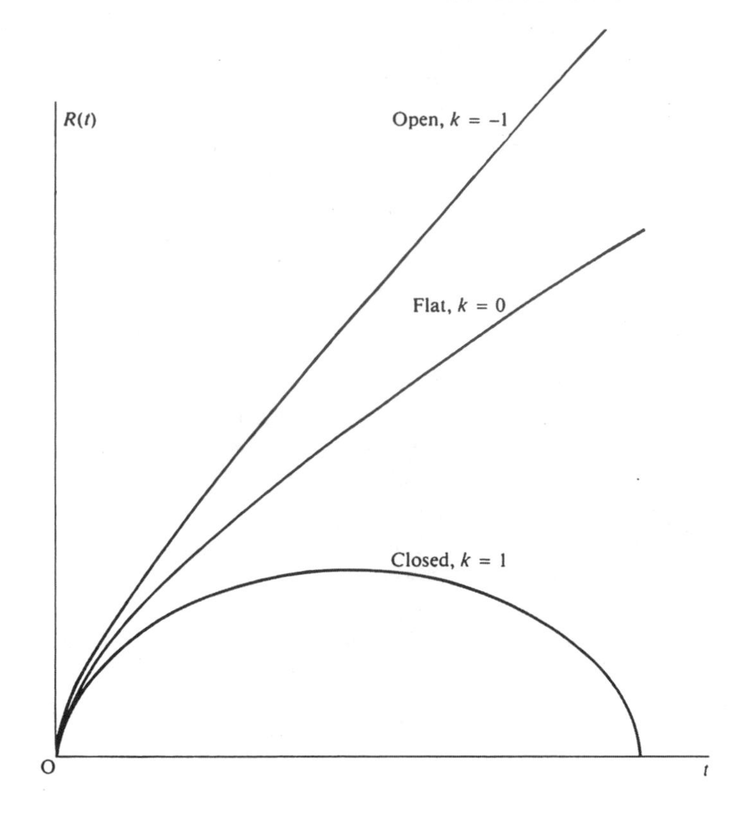

Fig. 6.2. Zero-pressure models of the universe.

$$t = \int_0^R \left(\frac{R}{A^2 + R} \right)^{1/2} dR.$$

Putting $R \equiv A^2 \sinh^2(\psi/2)$ gives

$$t = A^2 \int_0^\psi \sinh^2(\psi/2)\, d\psi = \tfrac{1}{2} A^2 \int_0^\psi (\cosh\psi - 1)\, d\psi = \tfrac{1}{2} A^2 (\sinh\psi - \psi).$$

So

$$R = \tfrac{1}{2} A^2 (\cosh\psi - 1), \quad t = \tfrac{1}{2} A^2 (\sinh\psi - \psi), \tag{6.20}$$

and these give $R(t)$ via the parameter ψ. Its graph is also shown in Figure 6.2. Note that $\dot{R} \to 1$ as ψ (and hence t) $\to \infty$.

We see that $k = 0$ and $k = -1$ give models that continually expand, while $k = 1$ gives a model that expands to a maximum value of R, and then contracts, so the latter is not only spatially but also temporally closed. The significance of the value of k is explained in Newtonian terms in Section 6.7.

The question naturally arises as to which (if any, bearing in mind our assumptions) represents our own universe. Astronomical observations over recent years have yielded estimates for H_0, q_0, and ρ_0. The methods employed are extremely ingenious and complex, and we shall not attempt a review of them.[8] Recent estimates for H_0 put it at about $(18 \times 10^9)^{-1}\,\text{yrs}^{-1}$, and using equation (6.16) gives a value for the critical density of about $5.7 \times 10^{-27}\,\text{kg}\,\text{m}^{-3}$. Estimates for the deceleration parameter q_0, based on masses and distributions of galaxies, are around 0.025, but are extremely difficult to be sure about. With $q_0 \approx 0.025$, equation (6.17) would give

$$\rho_0 \approx 2\rho_c q_0 \approx 0.28 \times 10^{-27}\,\text{kg}\,\text{m}^{-3},$$

which is greater than the observed density of the universe estimated on the assumption that its matter content is visible and lies principally in the galaxies. This discrepancy leads to the so-called "problem of missing matter" whose existence is postulated for many reasons in modern astronomy. Not the least of these is Oort's analysis of the motions of stars in our galaxy, which shows that much more matter must exist gravitationally than can be accounted for by visible matter. It is suggested that this matter might take the form of an intergalactic gas, which might include small dark stars, small primeval black holes and neutrinos, and it is thought that by itself this dark matter has a density of about $1 \times 10^{-27}\,\text{kg}\,\text{m}^{-3}$. So we do not really know yet whether the universe is open or closed.

One final connection we can make between theory and observation concerns the present age t_0 of the universe. For the flat model, equation (6.18) gives

$$H(t) \equiv \dot{R}(t)/R(t) = 2/(3t), \tag{6.21}$$

so

$$t_0 = 2/(3H_0) \approx 12 \times 10^9\,\text{yrs}.$$

To find the values of t_0 given by the other two models is somewhat complicated, and is left as an exercise (Problem 6.2).

None of these estimates of age from the Friedmann models conflicts with, nor is especially supported by, the meteorite data.

Exercise 6.3

1. Check the integrations leading to the parametric equations (6.19) and (6.20).

6.4 Redshift, distance, and speed of recession

We return in this section to the implications of something mentioned in the introduction to this chapter: the observed redshift $z \equiv \Delta\lambda/\lambda_0$ for the wavelength of light emitted by distant galaxies. Here λ_0 is the proper wavelength

[8]See, for example, Weinberg, 1972, Chap. 14, and Ohanian, 1989, Chap. 9.

and $\Delta\lambda$ is the difference between the observed and the proper wavelengths. As noted in Exercise 6.0.1, if this is interpreted as the Doppler shift due to a speed of recession v (small compared with c), then we have the approximate result that $z = v/c$. We shall replace this rough argument by one more appropriate to the context of the Friedmann cosmological models and improve on the approximate result. We shall also establish the fundamental relation between the speed of recession of a galaxy and its distance from the observer, a relation generally known as *Hubble's law.*

In this section (and the next) we shall make it clear where the speed of light c is involved, and not have $c = 1$, as in the earlier sections of this chapter. The form that the Robertson–Walker line-element then takes is

$$c^2 d\tau^2 = c^2 dt^2 - (R(t))^2 \left((1 - kr^2)^{-1} dr^2 + r^2 d\theta^2 + r^2 \sin^2\theta \, d\phi^2\right). \quad (6.22)$$

When we refer to the *distance* of a galaxy (or other remote object) from the observer, we shall mean the length of the spatial geodesic connecting the two at the same cosmic time t. It is the distance we would obtain if we could stop the expansion of the universe at time t and measure it using measuring rods. In mathematical terms, we put $t = \text{const}$ in the spacetime line element above and work with the line element

$$ds^2 = (R(t))^2 \left((1 - kr^2)^{-1} dr^2 + r^2 d\theta^2 + r^2 \sin^2\theta \, d\phi^2\right) \quad (6.23)$$

of the corresponding three-dimensional space. By the *speed of recession* of the galaxy, we mean the magnitude of the rate of change of this distance with respect to t.

The only observers we can realistically consider are situated on, or near, the Earth: they inhabit our own galaxy, the Milky Way. Because of the assumed spatial homogeneity of the universe, there is no loss of generality in arranging the coordinate system so that $r = 0$ for our own galaxy and the observers who inhabit it. If the remote galaxy has (spatial) coordinates (r_G, θ_G, ϕ_G), then the spatial geodesic[9] whose length we need is given by

$$\theta = \theta_G, \quad \phi = \phi_G, \quad 0 \leq r \leq r_G. \quad (6.24)$$

(Here r_G is constant, corresponding to the fact that the galaxy takes part in the general expansion of the universe.) The length of this geodesic at time t is given by

$$L_G(t) = \int_0^{r_G} \frac{R(t)}{\sqrt{1 - kr^2}} \, dr = R(t) \int_0^{r_G} \frac{1}{\sqrt{1 - kr^2}} \, dr. \quad (6.25)$$

Consider now a pulse of light (or photon) emitted by the galaxy at time t_E and arriving at the observer at time t_R. It will follow a null geodesic[10] given

[9] The spherical symmetry of the line element suggests that the equations (6.24) give a geodesic, but it can be checked. See Exercise 6.4.1.

[10] Again, the spherical symmetry suggests that null curves with θ and ϕ constant are geodesics. Exercise 6.4.2 gives a verification.

by $\theta = \theta_G$, $\phi = \phi_G$. Since the geodesic is null $c^2 d\tau^2 = 0$, so the *coordinate* velocity of the pulse is

$$\frac{dr}{dt} = -\frac{c\sqrt{1-kr^2}}{R(t)}, \tag{6.26}$$

on taking the negative square root, as r is decreasing. So

$$\int_{t_E}^{t_R} \frac{dt}{R(t)} = -\int_{r_G}^{0} \frac{dr}{c\sqrt{1-kr^2}}, \tag{6.27}$$

showing that the integral on the left does not depend on t_E and t_R. So if $t_E + \delta t_E$ and $t_R + \delta t_R$ are the times of emission and reception for a later pulse of light, then

$$\int_{t_E}^{t_R} \frac{dt}{R(t)} = \int_{t_E+\delta t_E}^{t_R+\delta t_R} \frac{dt}{R(t)}.$$

For coordinate differences δt_E and δt_R that are small, this gives $\delta t_E / R(t_E) = \delta t_R / R(t_R)$. If these time differences correspond to the proper and observed periods respectively of the emitted light, then for the corresponding wavelengths we have

$$\frac{\lambda_R}{\lambda_E} = \frac{\delta\tau_R}{\delta\tau_E} = \frac{\delta t_R}{\delta t_E} = \frac{R(t_R)}{R(t_E)}, \tag{6.28}$$

since $\delta\tau = \delta t$ for a spatially fixed emitter or receiver. But λ_E is the proper wavelength λ_0 and λ_R is the observed wavelength, which is usually denoted simply by λ, so for the redshift z we have

$$z = \frac{\lambda - \lambda_0}{\lambda_0} = \frac{R(t_R) - R(t_E)}{R(t_E)}. \tag{6.29}$$

This last equation is the key to the relations between redshift, distance, speed of recession, the deceleration parameter, and Hubble's constant, but it needs careful handling to yield the best form of results. Our analysis makes use of the Taylor expansion of $R(t)$ about t_R:

$$R(t) = R(t_R) + \dot{R}(t_R)(t - t_R) + \tfrac{1}{2}\ddot{R}(t_R)(t - t_R)^2 + \cdots . \tag{6.30}$$

We shall assume that $t_R - t_E$ is small[11] and retain explicitly terms involving $(t_R - t_E)^2$.

Putting $t = t_E$ in equation (6.30) gives

$$\begin{aligned} R(t_E) &= R(t_R) - \dot{R}(t_R)(t_R - t_E) + \tfrac{1}{2}\ddot{R}(t_R)(t_R - t_E)^2 + \cdots \\ &= R(t_R)\left(1 - H(t_R)(t_R - t_E) - \tfrac{1}{2}q(t_R)H(t_R)^2(t_R - t_E)^2 + \cdots\right), \end{aligned}$$

where $H = \dot{R}/R$ is and $q = -R\ddot{R}/\dot{R}^2$. If we take t_R to be its present-day value, then $H(t_R)$ is Hubble's "constant" H_0 and $q(t_R)$ is the deceleration parameter q_0, and on setting $t_R - t_E = \Delta t$ we get

[11] Small compared to what? This question will be answered at the end of the section.

$$R(t_E) = R(t_R)\left(1 - H_0\Delta t - \tfrac{1}{2}q_0(\Delta t)^2 + \cdots\right). \tag{6.31}$$

Equation (6.29) then gives

$$\begin{aligned} z &= \left(H_0\Delta t + \tfrac{1}{2}q_0 H_0^2(\Delta t)^2 + \cdots\right)\left(1 - H_0\Delta t - \tfrac{1}{2}q_0(\Delta t)^2 + \cdots\right)^{-1} \\ &= \left(H_0\Delta t + \tfrac{1}{2}q_0 H_0^2(\Delta t)^2 + \cdots\right)\left(1 + H_0\Delta t + \cdots\right) \end{aligned}$$

So

$$z = H_0\Delta t + H_0^2(1 + \tfrac{1}{2}q_0)(\Delta t)^2 + \cdots. \tag{6.32}$$

We now need an expression for Δt.

From equations (6.25) and (6.27) we have

$$c\int_{t_E}^{t_R} \frac{dt}{R(t)} = \frac{L_G(t_R)}{R(t_R)}, \tag{6.33}$$

and from the expansion (6.30) we have

$$\begin{aligned} \frac{1}{R(t)} &= \frac{1}{R(t_R)}\left(1 + H(t_R)(t - t_R) - \tfrac{1}{2}q(t_R)(t - t_R)^2 + \cdots\right)^{-1} \\ &= \frac{1}{R(t_R)}\left(1 - H_0(t - t_R) + (H_0^2 - \tfrac{1}{2}q_0)(t - t_R)^2 + \cdots\right), \end{aligned}$$

on expanding and collecting together terms up to those in $(t - t_R)^2$. So, from equation (6.33), it follows that

$$\begin{aligned} \frac{L_G(t_R)}{c} &= \int_{t_E}^{t_R} \left(1 - H_0(t - t_R) + (H_0^2 - \tfrac{1}{2}q_0)(t - t_R)^2 + \cdots\right) dt \\ &= \Delta t + \tfrac{1}{2}H_0(\Delta t)^2 + \cdots. \end{aligned}$$

This expression can be inverted[12] to get

$$\Delta t = \frac{L_G(t_R)}{c} - \frac{H_0}{2}\left(\frac{L_G(t_R)}{c}\right)^2 + \cdots. \tag{6.34}$$

Substituting into (6.32) gives

$$\boxed{z = \frac{H_0 L}{c} + \frac{1 + q_0}{2}\left(\frac{H_0 L}{c}\right)^2 + \cdots,} \tag{6.35}$$

where $L = L_G(t_R)$ is the distance of the observed galaxy from the observer.

This last equation is the *distance–redshift* relation for Friedmann models of the universe. In its simplest form (i.e., neglecting squares and higher powers

[12]See Exercise 6.4.3

of H_0L) it states that the redshift of an observed galaxy is proportional to its distance from the observer.

By differentiating equation (6.25) with respect to t, we can deduce that

$$\dot{L}_G(t) = \frac{\dot{R}(t)}{R(t)} L_G(t) = H(t) L_G(t), \tag{6.36}$$

so on evaluating at $t = t_R$ and putting $v = \dot{L}_G(t_R)$ for the speed of recession of the observed galaxy we have the relation $v = H_0 L$ connecting the speed of recession with distance. We can then write equation (6.35) in the equivalent form

$$\boxed{z = \frac{v}{c} + \frac{1+q_0}{2}\left(\frac{v}{c}\right)^2 + \cdots,} \tag{6.37}$$

relating the redshift z to the speed of recession v.

We can now answer the question posed in footnote 11. For the expansions (6.35) and (6.37) to be valid, Δt must be such that $v = H_0 L < c$. If $v \ll c$, then dropping the cubic and higher-power terms of the expansions gives valid approximations.

The redshift–distance relation (6.35) is used to get estimates for Hubble's constant, as given in the discussion at the end of Section 6.3.

Exercises 6.4

1. Verify that equations (6.24) give a geodesic of the space with line element (6.23).

2. Verify that a null curve with θ and ϕ constant is a null geodesic of the Robertson–Walker line element.

3. Verify the expression (6.34) for Δt.

6.5 Objects with large redshifts

Objects with high redshifts are continually being detected with values considerably greater than those originally observed by Hubble in the 1920s. Redshifts as large as 4.7 have been found for quasars[13] and as large as 3.8 for galaxies. So the questions arise: are some galaxy and quasar recession speeds greater than the speed of light, and if so, can they be observed? The results of the previous section cannot provide the answers, as they are valid only for $v < c$. We shall not attempt to provide a general answer to these questions, but we shall illustrate the approach that might be taken by working through a particular example.

[13] See Stuckey, 1992.

Consider, then, a galaxy receding from us with a redshift $z = 4$. If we take (for mathematical tractability only) the flat model of the universe, then, as a function of the cosmic time t,

$$R(t) = Bt^{2/3}, \quad \text{where } B \equiv (3A/2)^{2/3} = (6\pi G\rho_0)^{1/3}, \tag{6.38}$$

as in Section 6.3. It follows from equation (6.29) that

$$1 + z = \left(\frac{t_R}{t_E}\right)^{2/3}, \tag{6.39}$$

where t_E is the time of emission of a photon from a galaxy and t_R is the time of reception. As in the previous section, we shall have $r = r_G$ for the galaxy and $r = 0$ for our observers, and we shall take t_R to be its present-day value t_0. Thus, with $z = 4$ and 12×10^9 years as a typical value for the age of the universe, we get $t_E \approx 1.07 \times 10^9$ years, so the light was emitted about 10.9×10^9 years ago.

We shall see below that redshifts of $z > 3$ correspond (in the flat model) to speeds of recession greater than c. Such speeds are permitted because there is no *single inertial frame* that can accommodate *both* galaxies: we are working in a curved spacetime.

Consider now a photon leaving the galaxy at cosmic time t_E, as in the Figure 6.3. If the galaxy is receding faster than light, can the photon ever reach us? It turns out that it *is possible*, but only because in this model the expansion is slowing.

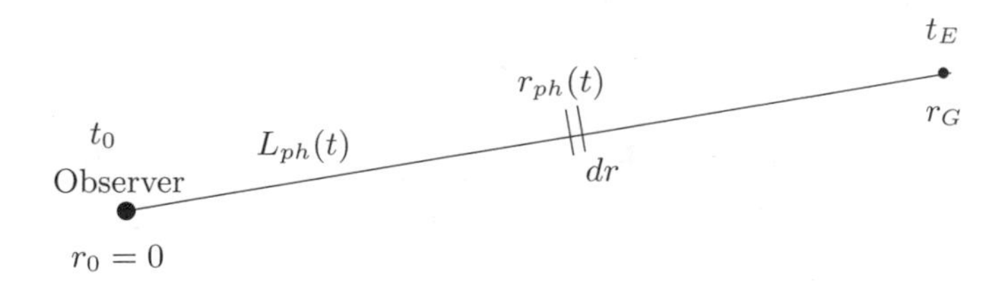

Fig. 6.3. Light from a receding galaxy.

Let $r_{ph}(t)$ be the r-coordinate of a photon which was originally emitted from the galaxy at time t_E. With $k = 0$, equation (6.26) gives

$$\frac{dr}{dt} = -\frac{c}{R(t)} = -\frac{c}{B^{2/3}}.$$

So

$$\int_{r_G}^{r_{ph}(t)} dr = -c \int_{t_E}^{t} \frac{dt}{Bt^{2/3}},$$

which gives

$$r_{ph}(t) = r_G - \frac{3c}{B}\left(t^{1/3} - t_E^{1/3}\right). \tag{6.40}$$

With $k = 0$, equation (6.25) gives

$$L_G(t) = R(t)\, r_G = Bt^{2/3}\, r_G$$

for the distance of the galaxy from the observer at time t. Similarly, we have

$$L_{ph}(t) = R(t)\, r_{ph}(t) = Bt^{2/3}\, r_{ph}(t)$$

for the distance of the photon from the observer at time t. So on multiplying equation (6.40) by $Bt^{2/3}$, we get

$$\begin{aligned} L_{ph}(t) &= L_G(t) - \frac{3c}{B}(Bt^{2/3})\left(t^{1/3} - t_E^{1/3}\right) \\ &= L_G(t) - 3ct\left(1 - (t_E/t)^{1/3}\right). \end{aligned} \tag{6.41}$$

Since $r_{ph}(t_0) = 0$, equation (6.40) gives

$$r_G = \frac{3c}{B}\left(t_0^{1/3} - t_E^{1/3}\right), \tag{6.42}$$

so

$$L_G(t) = R(t) r_G = 3ct^{2/3}\left(t_0^{1/3} - t_E^{1/3}\right). \tag{6.43}$$

Substituting this last expression in equation (6.41) gives

$$L_{ph}(t) = 3ct^{2/3}\left(t_0^{1/3} - t^{1/3}\right) \tag{6.44}$$

for the distance of the photon from the observer at time t.

Writing $L \equiv L_G(t_0)$ for the present-day distance of the remote galaxy from our galaxy, we have

$$L = 3ct_0^{2/3}\left(t_0^{1/3} - t_E^{1/3}\right) = 3ct_0\left(1 - (t_E/t_0)^{1/3}\right),$$

so, from equation (6.39) (with $t_R = t_0$),

$$L = 3ct_0\left(1 - \frac{1}{\sqrt{1+z}}\right). \tag{6.45}$$

Hence for our example of $z = 4$ the present distance to the galaxy is about $1.7ct_0$, or 20.4×10^9 light years.

Differentiating equations (6.41) and (6.44) with respect to t yields two expressions for the photon's "velocity":

$$\dot{L}_{ph}(t) = \dot{L}_G(t) - 3c\left(1 - \tfrac{2}{3}(t_E/t)^{1/3}\right), \tag{6.46}$$

and

$$\dot{L}_{ph}(t) = c\left(2(t_0/t)^{1/3} - 3\right). \tag{6.47}$$

At the time of emission t_E, the first of these gives $\dot{L}_{ph}(t_E) = \dot{L}_G(t_E) - c$, which is simply the difference between the galaxy's speed of recession and the speed of light, and at t_0 the second one gives $\dot{L}_{ph}(t_0) = -c$, as it should. Of course, $\dot{L}_{ph}(t)$ is not really the photon's velocity as measured in a local inertial frame (except at the time of reception t_0), but it does tell us whether the distance $L_{ph}(t)$ is increasing, or decreasing.

Figure 6.4 shows how the photon's distance depends on t for an object with $z = 4$ and $t_0 = 12 \times 10^9$ years as a typical value for the age of the universe. As noted above, this implies that $t_E \approx 1.07 \times 10^9$ years. At the time of emission $\dot{L}_{ph}(t_E) > 0$ and the photon is receding. It continues to recede (being carried away by the general expansion of the universe) until $\dot{L}_{ph}(t) = 0$, which occurs when

$$t = 8t_0/27 \approx 3.56 \times 10^9 \text{years}.$$

After this, $\dot{L}_{ph}(t) < 0$ and the photon is approaching, eventually arriving to be observed at time $t_0 = 12 \times 10^9$ years.

In the flat-space model, equation (6.36) gives

$$\dot{L}_G(t) = \frac{\dot{R}(t)}{R(t)} L_G(t) = \frac{2}{3t} L_G(t),$$

so, making use of equation (6.45) (in which $L = L_G(t_0)$), we get

$$\dot{L}_G(t_0) = \frac{2}{3t_0} L_G(t_0) = 2c\left(1 - \frac{1}{\sqrt{1+z}}\right).$$

That is,

$$v = 2c\left(1 - \frac{1}{\sqrt{1+z}}\right), \tag{6.48}$$

where $v \equiv \dot{L}_G(t_0)$ is the speed of recession. Clearly (in this model), the speed of recession exceeds c for all redshifts $z > 3$.

In the flat model there is also a boundary, known as a *particle horizon* that divides those objects that can be seen from those that cannot. This follows from equation (6.42), which can be written as

$$r_G = \frac{3c\,t_0^{1/3}}{B}\left(1 - \left(\frac{t_E}{t_0}\right)^{1/3}\right),$$

showing that $r_G < 3c\,t_0^{1/3}/B$. Here $B = (3A/2)^{2/3}$ and from equation (6.14) (with $k = 0$) we have that $A^2 = R_0^3 H_0^2$. Also, from equation (6.21), $t_0 = 2/3H_0$, so after substitution and simplification we arrive at

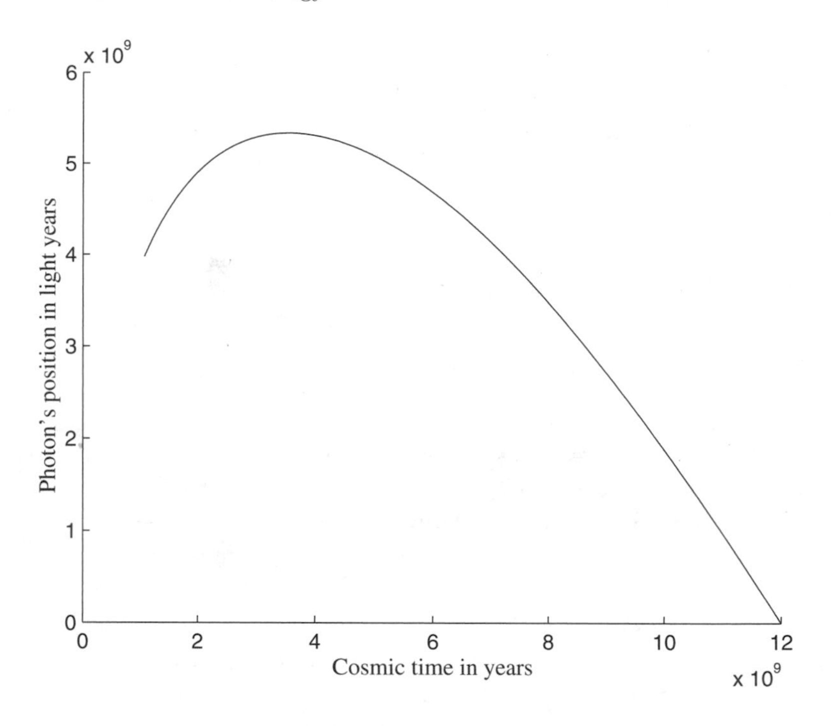

Fig. 6.4. Photon's position as a function of cosmic time. The photon is dragged away, but ultimately arrives in our galaxy.

$$r_G < \frac{2c}{R_0 H_0} \tag{6.49}$$

for a visible galaxy. This converts to $L < 2c/H_0$ as the limit on the distance of a visible galaxy and gives $v < 2c$ as a limit on its speed of recession. This last result can also be inferred from equation (6.48).

Exercises 6.5

1. Work through the substitutions and simplification leading to the inequality $r_G < 2c/R_0 H_0$ that defines the particle horizon.

2. Does the existence of a particle horizon in the flat model place an upper bound on the redshifts that can be observed?

6.6 Comment on Einstein's models; inflation

As early as 1917 Einstein applied his field equations to a "cosmic gas" of the kind we have been discussing. He was strongly drawn, on philosophical rather than mathematical grounds, to the idea of a stable universe, with $k = 1$, that did not change with time, that is, with $H(t) = \dot{R}(t)/R(t) = 0$. In order to counteract the obvious gravitational collapse of such a gas he introduced into the field equations a cosmological term to act as a repulsion mechanism (possibly due to negative matter, if it existed in the universe). The revised field equations were

$$R^{\mu\nu} - \tfrac{1}{2}Rg^{\mu\nu} + \Lambda g^{\mu\nu} = \kappa T^{\mu\nu}, \tag{6.50}$$

where Λ was a constant known as the *cosmological constant.*

Since ${g^{\mu\nu}}_{;\mu} = 0$, this did not alter the divergence property ${T^{\mu\nu}}_{;\mu} = 0$. The constant Λ had to be extremely small, so as not to interfere with the general-relativistic predictions of the solar system (see Chap. 4).

After Hubble's detection of the redshift in the 1920s and 1930s, interpreted now by almost all astronomers and physicists as a Doppler-type shift due to expansion, (but with the subtle difference that the wavelength lengthens *during* its aeons of travel) Einstein came to believe that the universe was flying apart with considerable kinetic energy, so that a repulsion mechanism was no longer required. He withdrew the cosmological term, later referring to it as the greatest blunder of his life. In 1932 he proposed with de Sitter a model in which $k = 0$ and $p = 0$, which is the flat model of Section 6.3. This leads to a present-day density of $\rho_0 = \rho_c \approx 5.7 \times 10^{-27}\,\text{kg}\,\text{m}^{-3}$ and to an age $t_0 \approx 12 \times 10^9$ yrs.

This Einstein–de Sitter model, which permitted expansion, and which incorporated *homogeneity* and *isotropy* (the Cosmological Principle) was modified (in 1948) by Hoyle, Bondi and Gold,[14] who made the added assumption of *temporal homogeneity* (the Perfect Cosmological Principle). In that model (the so-called 'steady-state model'), the universe did not come from a big bang, and is the same at all times—because they postulated a continuous creation of matter (of about one hydrogen atom per year per 6 km^3) in the intergalactic vacuum, to balance the reduction in density due to expansion. While this violated energy conservation, the real blow to the theory was the discovery in 1965, by Penzias and Wilson[15] of a black-body radiation (see Fig. 6.1) coming in from all directions, interpreted to be 'relic radiation' from the big bang.

Although the model with $k = 0$ is the most favored, it would be wrong to believe that all is clear-cut, and there are physicists (for example, the Nobel laureate, Hannes Alfven) who believe evidence for the big-bang is lacking[16]. Further, if energy conservation—a pillar of physics, and violated in steady

[14]See Hoyle, 1948, and Bondi and Gold, 1948.

[15]See Penzias and Wilson, 1965.

[16]There have been articles—too numerous to list here—by physicists H. Alfven, E.J. Lerner, and others, and astronomers such as H. Arp and J.V.Narlikar, which query big-bang cosmologies. The bases for many of these arguments involve paucity

state cosmologies—is to be preserved, it must also be satisfactorily explained how the entire universe came from nothing at $t = 0$.

Be that as it may, we should also mention difficulties or 'puzzles' that have been encountered when considering the very early universe of the Friedmann models. While it may seem odd to single out the first fraction of a second in a proposed expansion lasting billions of years, much is going on in the first few moments—the temperatures will be extremely high (because the temperature scales as $1/R$—see Problem 8); and particle physicists tell us that the strong interaction is then unified with the electroweak as in GUTs (Grand Unified Theories). Thus, quantum field theory must play a role[17], making a blend of general relativity with quantum physics more appropriate. However, since our short introduction confines itself to a study of general relativity, we mention here just two 'classical' puzzles: (i) the *flatness* problem, and (ii) the *causal* (or so-called *horizon*) problem.

The critical density as defined by equation (6.16) is its present-day value (as the present-day value H_0 of Hubble's constant is used), but (like Hubble's constant) it can be given a value at any cosmic time t by replacing H_0 by $H(t)$ in its defining equation so that ρ_c becomes a time-dependent quantity. We can now introduce the quantity $\Omega \equiv \rho/\rho_c$, where both ρ and ρ_c are evaluated at a general time t.

(i) The *flatness* problem can then be put in the form of a question: why is Ω so close to unity throughout the expansion, even in the radiation-dominated era? The Friedmann equation (6.14) may be written in the form

$$k = \frac{8\pi G R^2 \rho_c}{3}\left(\frac{\rho}{\rho_c} - 1\right) = \frac{8\pi G R^2 \rho_c}{3}\left(\Omega - 1\right), \tag{6.51}$$

which gives

$$\Omega = 1 + k/\dot{R}^2, \tag{6.52}$$

showing that Ω would rapidly increase, or decrease, for $k = +1$, or -1, respectively. Also, in case we suspect that the constant k can change during the evolution, this seems unlikely, because then the whole topology of the universe would change.

(ii) The *causal* (or *horizon*) problem may be stated in the following way. How did early parts of the proto-universe, flying apart at speeds greater than the speed of light (see Section 6.5) ever come to be in thermal equilibrium? Causal communication is *needed* for establishing homogeneity across a large region, and it is homogeneity that, to the best of our measurements, we appear to see today.

---

of observational evidence, ongoing lack of an exact value for H, difficulties with relative abundances of hydrogen and helium, large-scale inhomogeneities in the universe's structure in conflict with the uniformity of the cosmic background radiation, and so on. See Horgan, 1987, for a review of some of these dissident views.

[17] See Narlikar and Padmanabhan, 1991.

These two difficulties (and others concerning monopoles, domain walls, and symmetries in the early universe) were addressed with some success by Alan Guth in 1981 in an 'inflationary' scenario.[18] The basic picture was that for a very short time (about 10^{-33} seconds) the universe expanded exponentially (somewhat like an airbag), following a curve $R(t) \approx R_I e^{\alpha(t-t_I)}$, where R_I is the scale factor at time t_I when inflation starts, and α is a positive constant.

The causal problem is then 'solved' by noting that $\dot{R}$ is proportional to $R(t)$, so that when R is close to zero, so is its time derivative (unlike the initial rapid rate of change of R in the untampered-with Friedmann models). The significance of this is that running the expansion backwards through a change ΔR takes quite a long time at the beginning of the inflation—enough time, it is claimed, to allow for thermal connection.

The flatness problem is 'solved' if we rule out any values for k that are not *extremely* close to 0, for this means there would be no increasing or decreasing of the ratio ρ/ρ_c, and thus no precipitous collapse or rapid expansion to infinity. Inflation does this for us, because the curvature term k/R^2 in equation (6.8), which may initially be of moderate value, is clearly negligible after inflation. This has the same effect as setting $k = 0$.

So it looks as if the solution to the flatness problem has to be $\Omega = 1$, (i.e., $k = 0$, and $\rho = \rho_c$), although observations do not yet find enough density of matter to bear this out. This is the reason astronomers are today searching for 'missing matter'.

Characteristics of the early universe presently occupy the interests of high-energy particle physicists, as well as string theorists, but the basic models (after those few first fractions of a second) are nevertheless thought to be the Friedmann models of general relativity.

Finally, we offer, for comparison with our general relativistic results, a discussion of the Newtonian view of the universe.

Exercises 6.6

1. Show that the empty spacetime field equations derived from equations (6.50) are $R^{\mu\nu} = \Lambda g^{\mu\nu}$.

2. Carry out the working that leads to equation (6.52).

6.7 Newtonian dust

Suppose we have a Newtonian dust (i.e., a fluid with zero pressure moving according to Newton's laws of motion and gravitation) of uniform density $\rho(t)$, which is in a state of uniform expansion, the only force on it being gravity.

---

[18] See Guth, 1981.

This means that the position vector $\mathbf{r}$ of a fluid particle at any time t is given by a relation of the form

$$\mathbf{r} = R(t)\mathbf{c}, \tag{6.53}$$

where $\mathbf{c}$ is a constant vector which is determined by the initial position of the fluid particle. Differentiation gives

$$\dot{\mathbf{r}} = H(t)\mathbf{r}, \tag{6.54}$$

where $H(t) = \dot{R}(t)/R(t)$, and in this way our dust has a Hubble constant $H(t)$ associated with the scale factor $R(t)$.

The Newtonian continuity equation

$$\partial\rho/\partial t + \nabla \cdot (\rho\dot{\mathbf{r}}) = 0$$

gives $\dot{\rho} + 3\rho H = 0$, or $R\dot{\rho} + 3\dot{R}\rho = 0$, which integrates to give

$$\rho R^3 = \text{constant}.$$

As in Section 6.3, let us indicate present-day values with a subscript zero, and write the above equation as

$$\rho R^3 = \rho_0 R_0^3. \tag{6.55}$$

Euler's equation of motion for such a fluid takes the form[19]

$$(\partial/\partial t + \dot{\mathbf{r}} \cdot \nabla)\dot{\mathbf{r}} = \mathbf{F},$$

where $\mathbf{F}$ is the body force per unit mass. With $\dot{\mathbf{r}} = H(t)\mathbf{r}$, this reduces to

$$(\dot{H} + H^2)\mathbf{r} = \mathbf{F}. \tag{6.56}$$

The body force $\mathbf{F}$ is due to gravity, and satisfies $\nabla \cdot \mathbf{F} = -4\pi G\rho$, and on taking the divergence of equation (6.56), we have

$$3(\dot{H} + H^2) = -4\pi G\rho.$$

Putting $H = \dot{R}/R$ results in

$$3\ddot{R}/R = -4\pi G\rho, \tag{6.57}$$

which is exactly the same as the relativistic equation (6.6) with p put equal to zero. Substitution for ρ from equation (6.55) and rearrangement gives

$$2\ddot{R} + A^2/R^2 = 0,$$

where, as before, $A^2 = 8\pi G\rho_0 R_0^3/3$. Multiplying by $\dot{R}$ and integrating gives

[19]See, for example, Landau and Lifshitz, 1987, §2.

$$\boxed{\dot{R}^2 + k = A^2/R,} \tag{6.58}$$

where k is a constant of integration. This is exactly the same as the Friedmann equation (6.14), but there k is either ± 1 or 0. In fact if $k \neq 0$, there is no loss of generality in taking it to be ± 1 in the Newtonian case (see Exercise 6.7.1), and we are therefore led to exactly the same three models for the evolution of the universe as we obtained in Section 6.3.

Equation (6.58) was obtained by integrating equation (6.57), which is effectively the equation of motion of the whole dust-filled Newtonian universe. There is therefore a sense in which we may regard equation (6.58) as the energy equation of the whole universe. Writing it as

$$\dot{R}^2 - A^2/R = -k,$$

we may regard $\dot{R}^2$ as a measure of its kinetic energy, $-A^2/R$ as a measure of its gravitational potential energy, and $-k$ *as a measure of its total energy.* If $k = -1$, the total energy is positive, and the universe has an excess of kinetic energy that allows it to keep expanding at an ultimately constant rate ($\dot{R} \to 1$ and $R \to \infty$ as $t \to \infty$). If $k = 0$, the total energy is zero, and the kinetic energy is just sufficient to allow the universe to keep expanding, but at a decreasing rate ($\dot{R} \to 0$ and $R \to \infty$ as $t \to \infty$). If $k = 1$, then the universe has insufficient kinetic energy for continued expansion. It expands until $\dot{R} = 0$ (when $R = A^2$) and subsequently contracts. This simplistic treatment of the universe as modeled by a Newtonian dust affords insight into the meaning of the curvature constant k appearing in the relativistic models.

Since the Newtonian analysis leads to the same differential equation and hence to the same results as in the relativistic case, we may ask why we bother with a relativistic treatment. Our answer must include the following. In the first place, there are difficulties with Newtonian cosmology that our simple treatment obscures.[20] Second, in relativity pressure contributes to the total energy, and hence to the gravitational field, and Newtonian gravity is deficient in this respect. Third, if the fluid contains particles (stars, etc.) having relativistic speeds, then Newtonian physics is inadequate. Finally, the problem of light propagation throughout the universe should be handled from a relativistic viewpoint.

Exercise 6.7

1. The relationship (6.53) is preserved if we replace $R(t)$ and $\mathbf{c}$ by $\tilde{R}(t)$ and $\tilde{\mathbf{c}}$, defined by

$$R(t) = \lambda \tilde{R}(t), \quad \mathbf{c} = \lambda^{-1}\tilde{\mathbf{c}},$$

where λ is constant. Show that this leads to $\tilde{H}(t) = H(t)$, $\tilde{A}^2 = A^2/\lambda^3$, and

[20] See Bondi, 1960, §9.3.

$$\dot{\tilde{R}}^2 + k/\lambda^2 = \tilde{A}^2/\tilde{R}$$

in place of equation (6.58).
Hence if $k \neq 0$, by a suitable choice of λ we can make $\tilde{k} \equiv k/\lambda^2 = \pm 1$.

Problems 6

1. Show that the three-dimensional manifold with line element (6.3) has a curvature scalar equal to $-6k$.

2. Show that for the cases $k = \pm 1$ the constant of integration A^2 occurring in equation (6.14) is given by either

$$\frac{A^2}{2} = R_0 q_0 \left(\frac{k}{2q_0 - 1}\right) \quad \text{or} \quad \frac{A^2}{2} = \frac{q_0}{H_0}\left(\frac{k}{2q_0 - 1}\right)^{3/2}.$$

   Hence show that for the closed model the present-day value ψ_0 of the parameter ψ (see equations (6.19)) is given by $\cos\psi_0 = (1 - q_0)/q_0$, while for the open model (see equations (6.20)) it is given by $\cosh\psi_0 = (1 - q_0)/q_0$.
   Use these results to show that for a closed model with $q_0 \approx 1$ (say) and $H_0 \approx (13 \times 10^9\,\text{yrs})^{-1}$ we have $t_0 \approx 7.4 \times 10^9\,\text{yrs}$, while for an open model with $q_0 \approx 0.014$ (say) and the same value for H_0, we have $t_0 \approx 12.4 \times 10^9\,\text{yrs}$.
   Repeat the calculations using a Hubble constant of $(18 \times 10^9\,\text{yrs})^{-1}$.

3. A galaxy on the horizon to-day is receding with a speed v just under the limiting value of $2c$ for the flat model of the universe and is now a distance L_0 away from us. In the flat model (with $k = 0$) it is, of course, slowing down. Taking H_0 to be $(18 \times 10^9)^{-1}\,\text{yrs}^{-1}$ (so that $t_0 = 12 \times 10^9\,\text{yrs}$), find
   (a) its present-day distance L_0 from us;
   (b) the future cosmic time t when its speed of recession will have slowed to c, and hence its distance from us then;
   (c) its redshift now, and its redshift when its speed is c.

4. Sketch the graph of the velocity $v_{ph}(t)$ of a photon in the flat model as a function of cosmic time t, given that it arrives here at t_0, having left the galaxy G at $t_E = t_0/5$. Verify that the photon is stationary when $t = 8t_0/27$.

5. A swimmer who can maintain a speed of $3\,\text{m}\,\text{s}^{-1}$ in static water jumps off a dock into a fiord while the tide is going out. The speed of the outgoing tide is $4(1 - 0.001t^{2/3})$, where t is the time in seconds after the jump is made. Find
   (a) how long it takes her to swim back to the dock;

(b) the maximum distance she is swept out from the dock.
Compare and contrast this Galilean calculation with the results obtained in Section 6.5.

6. The time $t_0 - t_E$ for light to travel from a galaxy to a present-day observer is sometimes called the *look-back* time.
   (a) Show that in the flat model the look-back time for a galaxy with redshift z is $(1 - (1+z)^{-3/2})t_0$.
   (b) An astronomer is observing a galaxy for which the redshift is $z = 2$. Taking the present age of the universe to be 12×10^9 yrs, calculate how long ago the light was emitted. Compare this result with that obtained for the case $z = 4$.

7. The value of z for the cosmic background radiation is about 1100. Use this value to obtain an estimate for the age of the universe in the flat model by assuming that the photons were discharged *not* at $t_E = 0$, but rather at $t_E \approx 340,000$ years. (This is the so-called *recombination* or *last-scattering* time—the epoch after which matter and radiation were decoupled, and photons were able to travel freely without being further scattered by the hot plasma of the very early universe.)

8. For the flat model, make an estimate of the absolute temperature of the universe at the recombination time of 340,000 years, given that the temperature T scales as $1/R$.
   Give a plausibility argument for the latter statement, using that $h\nu$ and kT are both measures of energy. (Here h is Planck's constant and k is Boltzmann's constant.)

Appendices

A

Special relativity review

A.0 Introduction

Newton believed that time and space were completely separate entities. Time flowed evenly, the same for everyone, and fixed spatial distances were identical, whoever did the observing. These ideas are still tenable, even for projects like manned rocket travel to the Moon, and almost all the calculations of everyday life in engineering and science rest on Newton's very reasonable tenets. Einstein's 1905 discovery that space and time were just two parts of a single higher entity, *spacetime*, alters only slightly the well-established Newtonian physics with which we are familiar. The new theory is known as *special relativity*, and gives a satisfactory description of all physical phenomena (when allied with quantum theory), with the exception of gravitation. It is of importance in the realm of high relative velocities, and is checked out by experiments performed every day, particularly in high-energy physics. For example, the Stanford linear accelerator, which accelerates electrons close to the speed of light, is about two miles long and cost $\$10^8$; if Newtonian physics were the correct theory, it need only have been about one inch long.

The fundamental postulates of the theory concern *inertial reference systems* or *inertial frames*. Such a reference system is a coordinate system based on three mutually orthogonal axes, which give coordinates x, y, z in space, and an associated system of synchronized clocks at rest in the system, which gives a time coordinate t, and which is such that when particle motion is formulated in terms of this reference system *Newton's first law holds*. It follows that if K and K' are inertial frames, then K' is moving relative to K without rotation and with constant velocity. The four coordinates (t, x, y, z) label points in spacetime, and such a point is called an *event*.

The fundamental postulates are:

1. **The speed of light c is the same in all inertial frames.**
2. **The laws of nature are the same in all inertial frames.**

Postulate 1 is clearly at variance with Newtonian ideas on light propagation. If the same system of units is used in two inertial frames K and K', then it implies that

$$c = dr/dt = dr'/dt', \tag{A.1}$$

where (in Cartesians) $dr^2 = dx^2 + dy^2 + dz^2$, and primed quantities refer to the frame K'. Equation (A.1) may be written as

$$c^2 dt^2 - dx^2 - dy^2 - dz^2 = c^2(dt')^2 - (dx')^2 - (dy')^2 - (dz')^2 = 0,$$

and is consistent with the assumption that there is an invariant *interval* ds between neighboring events given by

$$\pm ds^2 = c^2 dt^2 - dx^2 - dy^2 - dz^2 = c^2(dt')^2 - (dx')^2 - (dy')^2 - (dz')^2, \tag{A.2}$$

which is such that $ds = 0$ for neighboring events on the spacetime curve representing a photon's history. It is convenient to introduce indexed coordinates x^μ $(\mu = 0, 1, 2, 3)$ defined by

$$x^0 \equiv ct, \quad x^1 \equiv x, \quad x^2 \equiv y, \quad x^3 \equiv z, \tag{A.3}$$

and to write the invariance of the interval as

$$\boxed{ds^2 = \eta_{\mu\nu} dx^\mu dx^\nu = \eta_{\mu\nu} dx^{\mu'} dx^{\nu'},} \tag{A.4}$$

where

$$[\eta_{\mu\nu}] = \begin{bmatrix} 1 & 0 & 0 & 0 \\ 0 & -1 & 0 & 0 \\ 0 & 0 & -1 & 0 \\ 0 & 0 & 0 & -1 \end{bmatrix}.$$

in Cartesian coordinates, and Einstein's summation convention has been employed (see Sec. 1.2). In the language of Section 1.9, we are asserting that the spacetime of special relativity is a four-dimensional pseudo-Riemannian manifold[1] with the property that, provided *Cartesian coordinate systems* based on inertial frames are used, the metric tensor components $g_{\mu\nu}$ take the form $\eta_{\mu\nu}$ given above.

Although special relativity may be formulated in arbitrary inertial coordinate systems, we shall stick to Cartesian systems, and raise and lower tensor suffixes using $\eta_{\mu\nu}$ or $\eta^{\mu\nu}$, where the latter are the components of the contravariant metric tensor (see Sec. 1.8). In terms of matrices (see remarks in Sec. 1.2), $[\eta^{\mu\nu}] = [\eta_{\mu\nu}]$ and associated tensors differ only in the signs of some of their components.

---

[1]Roughly speaking, a Riemannian *manifold* is the N-dimensional generalization of a surface. What makes spacetime *pseudo*-Riemannian is the presence of the minus signs in the expression for ds^2. See Sec. 1.9 for details.

Example A.0.1
If $\lambda^\mu = (\lambda^0, \lambda^1, \lambda^2, \lambda^3)$, then

$$\lambda_\mu = \eta_{\mu\nu}\lambda^\nu = (\lambda^0, -\lambda^1, -\lambda^2, -\lambda^3).$$

In a Cartesian coordinate system, the inner product $g_{\mu\nu}\lambda^\mu\sigma^\nu = \lambda^\mu\sigma_\mu = \lambda_\mu\sigma^\mu$ (see Sec. 1.9) takes the simple form

$$\eta_{\mu\nu}\lambda^\mu\sigma^\nu = \lambda^0\sigma^0 - \lambda^1\sigma^1 - \lambda^2\sigma^2 - \lambda^3\sigma^3.$$

The frame-independence contained in the second postulate is incorporated into the theory by expressing the laws of nature as tensor equations which are invariant under a change of coordinates from one inertial reference system to another.

We conclude this introduction with some remarks about time. Each inertial frame has its own coordinate time, and we shall see in the next section how these different coordinate times are related. However, it is possible to introduce an invariantly defined time associated with any given particle (or an idealized observer whose position in space may be represented by a point). The path through spacetime which represents the particle's history is called its *world line*,[2] and the *proper time interval* $d\tau$ between points on its world line, whose coordinate differences relative to some frame K are dt, dx, dy, dz, is defined by

$$c^2 d\tau^2 \equiv c^2 dt^2 - dx^2 - dy^2 - dz^2$$

or

$$\boxed{c^2 d\tau^2 \equiv \eta_{\mu\nu} dx^\mu dx^\nu.} \tag{A.5}$$

So

$$\boxed{d\tau = (1 - v^2/c^2)^{1/2} dt,} \tag{A.6}$$

where v is the particle's speed. Finite proper time intervals are obtained by integrating equation (A.6) along portions of the particle's world line.

Equation (A.6) shows that for a particle at rest in K the proper time τ is nothing other than the coordinate time t (up to an additive constant) measured by stationary clocks in K. If at any instant of the history of a moving particle we introduce an *instantaneous rest frame* K_0, such that the particle is momentarily at rest in K_0, then we see that the proper time τ is the *time recorded by a clock which moves along with the particle.* It is therefore an invariantly defined quantity, a fact which is clear from equation (A.5).

---

[2]For an extended object we have a *world tube.*

A.1 Lorentz transformations

A *Lorentz transformation* is a coordinate transformation connecting two inertial frames K and K'. We observed in the previous section that K' moves relative to K without rotation and with constant velocity, and it is fairly clear that this implies that the primed coordinates $x^{\mu'}$ of K' are given in terms of the unprimed coordinates x^μ of K via a linear (or, strictly speaking, an affine[3]) transformation

$$x^{\mu'} = \Lambda^{\mu'}_{\nu} x^\nu + a^\mu, \tag{A.7}$$

where $\Lambda^{\mu'}_{\nu}$ and a^μ are constants. (This result also follows from the transformation formula for connection coefficients given in Exercise 2.2.5, since, as a consequence of $g_{\mu\nu} = g_{\mu'\nu'} = \eta_{\mu\nu}$, $\Gamma^{\mu}_{\nu\sigma} = \Gamma^{\mu'}_{\nu'\sigma'} = 0$, and hence $X^{\mu'}_{\nu\sigma} \equiv \partial^2 x^{\mu'}/\partial x^\nu \partial x^\sigma = 0$, which integrates to give equation (A.7).) Differentiation of equation (A.7) and substitution into equation (A.4) yields

$$\boxed{\eta_{\mu\nu} = \Lambda^{\rho'}_{\mu} \Lambda^{\sigma'}_{\nu} \eta_{\rho\sigma}} \tag{A.8}$$

as the necessary and sufficient condition for $\Lambda^{\mu'}_{\nu}$ to represent a Lorentz transformation.

If in the transformation (A.7) $a^\mu = 0$, so that the spatial origins of K and K' coincide when $t = t' = 0$, then the Lorentz transformation is called *homogeneous*, while if $a^\mu \neq 0$ (i.e., not all the a^μ are zero) then it is called *inhomogeneous*. (Inhomogeneous transformations are often referred to as *Poincaré transformations*, in which case homogeneous transformations are referred to simply as Lorentz transformations.)

To gain some insight into the meaning of a Lorentz transformation, let us consider the special case of a *boost* in the x direction. This is the situation where the spatial origin O$'$ of K' is moving along the x axis of K in the positive direction with constant speed v relative to K, the axes of K and K' coinciding when $t = t' = 0$ (see Fig. A.1). The transformation is homogeneous and could take the form

$$\begin{aligned} t' &= Bt + Cx, \\ x' &= A(x - vt), \\ y' &= y, \\ z' &= z, \end{aligned} \tag{A.9}$$

the last three equations being consistent with the requirement that O$'$ moves along the x axis of K with speed v relative to K. Adopting this as a "trial solution" and substituting in equation (A.2) gives

$$B^2c^2 - A^2v^2 = c^2, \quad BCc^2 + A^2v = 0, \quad C^2c^2 - A^2 = -1.$$

[3] An affine transformation is a linear transformation that includes a shift of origin.

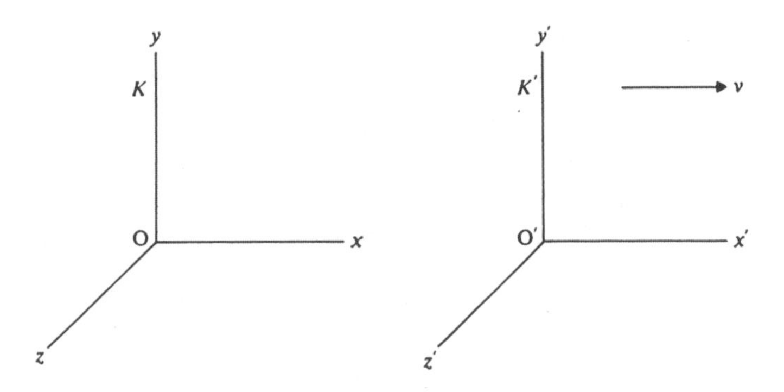

Fig. A.1. A boost in the x direction.

These imply that (see Exercise A.1.1)

$$A = B = (1 - v^2/c^2)^{-1/2}, \quad C = -(v/c^2)(1 - v^2/c^2)^{-1/2}. \tag{A.10}$$

If we put

$$\gamma \equiv (1 - v^2/c^2)^{-1/2}, \tag{A.11}$$

then our boost may be written as

$$\begin{aligned} t' &= \gamma(t - xv/c^2), \\ x' &= \gamma(x - vt), \\ y' &= y, \\ z' &= z, \end{aligned} \tag{A.12}$$

or (in matrix form)

$$\begin{bmatrix} ct' \\ x' \\ y' \\ z' \end{bmatrix} = \begin{bmatrix} \gamma & -v\gamma/c & 0 & 0 \\ -v\gamma/c & \gamma & 0 & 0 \\ 0 & 0 & 1 & 0 \\ 0 & 0 & 0 & 1 \end{bmatrix} \begin{bmatrix} ct \\ x \\ y \\ z \end{bmatrix}. \tag{A.13}$$

Putting $\tanh\psi \equiv v/c$ gives (from equation (A.11)) $\gamma = \cosh\psi$, so the boost may also be written as

$$\begin{aligned} ct' &= ct\cosh\psi - x\sinh\psi, \\ x' &= x\cosh\psi - ct\sinh\psi, \\ y' &= y, \\ z' &= z. \end{aligned} \tag{A.14}$$

It may be shown that a general homogeneous Lorentz transformation is equivalent to a boost in some direction followed by a spatial rotation. The general inhomogeneous transformation requires an additional translation (i.e., a shift of spacetime origin).

Since $X^{\mu'}_{\nu} \equiv \partial x^{\mu'}/\partial x^{\nu} = \Lambda^{\mu'}_{\nu}$, a contravariant vector has components λ^{μ} relative to inertial frames which transform according to

$$\lambda^{\mu'} = \Lambda^{\mu'}_{\nu} \lambda^{\nu},$$

while a covariant vector has components λ_{μ} which transform according to

$$\lambda_{\mu'} = \Lambda^{\nu}_{\mu'} \lambda_{\nu},$$

where $\Lambda^{\nu}_{\mu'}$ is such that $\Lambda^{\nu}_{\mu'} \Lambda^{\mu'}_{\sigma} = \delta^{\nu}_{\sigma}$. These transformation rules extend to tensors. For example, a mixed tensor of rank two has components τ^{μ}_{ν} which transform according to

$$\tau^{\mu'}_{\nu'} = \Lambda^{\mu'}_{\rho} \Lambda^{\sigma}_{\nu'} \tau^{\rho}_{\sigma}.$$

The equations of electromagnetism are invariant under Lorentz transformations, and in Section A.8 we present them in tensor form which brings out this invariance. However, the equations of Newtonian mechanics are not invariant under Lorentz transformations, and some modifications are necessary (see Sec. A.6). The transformations which leave the equations of Newtonian mechanics invariant are Galilean transformations, to which Lorentz transformations reduce when v/c is negligible (see Exercise A.1.3).

Exercises A.1

1. Verify that A, B, C are as given by equations (A.10).

2. Equation A.13 gives the matrix $[\Lambda^{\mu'}_{\nu}]$ for the boost in the x direction. What form does the inverse matrix $[\Lambda^{\nu}_{\mu'}]$ take? What is the velocity of K relative to K'?

3. Show that when v/c is negligible, equations (A.12) of a Lorentz boost reduce to those of a Galilean boost:

$$t' = t, \quad x' = x - vt, \quad y' = y, \quad z' = z.$$

A.2 Relativistic addition of velocities

Suppose we have three inertial frames K, K', and K'', with K' connected to K by a boost in the x direction, and K'' connected to K' by a boost in the x' direction. If the speed of K' relative to K is v, then equations (A.14) hold, where $\tanh\psi = v/c$ and if the speed of K'' relative to K' is w, then we have analogously

$$\begin{aligned} ct'' &= ct' \cosh\phi - x' \sinh\phi, \\ x'' &= x' \cosh\phi - ct' \sinh\phi, \\ y'' &= y', \\ z'' &= z', \end{aligned} \tag{A.15}$$

where $\tanh\phi = w/c$. Substituting for ct', x', y', z' from equations (A.14) into the above gives

$$\begin{aligned} ct'' &= ct \cosh(\psi + \phi) - x \sinh(\psi + \phi), \\ x'' &= x \cosh(\psi + \phi) - ct \sinh(\psi + \phi), \\ y'' &= y, \\ z'' &= z. \end{aligned} \tag{A.16}$$

This shows that K'' is connected to K by a boost, and that K'' is moving relative to K in the positive x direction with a speed u given by $u/c = \tanh(\psi + \phi)$. But

$$\tanh(\psi + \phi) = \frac{\tanh\psi + \tanh\phi}{1 + \tanh\psi \tanh\phi},$$

so

$$\boxed{u = \frac{v + w}{1 + vw/c^2}.} \tag{A.17}$$

This is the relativistic formula for the addition of velocities, and replaces the Newtonian formula $u = v + w$.

Note that $v < c$ and $w < c$ implies that $u < c$, so that by compounding speeds less than c one can never exceed c. For example, if $v = w = 0.75c$, then $u = 0.96c$.

Exercises A.2

1. Verify equations (A.16).

2. Verify that if $v < c$ and $w < c$ then the addition formula (A.17) implies that $u < c$.

A.3 Simultaneity

Many of the differences between Newtonian and relativistic physics are due to the concept of simultaneity. In Newtonian physics this is a frame-independent concept, whereas in relativity it is not. To see this, consider two inertial frames K and K' connected by a boost, as in Section A.1. Events which are simultaneous in K are given by $t = t_0$, where t_0 is constant. Equations (A.12) show that for these events

$$t' = \gamma(t_0 - xv/c^2),$$

so t' depends on x, and is not constant. The events are therefore not simultaneous in K'. (See also Fig. A.5.)

A.4 Time dilation, length contraction

Since a moving clock records its own proper time τ, equation (A.6) shows that the proper time interval $\Delta\tau$ recorded by a clock moving with constant speed v relative to an inertial frame K is given by

$$\boxed{\Delta\tau = (1 - v^2/c^2)^{1/2}\Delta t,} \tag{A.18}$$

where Δt is the coordinate time interval recorded by stationary clocks in K. Hence $\Delta t > \Delta\tau$ and the moving clock "runs slow." This is the phenomenon of *time dilation*. The related phenomenon of *length contraction* (also known as *Lorentz contraction*) arises in the following way.

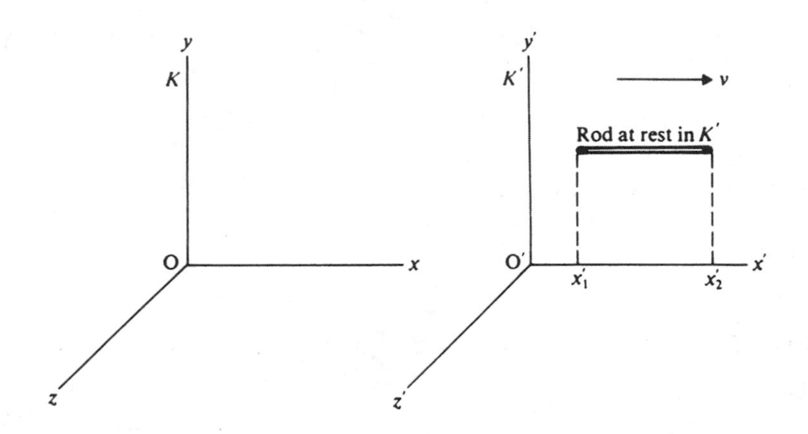

Fig. A.2. Length contraction.

Suppose that we have a rod moving in the direction of its own length with constant speed v relative to an inertial frame K. There is no loss of generality in choosing this direction to be the positive x direction of K. If K' is a frame moving in the same direction as the rod with speed v relative to K, so that K' is connected to K by a boost as in Section A.1, then the rod will be at rest in K', which is therefore a rest frame for it (see Fig. A.2). The *proper length* or *rest length* l_0 of the rod is the length as measured in the rest frame K', so

$$l_0 = x_2' - x_1',$$

where x'_2 and x'_1 are the x' coordinates of its endpoints in K'. According to equations (A.12), the x coordinates x_1, x_2 of its endpoints at any time t in K are given by

$$x'_1 = \gamma(x_1 - vt),$$
$$x'_2 = \gamma(x_2 - vt).$$

Hence if we take the difference between the endpoints *at the same time* t in K, we get

$$x'_2 - x'_1 = \gamma(x_2 - x_1).$$

The length l of the rod, as measured by noting the simultaneous positions of its endpoints in K, is therefore given by

$$\boxed{l_0 = \gamma l \quad \text{or} \quad l = l_0(1 - v^2/c^2)^{1/2}.} \tag{A.19}$$

So $l < l_0$ and the moving rod is contracted.

A straightforward calculation shows that if the rod is moving relative to K in a direction perpendicular to its length, then it suffers no contraction. It follows that the volume V of a moving object, as measured by simultaneously noting the positions of its boundary points in K, is related to its rest volume V_0 by $V = V_0(1 - v^2/c^2)^{1/2}$. This fact must be taken into account when considering densities.

A.5 Spacetime diagrams

Spacetime diagrams are either three- or two-dimensional representations of spacetime, having either one or two spatial dimensions suppressed. When events are referred to an inertial reference system, it is conventional to orient the diagrams so that the t axis points vertically upwards and the spatial axes are horizontal. It is also conventional to scale things so that the straight-line paths of photons are inclined at 45°; this is equivalent to using so-called relativistic units in which $c = 1$, or using the coordinates x^μ defined by equations (A.3).

If we consider all the photon paths passing through an event O then these constitute the *null cone* at O (see Fig. A.3). The region of spacetime contained within the upper half of the null cone is the *future* of O, while that contained within the lower half is its *past*. The region outside the null cone contains events which may either come before or after the event O in time, depending on the reference system used, but there is no such ambiguity about the events in the future and in the past. This follows from the fact that the null cone at O is invariantly defined. If the event O is taken as the origin of an inertial reference system, then the equation of the null cone is

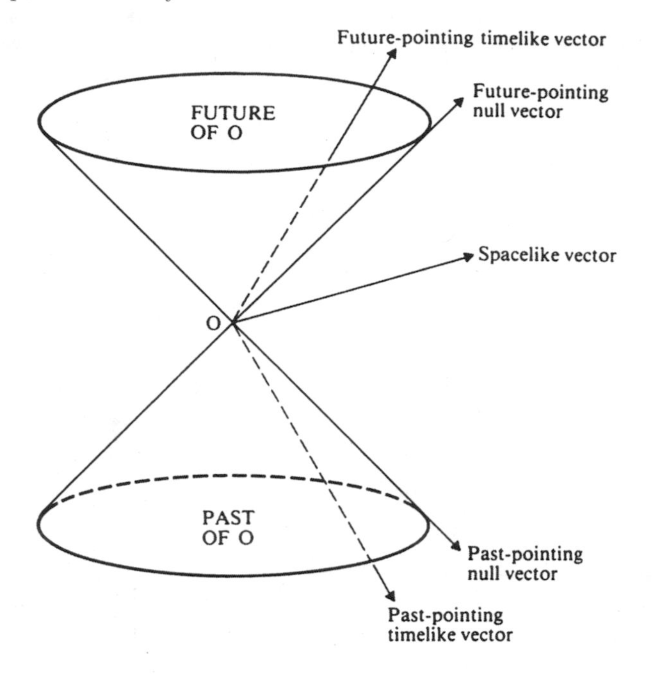

Fig. A.3. Null cone and vectors at an event O.

$$x^2 + y^2 + z^2 = c^2t^2. \tag{A.20}$$

If we have a vector λ^μ localized at O, then λ^μ is called *timelike* if it lies within the null cone, *null* if it is tangential to the null cone, and *spacelike* if it lies outside the null cone. That is, λ^μ is

$$\left\{\begin{array}{l}\text{timelike}\\ \text{null}\\ \text{spacelike}\end{array}\right. \quad \text{if } \eta_{\mu\nu}\lambda^\mu\lambda^\nu \left\{\begin{array}{l} >0\\ =0\\ <0\end{array}\right. . \tag{A.21}$$

Timelike and null vectors may be characterized further as *future-pointing* or *past-pointing* (see Fig. A.3).

Consider now the world line of a particle with mass. Relativistic mechanics prohibits the acceleration of such a particle to speeds up to c (a fact suggested by the formula (A.17) for the addition of velocities),[4] which implies that its world line must lie within the null cone at each event on it, as the following remarks show. With the speed $v < c$ the proper time τ as defined by equation (A.6) is real, and may be used to parameterize the world line: $x^\mu = x^\mu(\tau)$. Its tangent vector $u^\mu \equiv dx^\mu/d\tau$ (see Sec. 1.7) is called the *world velocity* of the particle, and equation (A.5) shows that

[4]Particles having speeds in excess of c, called tachyons, have been postulated, but attempts to detect them have (to date) been unsuccessful. They cannot be decelerated to speeds below c.

$$\eta_{\mu\nu} u^\mu u^\nu = c^2,$$

so u^μ is timelike and lies within the null cone at each event on the world line (see Fig. A.4 (a)). The tangent vector at each event on the world line of a photon is clearly null (see Fig. A.4 (b)).

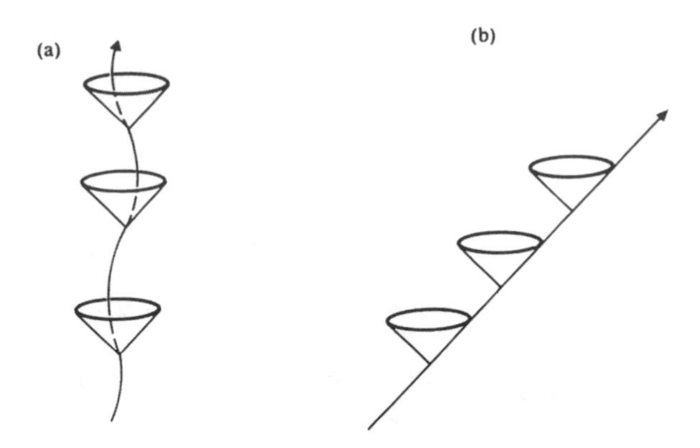

Fig. A.4. World lines of (a) a particle with mass, and (b) a photon.

Spacetime diagrams may be used to illustrate Lorentz transformations. A two-dimensional diagram suffices to illustrate the boost of Section A.1 connecting the frames K and K'. The x' axis of K' is given by $t' = 0$, that is, by $t = xv/c^2$, while the t' axis of K' is given by $x' = 0$, that is, by $x = vt$. So with $c = 1$, the slope of the x' axis relative to K is v, while that of the t' axis is $1/v$. So if the axes of K are drawn perpendicular as in Figure A.5, then those of K' are not perpendicular, but inclined as shown.

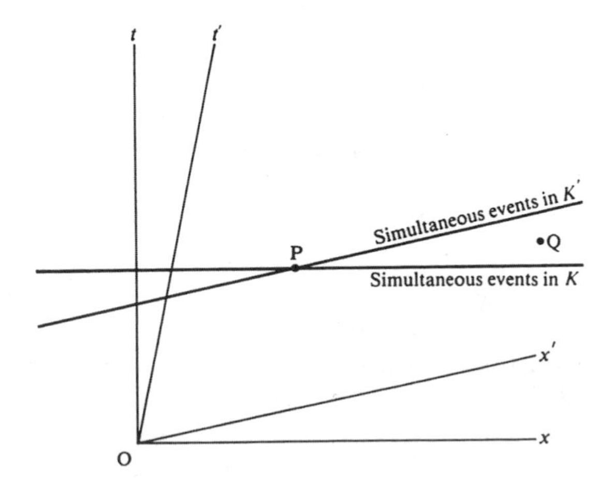

Fig. A.5. Spacetime diagram of a boost.

Events which are simultaneous in K are represented by a line parallel to the x-axis, while those which are simultaneous in K' are represented by a line parallel to the x' axis, and the frame dependence of the concept of simultaneity is clearly illustrated in a spacetime diagram. Note that the event Q of Figure A.5 occurs *after* the event P according to observers in K, while it occurs *before* the event P according to observers in K'.

Exercises A.5

1. Check the criterion (A.21).

2. Check the invariance of the light-cone under a boost, by showing that equation (A.20) transforms into the equation

$$(x')^2 + (y')^2 + (z')^2 = c^2(t')^2.$$

A.6 Some standard 4-vectors

Here we introduce some standard 4-vectors of special relativity, and comment briefly on their roles in relativistic mechanics. The prefix 4- serves to distinguish vectors in spacetime from those in space, which we shall call 3-vectors. It is useful to introduce the notation

$$\lambda^\mu = (\lambda^0, \lambda^1, \lambda^2, \lambda^3) = (\lambda^0, \boldsymbol{\lambda}), \tag{A.22}$$

so that bold-faced letters represent spatial parts.

We have already defined the world velocity $u^\mu \equiv dx^\mu/d\tau$ of a particle with mass. If we introduce the *coordinate velocity* v^μ (which is *not* a 4-vector) defined by

$$v^\mu \equiv dx^\mu/dt = (c, \mathbf{v}), \tag{A.23}$$

where $\mathbf{v}$ is the particle's 3-velocity, then

$$u^\mu = (dt/d\tau)v^\mu = (\gamma c, \gamma \mathbf{v}), \tag{A.24}$$

where $\gamma = (1 - v^2/c^2)^{-1/2}$. The particle's 4-*momentum* p^μ is defined in terms of u^μ by

$$\boxed{p^\mu \equiv m u^\mu,} \tag{A.25}$$

where m is the particle's rest mass.[5] The zeroth component p^0 is E/c, where E is the energy of the particle, and we can put

$$p^\mu = (E/c, \mathbf{p}). \tag{A.26}$$

[5] As in Chap. 3, we use m rather than the more emphatic m_0 for rest mass.

The wave aspect of light may be built into the particle approach by associating with a photon a *wave 4-vector* k^μ defined by

$$\boxed{k^\mu \equiv (2\pi/\lambda, \mathbf{k}),} \tag{A.27}$$

where λ is the wavelength and $\mathbf{k} = (2\pi/\lambda)\mathbf{n}$, $\mathbf{n}$ being a unit 3-vector in the direction of propagation.[6] It follows that $k^\mu k_\mu = 0$, so that k^μ is null. It is, of course, tangential to the photon's world line. The photon's 4-momentum p^μ is given by

$$p^\mu \equiv (h/2\pi)k^\mu, \tag{A.28}$$

where h is Planck's constant. Thus the photon's energy is

$$E = cp^0 = c(h/2\pi)k^0 = hc/\lambda = h\nu,$$

where ν is the frequency, in agreement with the quantum-mechanical result.

In relativistic mechanics, Newton's second law is modified to

$$dp^\mu/d\tau = f^\mu, \tag{A.29}$$

where f^μ is the 4-force on the particle. This is given in terms of the 3-force $\mathbf{F}$ by

$$f^\mu \equiv \gamma(\mathbf{F}\cdot\mathbf{v}/c, \mathbf{F}). \tag{A.30}$$

Example A.6.1

The invariance of the inner product gives

$$p^\mu p_\mu = p^{\mu'} p_{\mu'}. \tag{A.31}$$

If we take the primed frame K' to be an instantaneous rest frame, then $p^{\mu'} = (mc, \mathbf{0})$, and the right-hand side is m^2c^2. The left-hand side is $E^2/c^2 - \mathbf{p}\cdot\mathbf{p}$, so equation (A.31) gives

$$E = (p^2c^2 + m^2c^4)^{1/2}, \tag{A.32}$$

where $p^2 = \mathbf{p}\cdot\mathbf{p}$. This is the well-known result connecting the energy E of a particle with its momentum and rest mass.

From equations (A.24) and (A.26) we see that

$$\mathbf{p} = \gamma m\mathbf{v} \tag{A.33}$$

and that $E/c = p^0 = \gamma mc$, so

[6] The factor 2π, which seems to be an encumbrance, simplifies expressions in relativistic optics and wave theory.

$$E = \gamma mc^2 = mc^2(1 - v^2/c^2)^{-1/2} = mc^2 + \frac{1}{2}mv^2 + \cdots . \tag{A.34}$$

Equation (A.33) shows that the spatial part $\mathbf{p}$ of the relativistic 4-momentum p^μ reduces to the Newtonian 3-momentum $m\mathbf{v}$ when v is small compared to c (giving $\gamma \approx 1$). However, equation (A.34) shows that E reduces to $mc^2 + \frac{1}{2}mv^2$, and that the total energy includes not only the kinetic energy $\frac{1}{2}mv^2$, but also the *rest energy* mc^2, the latter being unsuspected in Newtonian physics. It should be noted that we have *not proved* the celebrated formula $E = \gamma mc^2$; it simply follows from our defining E by $p^0 = E/c$. Although this definition is standard in relativity, it is sensible to ask how it ever came about.

Conservation of momentum is an extremely useful principle, and if we wish to preserve it in relativity, then it turns out that we must define momentum $\mathbf{p}$ by $\mathbf{p} = \gamma m\mathbf{v}$ rather than $\mathbf{p} = m\mathbf{v}$. This follows from a consideration of simple collision problems in different inertial frames.[7] But $\gamma m\mathbf{v}$ is the spatial part of the 4-vector p^μ defined by equation (A.25), and it follows from equations (A.29) and (A.30) that $dp^0/d\tau = (\gamma/c)\mathbf{F} \cdot \mathbf{v}$, which implies that

$$\mathbf{F} \cdot \mathbf{v} = c\, dp^0/dt.$$

But $\mathbf{F} \cdot \mathbf{v}$ is the rate at which the 3-force $\mathbf{F}$ imparts energy to the particle, hence it is natural to define the energy E of the particle by $E = cp^0$. The conservation of energy and momentum of a free particle is then incorporated in the single 4-vector equation

$$p^\mu = \text{constant}.$$

This extends to a system of interacting particles with no external forces:

$$\boxed{\sum_{\text{all particles}} p^\mu = \text{constant}.} \tag{A.35}$$

Example A.6.2

Consider the Compton effect in which a photon collides with a stationary electron (see Fig. A.6). Initially the photon is traveling along the x^1 axis of our reference system and it collides with an electron at rest. After collision the electron and photon move off in the plane $x^3 = 0$, making angles θ and ϕ with the x^1 axis as shown. Remembering that the energy of a photon is $h\nu$, and that for a photon $p^\mu p_\mu = 0$, we have before collision:

$$p^\mu_{\text{ph}} = (h\nu/c, h\nu/c, 0, 0),$$
$$p^\mu_{\text{el}} = (mc, 0, 0, 0),$$

[7] See, for example, Rindler, 1982, §26. Note that in Rindler m_0 is proper mass and m is *relativistic mass*; in our notation these quantities are m and γm, respectively.

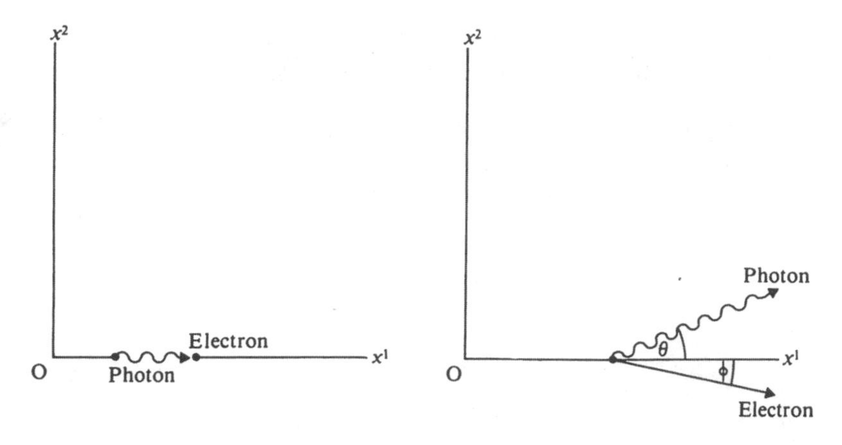

Fig. A.6. Geometry of the Compton effect.

and after collision

$$\bar{p}^{\mu}_{\text{ph}} = (h\bar{\nu}/c, (h\bar{\nu}/c)\cos\theta, (h\bar{\nu}/c)\sin\theta, 0),$$
$$\bar{p}^{\mu}_{\text{el}} = (\gamma mc, \gamma mv\cos\phi, -\gamma mv\sin\phi, 0),$$

where v is the electron's speed after collision. The conservation laws contained in

$$p^{\mu}_{\text{ph}} + p^{\mu}_{\text{el}} = \bar{p}^{\mu}_{\text{ph}} + \bar{p}^{\mu}_{\text{el}}$$

then give

$$h\nu/c + mc = h\bar{\nu}/c + \gamma mc,$$
$$h\nu/c = (h\bar{\nu}/c)\cos\theta + \gamma mv\cos\phi,$$
$$0 = (h\bar{\nu}/c)\sin\theta - \gamma mv\sin\phi.$$

Eliminating v and ϕ from these leads to the formula for Compton scattering (see Exercise A.6.3) giving the frequency of the photon after collision as

$$\boxed{\bar{\nu} = \frac{\nu}{1 + (h\nu/mc^2)(1 - \cos\theta)}.} \tag{A.36}$$

Exercises A.6

1. In a laboratory frame, write u^{μ} for (a) a stationary chair, (b) a speeding bullet.
   Is it possible to write u^{μ} for a photon?

2. Show that as a consequence of equation (A.29) we have $u_\mu f^\mu = 0$, and that f^μ as given by equation (A.30) satisfies this relation.

3. Check the derivation of the formula (A.36).

A.7 Doppler effect

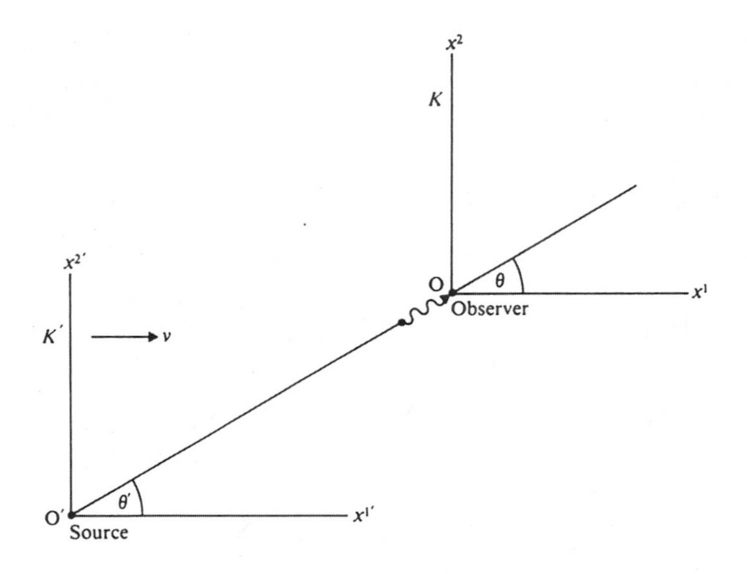

Fig. A.7. Photon arriving at an observer from a moving source.

Suppose that we have a source of radiation which is moving relative to an inertial frame K with speed v in the positive x^1 direction in the plane $x^3 = 0$, and that at some instant an observer fixed at the origin O of K receives a photon in a direction which makes an angle θ with the positive x^1 direction (see Fig. A.7). Let us attach to the source a frame K' whose axes are parallel to those of K, and which moves along with the source, so that it is at rest in K' at the origin O′. The frame K' is therefore connected to K by a Lorentz transformation comprising a boost in the x^1 direction and a translation (see Sec. A.1). It follows that the wave 4-vector k^μ of the photon transforms according to

$$k^{\mu'} = \Lambda^{\mu'}_{\nu} k^\nu, \tag{A.37}$$

where

$$[\Lambda^{\mu'}_{\nu}] = \begin{bmatrix} \gamma & -\gamma v/c & 0 & 0 \\ -\gamma v/c & \gamma & 0 & 0 \\ 0 & 0 & 1 & 0 \\ 0 & 0 & 0 & 1 \end{bmatrix}$$

(see Exercise A.1.2). The zeroth component gives $k^{0'} = \Lambda^{0'}_{\nu} k^{\nu}$, or

$$k^{0'} = \gamma k^0 - (\gamma v/c)k^1. \tag{A.38}$$

Now $k^{0'} = 2\pi/\lambda_0$, where λ_0 is the *proper wavelength* as observed in the frame K' in which the source is at rest, and

$$k^{\mu} = (2\pi/\lambda)(1, \cos\theta, \sin\theta, 0),$$

where λ is the wavelength as observed by the observer at the origin O of K. (Note that we are making use of the fact that k^{μ} is constant along the photon's world line.) Hence equation (A.38) gives

$$\frac{1}{\lambda_0} = \frac{1}{\lambda}(\gamma - \frac{\gamma v}{c}\cos\theta),$$

so

$$\lambda/\lambda_0 = \gamma[1 - (v/c)\cos\theta], \tag{A.39}$$

and the observed wavelength is different from the proper wavelength.

If the source is on the negative x^1 axis, so that it is approaching the observer, then $\theta = 0$ and equation (A.39) gives

$$\boxed{\frac{\lambda}{\lambda_0} = \gamma(1 - v/c) = \left(\frac{1 - v/c}{1 + v/c}\right)^{1/2}.} \tag{A.40}$$

Thus $\lambda < \lambda_0$ and the observed wavelength is *blueshifted*.

If the source is on the positive x^1 axis, so that it is receding from the observer, then $\theta = \pi$ and equation (A.39) gives

$$\boxed{\frac{\lambda}{\lambda_0} = \gamma(1 + v/c) = \left(\frac{1 + v/c}{1 - v/c}\right)^{1/2}.} \tag{A.41}$$

Thus $\lambda > \lambda_0$ and the observed wavelength is *redshifted.*

If the source is displaced away from the x^1 axis, then at some instant we will have $\theta = \pm\pi/2$ giving

$$\boxed{\lambda/\lambda_0 = \gamma = (1 - v^2/c^2)^{-1/2},} \tag{A.42}$$

which is also a redshift.

These shifts in the observed spectrum are examples of the *Doppler effect.* Formulae (A.40) and (A.41) refer to approach and recession, and have their counterparts in nonrelativistic physics. Formula (A.42) is that of the *transverse* Doppler effect, and has no such counterpart. The transverse effect was observed in 1938 by Ives and Stillwell, who examined the spectra of rapidly moving hydrogen atoms. Formula (A.42) may be used in discussions of the celebrated twin paradox.[8]

Exercise A.7

1. Using equation (A.37), show that the angle θ' (see Fig. A.7) at which the photon leaves the source, as measured in K', is given by

$$\tan\theta' = \frac{\tan\theta}{\gamma[1-(v/c)\sec\theta]}.$$

(This is essentially the *relativistic aberration formula.*)

A.8 Electromagnetism

The equations which govern the behavior of the electromagnetic field in free space are Maxwell's equations, which in SI units take the form

$$\nabla\cdot\mathbf{B} = 0, \tag{A.43}$$
$$\nabla\cdot\mathbf{E} = \rho/\varepsilon_0, \tag{A.44}$$
$$\nabla\times\mathbf{E} = -\partial\mathbf{B}/\partial t, \tag{A.45}$$
$$\nabla\times\mathbf{B} = \mu_0\mathbf{J} + \mu_0\varepsilon_0\partial\mathbf{E}/\partial t. \tag{A.46}$$

Here $\mathbf{E}$ is the *electric field intensity*, $\mathbf{B}$ is the *magnetic induction*, ρ is the charge density (charge per unit volume), $\mathbf{J}$ is the current density, μ_0 is the *permeability* of free space, and ε_0 is the *permittivity* of free space. The last two quantities satisfy

$$\mu_0\varepsilon_0 = 1/c^2. \tag{A.47}$$

The vector fields $\mathbf{B}$ and $\mathbf{E}$ may be expressed in terms of a *vector potential* $\mathbf{A}$ and a *scalar potential* ϕ:

$$\mathbf{B} = \nabla\times\mathbf{A}, \quad \mathbf{E} = -\nabla\phi - \partial\mathbf{A}/\partial t. \tag{A.48}$$

Equations (A.43) and (A.45) are then satisfied. These potentials are not uniquely determined by Maxwell's equations, and $\mathbf{A}$ may be replaced by $\mathbf{A}+\nabla\psi$ and ϕ by $\phi-\partial\psi/\partial t$, where ψ is arbitrary. Such transformations of the potentials are known as *gauge transformations*, and allow one to choose $\mathbf{A}$ and ϕ so that they satisfy the *Lorentz gauge condition*, which is

$$\nabla\cdot\mathbf{A} + \varepsilon_0\mu_0\partial\phi/\partial t = 0. \tag{A.49}$$

[8] See Feenberg, 1959.

The remaining two Maxwell equations (A.44) and (A.46) then imply that $\mathbf{A}$ and ϕ satisfy

$$\Box^2 \mathbf{A} = -\mu_0 \mathbf{J}, \quad \Box^2 \phi = -\rho/\varepsilon_0, \tag{A.50}$$

where $\Box^2$ is the d'Alembertian defined by[9]

$$\Box^2 \equiv \nabla^2 - c^{-2}\partial^2/\partial t^2. \tag{A.51}$$

Equations (A.50) may be solved in terms of retarded potentials, and the form of the solution shows that we may take ϕ/c and $\mathbf{A}$ as the temporal and spatial parts of a 4-vector:

$$A^\mu \equiv (\phi/c, \mathbf{A}), \tag{A.52}$$

which is known as the 4-*potential*.[10]

Maxwell's equations take on a particularly simple and elegant form if we introduce the *electromagnetic field tensor* $F_{\mu\nu}$ defined by

$$F_{\mu\nu} \equiv A_{\mu,\nu} - A_{\nu,\mu}, \tag{A.53}$$

where $A_\mu = (\phi/c, -\mathbf{A})$ is the covariant 4-potential, and commas denote partial derivatives. Equations (A.48) show that

$$[F_{\mu\nu}] = \begin{bmatrix} 0 & -E^1/c & -E^2/c & -E^3/c \\ E^1/c & 0 & B^3 & -B^2 \\ E^2/c & -B^3 & 0 & B^1 \\ E^3/c & B^2 & -B^1 & 0 \end{bmatrix}, \tag{A.54}$$

where $(E^1, E^2, E^3) \equiv \mathbf{E}$ and $(B^1, B^2, B^3) \equiv \mathbf{B}$. It is then a straightforward process (using the result of Exercise A.8.4) to check that Maxwell's equations are equivalent to

$$\boxed{F^{\mu\nu}{}_{,\nu} = \mu_0 j^\mu,} \tag{A.55}$$

$$\boxed{F_{\mu\nu,\sigma} + F_{\nu\sigma,\mu} + F_{\sigma\mu,\nu} = 0,} \tag{A.56}$$

where $j^\mu \equiv (\rho c, \mathbf{J})$ is the 4-*current density*. Note that $j^\mu = \rho v^\mu = (\gamma\rho_0)v^\mu = \rho_0 u^\mu$, where u^μ is the world velocity of the charged particles producing the current distribution, and ρ_0 is the *proper* charge density. That is, ρ_0 is the charge per unit *rest* volume, whereas ρ is the charge per unit volume (see remark at end of Sec. A.4).

The equation of motion of a particle of charge q moving in an electromagnetic field is

$$d\mathbf{p}/dt = q(\mathbf{E} + \mathbf{v} \times \mathbf{B}), \tag{A.57}$$

---

[9]We are using the more consistent-looking notation $\Box^2$, rather than $\Box$ used in some European texts.

[10]See, for example, Rindler, 1982, §38.

where $\mathbf{p}$ is its momentum and $\mathbf{v}$ its velocity. The right-hand side of this equation is known as the *Lorentz force*. It follows that the rate at which the electromagnetic field imparts energy E to the particle is given by

$$dE/dt = \mathbf{F} \cdot \mathbf{v} = q\mathbf{E} \cdot \mathbf{v}. \tag{A.58}$$

Equations (A.57) and (A.58) may be brought together in a single 4-vector equation (see Exercise A.8.5):

$$dp^\mu/d\tau = -qF^\mu{}_\nu u^\nu, \tag{A.59}$$

which gives

$$m\, d^2x^\mu/d\tau^2 = -qF^\mu{}_\nu u^\nu, \tag{A.60}$$

where m is the rest mass of the particle. The continuous version of equation (A.60) is

$$\mu\, d^2x^\mu/d\tau^2 = -F^\mu{}_\nu j^\nu, \tag{A.61}$$

where μ is the proper (mass) density of the charge distribution giving rise to the electromagnetic field, and this is the equation of motion of a charged unstressed fluid. That is, the only forces acting on the fluid particles arise from their electromagnetic interaction, there being no body forces nor mechanical stress forces such as pressure.

It is evident that Maxwell's equations and related equations may be formulated as 4-vector and tensor equations without modification. They are therefore invariant under Lorentz transformations, but not under Galilean transformations, and this observation played a leading role in the development of special relativity. By contrast, the equations of Newtonian mechanics are invariant under Galilean transformations, but not under Lorentz transformations, and therefore require modification to incorporate them into special relativity.

Exercises A.8

1. Show that the Lorentz gauge condition (A.49) may be written as $A^\mu{}_{,\mu} = 0$.

2. Check that the components $F_{\mu\nu}$ are as displayed in equation (A.54).

3. Show that the mixed and contravariant forms of the electromagnetic field tensor are given by

$$[F^\mu{}_\nu] = \begin{bmatrix} 0 & -E^1/c & -E^2/c & -E^3/c \\ -E^1/c & 0 & -B^3 & B^2 \\ -E^2/c & B^3 & 0 & -B^1 \\ -E^3/c & -B^2 & B^1 & 0 \end{bmatrix}, \tag{A.62}$$

$$[F^{\mu\nu}] = \begin{bmatrix} 0 & E^1/c & E^2/c & E^3/c \\ -E^1/c & 0 & B^3 & -B^2 \\ -E^2/c & -B^3 & 0 & B^1 \\ -E^3/c & B^2 & -B^1 & 0 \end{bmatrix}. \tag{A.63}$$

(Use caution in matrix multiplication.)

4. Verify that equations (A.55) and (A.56) are equivalent to Maxwell's equations.

5. Verify that equations (A.57) and (A.58) may be brought together in the single 4-vector equation (A.59).

Problems A

1. If an astronaut claims that a spaceflight took her 3 days, while a base station on Earth claims that she took 3.000 000 015 days, what kind of average rocket speed are we talking about?
(Assume only special-relativistic effects.)

2. Illustrate the phenomena of time dilation and length contraction using spacetime diagrams.
(Note that there is a scale difference between the inclined x' axis and the horizontal x axis: if 1 cm along the x axis represents 1 m, then along the x' axis it does *not* represent 1 m. There is also a scale difference between the time axes.)

3. If a laser in the laboratory has a wavelength of 632.8 nm, what wavelength would be observed by an observer approaching it directly at a speed $c/2$?

4. Show that under a boost in the x^1 direction the components of the electric field intensity **E** and the magnetic induction **B** transform according to

$$\begin{aligned} E^{1'} &= E^1, & B^{1'} &= B^1, \\ E^{2'} &= \gamma(E^2 - vB^3), & B^{2'} &= \gamma(B^2 + vE^3/c^2), \\ E^{3'} &= \gamma(E^3 + vB^2), & B^{3'} &= \gamma(B^3 - vE^2/c^2). \end{aligned}$$

5. Plot a graph of $\gamma(v) \equiv (1 - v^2/c^2)^{-1/2}$.

6. In a laboratory a certain switch is turned on, and then turned off 3 s later. In a "rocket frame" these events are found to be separated by 5 s. Show that in the rocket frame the spatial separation between the two events is 12×10^8 m, and that the rocket frame has a speed 2.4×10^8 m s^{-1} relative to the laboratory.
(Take $c = 3 \times 10^8$ m s^{-1}.)

7. A uniform charge distribution of proper density ρ_0 is at rest in an inertial frame K. Show that an observer moving with a velocity $\mathbf{v}$ relative to K sees a charge density $\gamma\rho_0$ and a current density $-\gamma\rho_0\mathbf{v}$.

8. A woman of mass 70 kg is at rest in the laboratory. Find her kinetic energy and momentum relative to an observer passing in the x direction at a speed $c/2$.

9. Show that the Doppler shift formula (A.39) may be expressed invariantly as

$$\lambda/\lambda_0 = (u^\mu_{\text{source}}\, k_\mu)/(u^\nu k_\nu),$$

where λ is the observed wavelength, λ_0 the proper wavelength, k^μ the wave 4-vector, u^μ the world velocity of the observer, and u^μ_{source} that of the source.
(Hint: In K', the wavelength is λ_0 and $u^{\mu'}_{\text{source}} = (c, \mathbf{0})$. In K, the wavelength is λ and $u^\mu_{\text{source}} = \gamma(c, \mathbf{v})$. For each frame of reference quantities like $k_\mu u^\mu$ are invariant, and thus may be evaluated in any reference frames we wish.)

10. It is found that a stationary "cupful" of radioactive pions has a half-life of 1.77×10^{-8} s. A collimated pion beam leaves an accelerator at a speed of $0.99c$, and it is found to drop to half its original intensity 37.3 m away. Are these results consistent?
(Look at the problem from two separate viewpoints, namely that of time dilation and that of path contraction. Take $c = 3 \times 10^8\,\mathrm{m\,s^{-1}}$.)

11. Verify that (in the notation of Sec. A.8) Ohm's law can be written as $j^\mu - u^\mu u_\nu j^\nu = \sigma u_\nu F^{\mu\nu}$, where σ is the conductivity of the material and u^μ is its 4-velocity.

12. Cesium-beam clocks have been taken at high speeds around the world in commercial jets. Show that for an equatorial circumnavigation at a height of 9 km (about 30,000 feet) and a speed of $250\,\mathrm{m\,s^{-1}}$ (about 600 m.p.h.) one would expect, *on the basis of special relativity alone*, the following time gains (or losses), when compared with a clock which remains fixed on Earth:

westbound flight	*eastbound Flight*
$+150 \times 10^{-9}$ s	-262×10^{-9} s.

(Begin by considering why there is a difference for westbound and eastbound flights, starting with a frame at the center of the Earth. Take $R_\oplus = 6378$ km, and the Earth's peripheral speed as 980 m.p.h. or $436\,\mathrm{m\,s^{-1}}$. Take $c = 3 \times 10^8\,\mathrm{m\,s^{-1}}$. In the early seventies Hafele and Keating[11] performed experiments along these lines, primarily to check the effect that the Earth's gravitational field had on the rate of clocks, which is to be ignored in this calculation.)

[11] Hafele and Keating, 1972.

B

The Chinese connection

B.0 Background

Accounts of a vehicle generally referred to as a *south-pointing carriage* are to be found in ancient Chinese writings.[1] Such a vehicle was equipped with a pointer, which always pointed south, no matter how the carriage was moved over the surface of the Earth. It thus acted like a compass, giving travelers a fixed direction from which to take their bearings. However, as is clear from their descriptions, these were mechanical and not magnetic devices: the direction of the pointer was maintained by some sort of gearing mechanism connecting the wheels of the carriage to the pointer. None of the descriptions of this mechanism that occurs in the literature is sufficiently detailed to serve as a blueprint for the construction of a south-pointing carriage, but they do contain clues which have led modern scholars to make conjectures and attempt reconstructions. The best-known and most elegant of these is that offered by the British engineer, George Lanchester, in 1947, and it forms the basis of the discussion in this appendix.

The way in which Lanchester's carriage attempts to maintain the direction of the pointer is by transporting it parallelly along the path taken by the carriage. The carriage has two wheels that can rotate independently on a common axle and the basic idea is to exploit the difference in rotation of the wheels that occurs when the carriage changes direction. The gearing uses this difference to adjust the angle of the pointer relative to the carriage, so that its direction relative to the piece of ground over which it is traveling is maintained. As a south-pointing device, Lanchester's carriage is flawed, for it only works on a flat Earth, but as a parallel-transporter it is perfect and

[1]See Needham, 1965, Vol. 4, §27(e)(5) for a thorough and detailed analysis of the descriptions of such vehicles and attempts at reconstructions by modern sinologists and others, and Cousins, 1955, for a popular account.

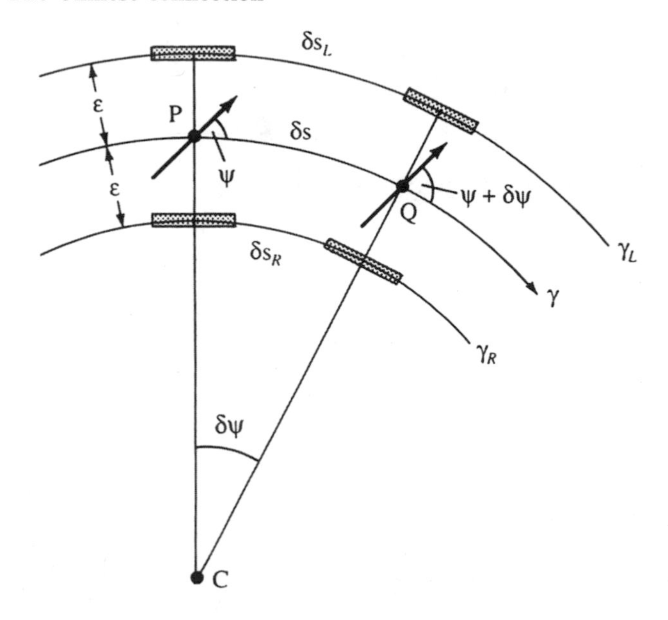

Fig. B.1. Plan view of the carriage rounding a bend.

yields a practical means[2] of transporting a vector parallelly along a curve on a surface.

As remarked above, Lanchester's transporter uses the difference in rotation of the wheels that arises when taking a bend, due to the inner wheel track being shorter than the outer wheel track. To see how the transporter works on a plane, we need to relate the path difference δp (due to a small change in direction of the carriage) to the required adjustment $\delta\psi$ in the direction of the pointer, and then show how the gearing maintains this relationship between δp and $\delta\psi$. This is done in the next section and it prepares the way for our discussion of using the transporter on a surface.

B.1 Lanchester's transporter on a plane

Figure B.1 shows the plan view of the carriage rounding a bend while being wheeled over a plane surface. The point P immediately below the midpoint of the axle follows the *base curve* γ, and to either side of this are the wheel tracks γ_L and γ_R of the left and right wheels. For each position of the carriage, the points of contact of the wheels with the ground define an *axle line* which is parallel to the direction of the axle and normal to the curves γ_L, γ, and γ_R. The figure shows two axle lines PC and QC meeting in C due to the carriage

[2]Provided it is miniaturized, so that its dimensions are small compared with the principal radii of curvature of the surface at points along its path. See Sec. B.2.

moving a short distance δs along the base curve while changing its direction by an amount $\delta\psi$ towards the right. The arrows in the figure represent the pointer on the carriage: for this to be parallelly transported, the angle that it[3] makes with γ (and therefore with an axis of the carriage at right angles to its axle) must increase from ψ to $\psi + \delta\psi$ in moving from P to Q. As explained in the previous section, we need to relate $\delta\psi$ to the path difference δp of the wheel tracks. For small δs, that part of γ between P and Q and the corresponding parts of γ_L and γ_R can be approximated by circles[4] with center C. If we let the track width be 2ε and put PC $= a$, then these circles have radii a and $a \pm \varepsilon$, giving

$$\delta s_L = (a + \varepsilon)\delta\psi, \qquad \delta s_R = (a - \varepsilon)\delta\psi$$

for the distances along γ_L and γ_R corresponding to δs along γ. Subtracting, we get

$$\boxed{\delta p \equiv \delta s_L - \delta s_R = 2\varepsilon\, \delta\psi} \tag{B.1}$$

for the path difference. This is the key equation that gives the relation between the adjustment $\delta\psi$ to the direction of the pointer (relative to the carriage) and the path difference δp in the wheel tracks in order that the pointer be parallelly transported along γ.

To appreciate how Lanchester's parallel transporter works, it is sufficient to consider the rear elevation shown in Figure B.2. The two wheels W_L and W_R have diameter 2ε, the same as the track width of the carriage; the wheel W_L is rigidly connected to a contrate gear wheel[5] A_L, with W_R and A_R similarly connected. The gear wheels B_L and B_R combine the functions of normal gear wheels and contrate gear wheels, having teeth round their edges and teeth set at right angles to these. Between B_L and B_R are two pinions[6] mounted on a stub axle, the whole assembly being similar to the differential gear box in the back axle of a truck. The rotation of W_L is transmitted to B_L via A_L and an intervening pinion, while the rotation of W_R is transmitted to B_R via A_R and a pair of rigidly connected pinions on a common axle. The pointer is mounted on a vertical axle which is *rigidly connected* to the stub axle, so that it turns with it; this axle also serves as the axle for B_L and B_R, which are *free* to turn about it. The gear wheels A_L, A_R, B_L, B_R have the same number of teeth; the number of teeth that the intervening pinions have is unimportant, but the two that transmit the rotation of W_R to B_R must have the same number of teeth.

---

[3]More correctly, its projection on the plane.

[4]The reader familiar with the notion of the *curvature* of a plane curve will recognize C as the center of curvature and PC as the radius of curvature of γ at the point P.

[5]That is, a gear wheel with teeth pointing in a direction parallel to its axis of rotation.

[6]That is, small cog wheels.

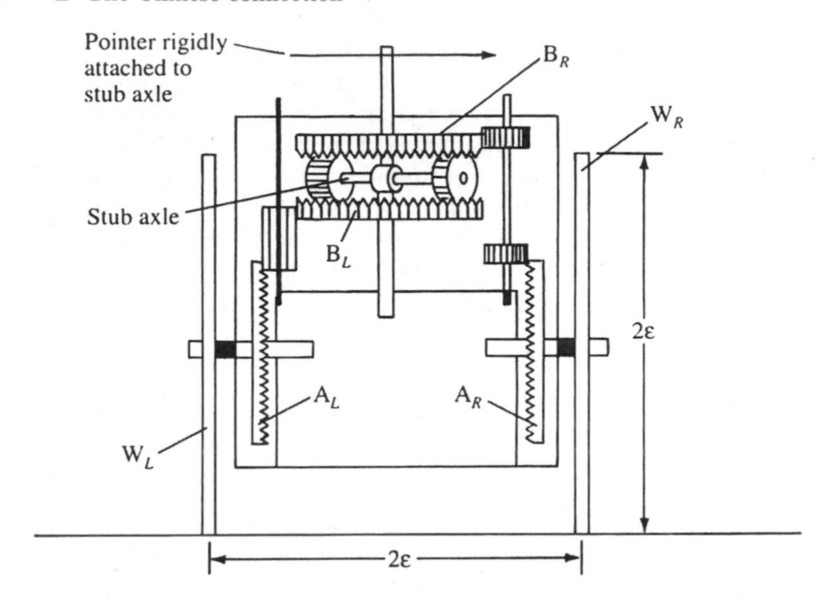

Fig. B.2. Rear elevation of Lanchester's transporter.

Having described the gearing mechanism, let us examine what happens when the transporter takes a right-hand bend as in Figure B.1. The wheel W_L travels δs_L and therefore turns through an angle $\delta s_L/\varepsilon$ (as its radius is ε), while W_R turns through an angle $\delta s_R/\varepsilon$. These rotations are faithfully transmitted to B_L and B_R (as all the gear wheels have the same number of teeth) and the result is that, when viewed from above, B_L turns *anticlockwise* through an angle $\delta s_L/\varepsilon$, while B_R turns *clockwise* through an angle $\delta s_R/\varepsilon$. The stub axle, and therefore the pointer, receives a rotation which is the *average* of the rotations of B_L and B_R. This amounts to an anticlockwise rotation of

$$\delta\psi = \frac{1}{2}\left(\frac{\delta s_L}{\varepsilon} - \frac{\delta s_R}{\varepsilon}\right) = \frac{\delta p}{2\varepsilon},$$

where $\delta p \equiv \delta s_L - \delta s_R$ is the path difference, in complete agreement with the requirement (B.1) for parallelly transporting the pointer.

The explanation above is based on Figure B.1, where both wheels are traveling in a forward direction with the center C to the right of both γ_L and γ_R. On a much tighter corner, the center C could lie between γ_L and γ_R, so that in turning the corner the inner wheel is traveling backwards. Some amendment to the explanation is then needed, but the outcome is the same: the pointer is still parallelly transported. In fact, if we let the two wheels turn at the same rate, but in opposite directions, then the carriage turns on the spot with no change in the direction of the pointer, as is easily checked. By this means we can move the carriage along a base curve γ that is *piecewise smooth*, by which we mean a curve having a number of vertices where the

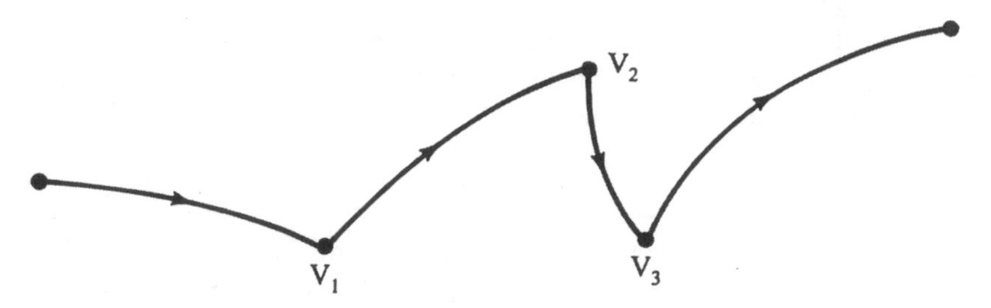

Fig. B.3. A piecewise smooth curve: in going along the curve the direction of the tangent changes discontinuously at the vertices V_1, V_2, V_3.

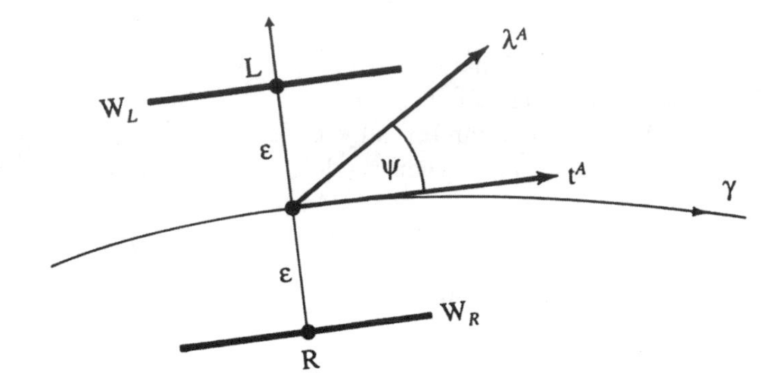

Fig. B.4. The basis vectors t^A, n^A, and the transported vector λ^A.

direction of its tangent suffers a discontinuity, as shown in Figure B.3. At a vertex V, the carriage can turn on the spot and then move off in a different direction, with the pointer parallelly transported in a purely automatic way, no matter how twisty the route.

B.2 Lanchester's transporter on a surface

The remarkable thing about Lanchester's south-pointing carriage is that it achieves parallel transport on any surface, as we shall verify in this section. To do this we work to first order in ε (where, as before, 2ε is the track width) and regard the carriage as having dimensions that are small, but small with respect to what? The answer to this question is that ε is small compared with the principal radii of curvature of the surface at all points of the route γ taken by the carriage, but to explain this fully requires too much digression.

Consider then Lanchester's carriage following a base-curve γ on a surface, as shown in Figure B.4. The vectors t^A and n^A are unit vectors, respectively tangential and normal to γ, so that t^A points in the direction of travel and n^A points along the axle line. The vector λ^A is a unit vector representing the pointer, which can be thought of as being obtained by lowering the pointer to ground level and adjusting its length as necessary. We use $\{t^A, n^A\}$ as a basis for the tangent plane at P and write

$$\lambda^A = \cos\psi\, t^A + \sin\psi\, n^A, \tag{B.2}$$

where ψ is the angle between λ^A and t^A, as shown in the figure. We wish to show that if the angle is adjusted according to equation (B.1), then the vector λ^A is parallelly transported along γ. The limiting version of equation (B.1), got by dividing by δs and letting $\delta s \to 0$, is

$$\dot{p} = 2\varepsilon\dot{\psi}, \tag{B.3}$$

where dots denote differentiation with respect to s, and it is sufficient to show that $D\lambda^A/ds = 0$ follows from equation (B.3).

Suppose that γ is given parametrically by $x^A(s)$, where s is arc-length along γ in the direction of travel. Then the left wheel track γ_L is given by

$$x_L^A(s) = x^A(s) + \varepsilon n^A. \tag{B.4}$$

As s increases by δs, the coordinates of the point L (see figure) change by

$$\delta x_L^A = \dot{x}^A \delta s + \varepsilon \dot{n}^A \delta s$$

(approximately) and the distance moved by L along γ_L is (again approximately)

$$\delta s_L = \{(g_{AB})_L \delta x_L^A \delta x_L^B\}^{1/2},$$

where the suffix L on g_{AB} indicates its value at L. To first order in ε this gives

$$\delta s_L = \{(g_{AB} + \partial_D g_{AB} \varepsilon n^D)(\dot{x}^A \delta s + \varepsilon \dot{n}^A \delta s)(\dot{x}^B \delta s + \varepsilon \dot{n}^B \delta s)\}^{1/2}, \tag{B.5}$$

where all quantities on the right are evaluated at P. On using the symmetry of g_{AB}, the binomial expansion and first-order approximation, equation (B.5) simplifies to

$$\delta s_L = (1 + \varepsilon g_{AB} \dot{x}^A \dot{n}^B + \tfrac{1}{2}\varepsilon \partial_D g_{AB} n^D \dot{x}^A \dot{x}^B)\delta s. \tag{B.6}$$

The corresponding expression for the distance moved by R along the right wheel track γ_R is got by changing the sign of ε:

$$\delta s_R = (1 - \varepsilon g_{AB} \dot{x}^A \dot{n}^B - \tfrac{1}{2}\varepsilon \partial_D g_{AB} n^D \dot{x}^A \dot{x}^B)\delta s. \tag{B.7}$$

Hence the path difference is (approximately)

$$\delta p \equiv \delta s_L - \delta s_R = 2\varepsilon(g_{AB} \dot{x}^A \dot{n}^B + \tfrac{1}{2}\partial_D g_{AB} n^D \dot{x}^A \dot{x}^B)\delta s,$$

and equation (B.3) is seen to be equivalent to

$$\dot{\psi} = g_{AB}\dot{x}^A\dot{n}^B + \tfrac{1}{2}\partial_D g_{AB} n^D \dot{x}^A \dot{x}^B, \tag{B.8}$$

which may be written as

$$\dot{\psi} = t_B Dn^B/ds, \tag{B.9}$$

as Exercise B.2.2 asks the reader to verify.

Returning now to equation (B.2), we can differentiate to obtain

$$\frac{D\lambda^A}{ds} = -\sin\psi\,\dot{\psi}\,t^A + \cos\psi\frac{Dt^A}{ds} + \cos\psi\,\dot{\psi}\,n^A + \sin\psi\frac{Dn^A}{ds}, \tag{B.10}$$

and we show that $D\lambda^A/ds = 0$ by using equation (B.9) to show that the two components $(D\lambda^A/ds)t_A$ and $(D\lambda^A/ds)n_A$ are both zero. For the first component, we have

$$\frac{D\lambda^A}{ds}t_A = -\sin\psi\,\dot{\psi} + \sin\psi\frac{Dn^A}{ds}t_A, \tag{B.11}$$

gotten by contracting equation (B.10) with t_A and using the orthonormality of t^A and n^A and the orthogonality of t^A and Dt^A/ds. It then follows from equation (B.9) that

$$\frac{D\lambda^A}{ds}t_A = -\sin\psi\left(\dot{\psi} - \frac{Dn^A}{ds}t_A\right) = 0,$$

as required. A similar argument gives

$$\frac{D\lambda^A}{ds}n_A = \cos\psi\left(\dot{\psi} + \frac{Dt^A}{ds}n_A\right)$$

for the second component and we can deduce that this is also zero by noting that differentiation of $t^A n_A = 0$ yields

$$\frac{Dt^A}{ds}n_A = -t_A\frac{Dn^A}{ds}.$$

Hence equation (B.3) implies that $D\lambda^A/ds = 0$, showing that Lanchester's carriage transports λ^A parallelly along γ, as claimed.

Exercises B.2

1. Working to first order in ε, show that equation (B.5) simplifies to give equation (B.6).

2. Show that equations (B.8) and (B.9) are equivalent.

3. Verify that equation (B.11) follows from equation (B.10), as claimed.

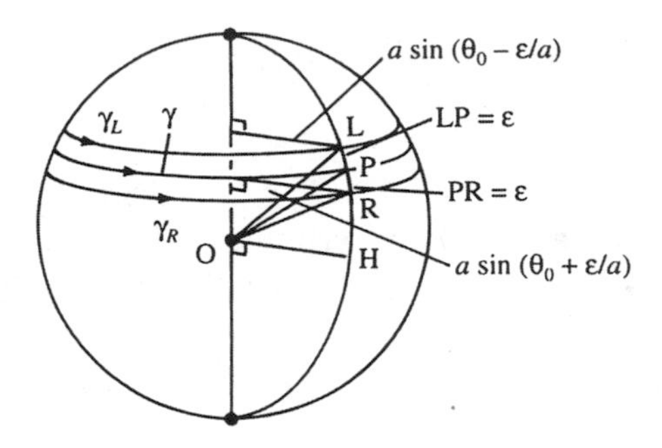

Fig. B.5. Wheel tracks on a sphere.

B.3 A trip at constant latitude

In Example 2.2.1, we showed the effect of parallel transport around a circle of latitude on a sphere. We can get the same result by using Lanchester's transporter. We note that, in traveling in an easterly direction (increasing ϕ) along a base curve γ given by $\theta = \theta_0$, the left wheel track γ_L has $\theta = \theta_0 - \varepsilon/a$ and the right wheel track γ_R has $\theta = \theta_0 + \varepsilon/a$. These are approximate expressions valid for ε small compared with the radius a of the sphere.[7] (See Fig. B.5.) It follows that the curves γ_L and γ_R are circles with radii equal to $a \sin(\theta_0 \mp \varepsilon/a)$ and that for a trip from $\theta = 0$ to $\theta = t$ the path difference is

$$\begin{aligned} \Delta p &= \left(a \sin\left(\theta_0 - \frac{\varepsilon}{a}\right) - a \sin\left(\theta_0 + \frac{\varepsilon}{a}\right)\right) t \\ &= -2at \cos\theta_0 \, \sin(\varepsilon/a) = -2\varepsilon t \cos\theta_0, \end{aligned}$$

on using the approximation $\sin(\varepsilon/a) = \varepsilon/a$. The corresponding adjustment to the direction of the pointer (got by integrating equation (B.3)) is

$$\Delta\psi = \Delta p/2\varepsilon = -t \cos\theta_0,$$

in agreement with equation (2.26) of Example 2.2.1.

[7] At every point of a sphere, its principal radii of curvature are both equal to its radius a.

C

Tensors and Manifolds

C.0 Introduction

In this appendix we present a more formal treatment of tensors and manifolds, enlarging on the concepts outlined in Section 1.10. The basic approach is to deal separately with the algebra of tensors and the coordinatization of manifolds, and then to bring these together to define tensor fields on manifolds.

We deal first with some algebraic preliminaries, namely the concepts of vector spaces, their duals, and spaces which may be derived from these by the process of tensor multiplication. The treatment here is quite general, though restricted to real finite-dimensional vector spaces.

We then give more formal definitions of a manifold and tensor fields than those given in Sections 1.7 and 1.8. The key notion here is that of the *tangent space* $T_P(M)$ at each point P of a manifold M, the *cotangent space* $T^*_P(M)$, and repeated tensor products of these. The result is the space $(T^r_s)_P(M)$ of tensors of type (r, s) at each point P of a manifold. A type (r, s) tensor field can then be defined as an assignment of a member of $(T^r_s)_P(M)$ to each point P of the manifold.

C.1 Vector spaces

We shall not attempt a formal definition of a vector space, but assume that the reader has some familiarity with the concept. The excellent text by Halmos[1] is a suitable introduction to those new to the concept.

The essential features of a vector space are that it is a set of *vectors* on which are defined two operations, namely addition of vectors and the multiplication of vectors by *scalars*; that there is a zero vector in the space; and that each vector in the space has an inverse such that the sum of a vector and its inverse equals the zero vector. It may be helpful to picture the set of vectors

[1]Halmos, 1974, Ch. I.

comprising a vector space as a set of arrows emanating from some origin, with addition of vectors given by the usual parallelogram law, and multiplication of vectors by a scalar as a scaling operation which changes its length but not its direction, with the proviso that if the scalar is negative, then the scaled vector will lie in the same line as the original, but point in the opposite direction. In this picture the zero vector is simply the point which is the origin (an arrow of zero length), and the inverse of a given vector is one of the same length as the given vector, but pointing in the opposite direction.

We shall restrict our treatment to *real* vector spaces whose scalars belong to the real numbers $\mathbb{R}$. As usual, we shall use bold type for vectors and non-bold type for scalars.

The notion of linear independence is of central importance in vector-space theory. If, for any scalars $\lambda^1, \ldots, \lambda^K$,

$$\lambda^1 \mathbf{v}_1 + \lambda^2 \mathbf{v}_2 + \cdots + \lambda^K \mathbf{v}_K = \mathbf{0} \tag{C.1}$$

implies that $\lambda_1 = \lambda_2 = \cdots = \lambda^K = 0$, then the set of vectors $\{\mathbf{v}_1, \mathbf{v}_2, \ldots, \mathbf{v}_K\}$ is said to be *linearly independent.* A set of vectors which is not linearly independent is said to be *linearly dependent.* Thus for a linearly dependent set $\{\mathbf{v}_1, \mathbf{v}_2, \ldots, \mathbf{v}_K\}$ there exists a non-trivial linear combination of the vectors which equals the zero vector. That is, there exists scalars $\lambda^1, \ldots, \lambda^K$, not all zero (though some may be) such that

$$\lambda^1 \mathbf{v}_1 + \lambda^2 \mathbf{v}_2 + \cdots + \lambda^K \mathbf{v}_K = \mathbf{0}. \tag{C.2}$$

Using Einstein's summation convention (as explained in Section 1.2), we can express the above more compactly as

$$\lambda^a \mathbf{v}_a = 0,$$

where the range of summation (in this case 1 to K) is gleaned from the context. We shall continue to use the summation convention in the rest of this appendix.

A set of vectors which has the property that every vector $\mathbf{v}$ in the vector space T may be written as a linear combination of its members is said to *span* the space T. Thus the set $\{\mathbf{v}_1, \mathbf{v}_2, \ldots, \mathbf{v}_K\}$ spans T if every vector $\mathbf{v} \in T$ may be expressed as

$$\mathbf{v} = \lambda^a \mathbf{v}_a, \tag{C.3}$$

for some scalars $\lambda^1, \ldots, \lambda^K$. (The symbol $\in$ is read as "belonging to", or as "belongs to", depending on the context.) If a set of vectors both spans T and is linearly independent, then it is a *basis* of T, and we shall restrict ourselves to vector spaces having finite bases. In this case it is possible to show that all bases of a given vector space T contain the same number of members[2] and this number is called the *dimension* of T.

[2]See Halmos, 1974, Ch. I, §8.

Let $\{\mathbf{e}_1, \mathbf{e}_2, \ldots, \mathbf{e}_N\}$ (or $\{\mathbf{e}_a\}$ for short) be a basis of an N-dimensional vector space T, so any $\boldsymbol{\lambda} \in T$ may be written as $\boldsymbol{\lambda} = \lambda^a \mathbf{e}_a$ for some scalars λ^a. This expression for $\boldsymbol{\lambda}$ is unique, for if $\boldsymbol{\lambda} = \tilde{\lambda}^a \mathbf{e}_a$, then subtraction gives $(\lambda^a - \tilde{\lambda}^a)\mathbf{e}_a = \mathbf{0}$, which implies that $\lambda^a = \tilde{\lambda}^a$ for all a, since basis vectors are independent. The scalars λ^a are the *components* of $\boldsymbol{\lambda}$ relative to the basis $\{\mathbf{e}_a\}$.

The last task of this section is to see how the components of a vector transform when a new basis is introduced. Let $\{\mathbf{e}_{a'}\}$ be a new basis for T, and let $\lambda^{a'}$ be the components of $\boldsymbol{\lambda}$ relative to the new basis. So

$$\boldsymbol{\lambda} = \lambda^{a'} \mathbf{e}_{a'}. \tag{C.4}$$

(As in Chapter 1, we use the same kernel for the vector and its components, and the basis to which the components are related is distinguished by the marks, or lack of them, on the superscript. In a similar way, the "unprimed" basis $\{\mathbf{e}_a\}$ is distinguished from the "primed" basis $\{\mathbf{e}_{a'}\}$. This notation is part of the *kernel–index method* initiated by Schouten and his co-workers.[3]) Each of the new basis vectors may be written as a linear combination of the old:

$$\mathbf{e}_{a'} = X^b_{a'} \mathbf{e}_b, \tag{C.5}$$

and conversely the old as a linear combination of the new:

$$\mathbf{e}_c = X^{a'}_c \mathbf{e}_{a'}. \tag{C.6}$$

(Although we use the same kernel letter X, the N^2 numbers $X^b_{a'}$ are different from the N^2 numbers $X^{a'}_c$, the positions of the primes indicating the difference.) Substitution for $\mathbf{e}_{a'}$ from equation (C.5) in (C.6) yields

$$\mathbf{e}_c = X^{a'}_c X^b_{a'} \mathbf{e}_b. \tag{C.7}$$

By the uniqueness of components we then have

$$X^{a'}_c X^b_{a'} = \delta^b_c, \tag{C.8}$$

where δ^b_c is the *Kronecker delta* introduced in Chapter 1. Similarly, by substituting for $\mathbf{e}_b$ in equation (C.5) from (C.6) and changing the lettering of suffixes, we also deduce that

$$X^b_{a'} X^{c'}_b = \delta^c_a. \tag{C.9}$$

Substitution for $\mathbf{e}_{a'}$ from equation (C.5) in (C.6) yields

$$\boldsymbol{\lambda} = \lambda^{a'} X^b_{a'} \mathbf{e}_b, \tag{C.10}$$

and by the uniqueness of components

---

[3]Schouten, 1954, p. 3, in particular footnote[1)].

$$\lambda^b = X^b_{a'}\lambda^{a'}. \tag{C.11}$$

Then

$$X^{c'}_a \lambda^a = X^{c'}_a X^a_{b'} \lambda^{b'}, \tag{C.12}$$

on changing the lettering of suffixes. (This change was to avoid a letter appearing more than twice, which would make a nonsense of the notation. See Section 1.2 for an explanation of *dummy suffixes*.)

To recap, if primed and unprimed bases are related by

$$\boxed{\mathbf{e}_{a'} = X^b_{a'}\mathbf{e}_b, \qquad \mathbf{e}_a = X^{b'}_a \mathbf{e}_{b'},} \tag{C.13}$$

then the components are related by

$$\boxed{\lambda^{a'} = X^{a'}_b \lambda^b, \qquad \lambda^a = X^a_{b'}\lambda^{b'},} \tag{C.14}$$

and

$$X^a_{b'} X^{b'}_c = \delta^a_c, \qquad X^{a'}_b X^b_{c'} = \delta^a_c. \tag{C.15}$$

We have thus reproduced the transformation formula (1.70) for a contravariant vector, but in this general algebraic approach the transformation matrix $[X^{a'}_b]$ is generated by a change of basis in the vector space rather than its being the Jacobian matrix arising from a change of coordinates.

Exercise C.1

1. Derive the result (C.9).

C.2 Dual spaces

The visualization of the vectors in a vector space as arrows emanating from an origin can be misleading, for sets of objects bearing no resemblance to arrows constitute vector spaces under suitable definitions of addition and multiplication by scalars. Among such objects are functions.

Let us confine our attention to real-valued functions defined on a real vector space T. In mathematical language such a function f would be written as $f : T \to \mathbb{R}$, indicating that it maps vectors of T into real numbers. The set of all such functions may be given a vector-space structure by defining:

(a) the sum of two functions f and g by
$(f + g)(\mathbf{v}) = f(\mathbf{v}) + g(\mathbf{v})$ for all $\mathbf{v} \in T$;

(b) the product αf of the scalar α and the function f by
$(\alpha f)(\mathbf{v}) = \alpha(f(\mathbf{v}))$ for all $\mathbf{v} \in T$;

(c) the zero function 0 by

$0(\mathbf{v}) = 0$ for all $\mathbf{v} \in T$

(where on the left 0 is a function, while on the right it is a number, there being no particular advantage in using different symbols);

(d) the inverse $-f$ by

$(-f)(\mathbf{v}) = -(f(\mathbf{v}))$ for all $\mathbf{v} \in T$.

That this does indeed define a vector space may be verified by checking the axioms given in Halmos.[4]

The space of all real-valued functions is too large for our purpose, and we shall restrict ourselves to those functions which are *linear*. That is, those functions which satisfy

$$f(\alpha\mathbf{u} + \beta\mathbf{v}) = \alpha f(\mathbf{u}) + \beta f(\mathbf{v}), \tag{C.16}$$

for $\alpha, \beta \in \mathbb{R}$ and all $\mathbf{u}, \mathbf{v} \in T$. Real-valued linear functions on a real vector space are usually called *linear functionals.* It is a simple matter to check that the sum of two linear functionals is itself a linear functional, and that the multiplication of a linear functional by a scalar yields a linear functional. These observations are sufficient to show that the set of linear functionals on a vector space T is itself a vector space. This space is the *dual* of T, and we denote it by T^*. Since linear functionals are vectors we shall use bold-face type for them also.

We now have two types of vectors, those in T and those in T^*. To distinguish them, those in T are called *contravariant vectors*, while those in T^* are called *covariant vectors.* As a further distinguishing feature, basis vectors of T^* will carry superscripts and components of vectors in T^* will carry subscripts. Thus if $\{\mathbf{e}^a\}$ is a basis of T^*, then $\boldsymbol{\lambda} \in T^*$ has a unique expression $\boldsymbol{\lambda} = \lambda_a \mathbf{e}^a$ in terms of components.

The use of the lower-case letter a in the implied summation above suggests that the range of summation is 1 to N, the dimension of T, i.e., that T^* has the same dimension as T. This is in fact the case, as we shall now prove by showing that a given basis $\{\mathbf{e}_a\}$ of T induces in a natural way a *dual basis* $\{\mathbf{e}^a\}$ of T^* having N members satisfying $\mathbf{e}^a(\mathbf{e}_b) = \delta^a_b$.

We start by defining $\mathbf{e}^a$ to be the real-valued function which maps $\boldsymbol{\lambda} \in T$ into the real number λ^a which is its ath component relative to $\{\mathbf{e}_a\}$, i.e., $\mathbf{e}^a(\boldsymbol{\lambda}) = \lambda^a$ for all $\boldsymbol{\lambda} \in T$. This gives us N real-valued functions which clearly satisfy $\mathbf{e}^a(\mathbf{e}_b) = \delta^a_b$, and it remains to show that they are linear and that they constitute a basis for T^*. The former is readily checked. As for the latter, we proceed as follows.

For any $\boldsymbol{\mu} \in T^*$ we can define N real numbers μ_a by $\boldsymbol{\mu}(\mathbf{e}_a) = \mu_a$. Then for any $\lambda \in T$,

[4] Halmos, 1974, Ch. I.

$$\begin{aligned}\boldsymbol{\mu}(\boldsymbol{\lambda}) &= \boldsymbol{\mu}(\lambda^a \mathbf{e}_a) = \lambda^a \boldsymbol{\mu}(\mathbf{e}_a) \qquad \text{(by the linearity of } \boldsymbol{\mu}) \\ &= \lambda^a \mu_a = \mu_a \mathbf{e}^a(\boldsymbol{\lambda}).\end{aligned}$$

Thus for any $\boldsymbol{\mu} \in T^*$ we have $\boldsymbol{\mu} = \mu_a \mathbf{e}^a$, showing that $\{\mathbf{e}^a\}$ spans T^*, and there remains the question of the independence of the $\{\mathbf{e}^a\}$. This is answered by noting that a relation $x_a \mathbf{e}^a = \mathbf{0}$, where $x_a \in \mathbb{R}$ and $\mathbf{0}$ is the zero functional, implies that

$$0 = x_a \mathbf{e}^a(\mathbf{e}_b) = x_a \delta^a_b = x_b \qquad \text{for all } b.$$

From the above it may be seen that given a basis $\{\mathbf{e}_a\}$ of T, the components μ_a of $\boldsymbol{\mu} \in T^*$ relative to the dual basis $\{\mathbf{e}^a\}$ are given by $\mu_a = \boldsymbol{\mu}(\mathbf{e}_a)$.

A change of basis (C.13) in T induces a change of the dual basis. Let us denote the dual of the primed basis $\{\mathbf{e}_{a'}\}$ by $\{\mathbf{e}^{a'}\}$, so by definition $\mathbf{e}^{a'}(\mathbf{e}_{b'}) = \delta^a_b$ and $\mathbf{e}^{a'} = Y^{a'}_b \mathbf{e}^b$ for some $Y^{a'}_b$. Then

$$\begin{aligned}\delta^a_b &= \mathbf{e}^{a'}(\mathbf{e}_{b'}) = Y^{a'}_d \mathbf{e}^d(X^c_{b'} \mathbf{e}_c) \\ &= Y^{a'}_d X^c_{b'} \mathbf{e}^d(\mathbf{e}_c) \qquad \text{(by the linearity of the } \mathbf{e}^d) \\ &= Y^{a'}_d X^c_{b'} \delta^d_c = Y^{a'}_c X^c_{b'}.\end{aligned}$$

Multiplying by $X^{b'}_d$ gives $X^{a'}_d = Y^{a'}_d$. Thus under a change of basis of T given by equations (C.13), the dual bases of T^* transform according to

$$\boxed{\mathbf{e}^{a'} = X^{a'}_b \mathbf{e}^b, \qquad \mathbf{e}^a = X^a_{b'} \mathbf{e}^{b'}.} \tag{C.17}$$

It is readily shown that the components of $\boldsymbol{\mu} \in T^*$ relative to the dual bases transform according to

$$\boxed{\mu_{a'} = X^b_{a'} \mu_b, \qquad \mu_a = X^{b'}_a \mu_{b'}.} \tag{C.18}$$

So the same matrix $[X^{a'}_b]$ and its inverse $[X^a_{b'}]$ are involved, but their roles relative to basis vectors and components are interchanged.

Given T and a basis $\{\mathbf{e}_a\}$ of it, we have seen how to construct its dual T^* with dual basis $\{\mathbf{e}^a\}$ satisfying $\mathbf{e}^a(\mathbf{e}_b) = \delta^a_b$. We can apply this process again to arrive at the dual T^{**} of T^*, with dual basis $\{\mathbf{f}_a\}$ say, satisfying $\mathbf{f}_a(\mathbf{e}^b) = \delta^b_a$, and vectors $\boldsymbol{\lambda} \in T^{**}$ may be expressed in terms of components as $\boldsymbol{\lambda} = \lambda^a \mathbf{f}_a$. Under a change of basis of T, components of vectors in T transform according to $\lambda^{a'} = X^{a'}_b \lambda^b$. This induces a change of dual basis of T^*, under which components of vectors in T^* transform according to $\mu_{a'} = X^b_{a'} \mu_b$. In turn, this induces a change of basis of T^{**}, under which it is readily seen that components of vectors in T^{**} transform according to $\lambda^{a'} = X^{a'}_b \lambda^b$ (because the inverse of the inverse of a matrix is the matrix itself). That is, the components of vectors in T^{**} transform in exactly the same way as the components of vectors in T. This means that if we set up a one-to-one correspondence between vectors in T and T^{**} by making $\lambda^a \mathbf{e}_a$ in T correspond to $\lambda^a \mathbf{f}_a$ in T^{**}, where $\{\mathbf{f}_a\}$ is the dual of the dual of $\{\mathbf{e}_a\}$, then this correspondence is *basis-independent.* A basis-independent one-to-one correspondence between vector spaces is called a *natural isomorphism*, and naturally isomorphic vector spaces are usually identified, by identifying corresponding vectors. Consequently, we shall identify T^{**} with T.

In this section we have given a more formal definition of a covariant vector as the dual of a contravariant vector, rather than as an object having components that transform in a certain way. Although we have reproduced the transformation formula (1.71) for the components, the transformation matrix $[X^a_{b'}]$ now arises from a change of the dual basis of T^* induced by a change of basis of T, rather than its being the Jacobian matrix arising from a change of coordinates.

Exercises C.2

1. Check that the sum of two linear functionals is itself a linear functional, and that multiplication of a linear functional by a scalar yields a linear functional.

2. Verify that the components $\boldsymbol{\mu} \in T^*$ relative to the dual bases transform according to equations (C.18), as asserted.

3. Identifying T^{**} with T means that a contravariant vector $\boldsymbol{\lambda}$ acts as a linear functional on a covariant vector $\boldsymbol{\mu}$. Show that in terms of components $\boldsymbol{\lambda}(\boldsymbol{\mu}) = \lambda^a \mu_a$.

C.3 Tensor products

Given a vector space T we have seen how to create a new vector space, namely its dual T^*, but here the process stops (on identifying T^{**} with T). However, it is possible to generate a new vector space from two vector spaces by forming what is called their tensor product. As a preliminary to this we need to define bilinear functionals on a pair of vector spaces.

Let T and U be two real finite-dimensional vector spaces. The *cartesian product* $T \times U$ is the set of all ordered pairs of the form $(\mathbf{v}, \mathbf{w})$, where $\mathbf{v} \in T$ and $\mathbf{w} \in U$. A *bilinear functional* f on $T \times U$ is a real-valued function $f : T \times U \to \mathbb{R}$, which is bilinear, i.e., satisfies

$$f(\alpha\mathbf{u} + \beta\mathbf{v}, \mathbf{w}) = \alpha f(\mathbf{u}, \mathbf{w}) + \beta f(\mathbf{v}, \mathbf{w}),$$
$$\text{for all } \alpha, \beta \in \mathbb{R},\ \mathbf{u}, \mathbf{v} \in T \text{ and } \mathbf{w} \in U,$$

and

$$f(\mathbf{v}, \gamma\mathbf{w} + \delta\mathbf{x}) = \gamma f(\mathbf{v}, \mathbf{w}) + \delta f(\mathbf{v}, \mathbf{x}),$$
$$\text{for all } \gamma, \delta \in \mathbb{R},\ \mathbf{v} \in T \text{ and } \mathbf{w}, \mathbf{x} \in U.$$

With definitions of addition, scalar multiplication, the zero function and inverses analogous to those given in the previous section, it is a straightforward matter to show that the set of bilinear functions on $T \times U$ is a vector space, so we shall now use bold-faced type for bilinear functionals.

We can now define the *tensor product* $T \otimes U$ of T and U as the vector space of all bilinear functionals on $T^* \times U^*$. Note that this definition uses the dual spaces T^* and U^*, and not T and U themselves.

The question naturally arises as to the dimension of $T \otimes U$. It is in fact NM, where N and M are the dimensions of T and U respectively, and we prove this by showing that from given bases of T and U we can define NM members of $T \otimes U$ which constitute a basis for it.

Let $\{\mathbf{e}^a\}$, $a = 1, \dots, N$, and $\{\mathbf{f}^\alpha\}$, $\alpha = 1, \dots, M$, be bases of T^* and U^*, dual to bases $\{\mathbf{e}_a\}$ and $\{\mathbf{f}_\alpha\}$ of T and U respectively. (Note that we use different alphabets for suffixes having different ranges.) Define NM functions $\mathbf{e}_{a\alpha} : T^* \times U^* \to \mathbb{R}$ by

$$\mathbf{e}_{a\alpha}(\boldsymbol{\lambda}, \boldsymbol{\mu}) = \lambda_a \mu_\alpha, \tag{C.19}$$

where λ_a are the components of $\boldsymbol{\lambda} \in T^*$ relative to $\{\mathbf{e}^a\}$ and μ_α are those of $\boldsymbol{\mu} \in U^*$ relative to $\{\mathbf{f}^\alpha\}$. In particular

$$\mathbf{e}_{a\alpha}(\mathbf{e}^b, \mathbf{f}^\beta) = \delta^b_a \delta^\beta_\alpha. \tag{C.20}$$

It is a simple matter to show that the $\mathbf{e}_{a\alpha}$ are bilinear and so belong to $T \otimes U$. To show that they constitute a basis we must show that they span $T \otimes U$ and are independent.

For any $\boldsymbol{\tau} \in T \otimes U$, define NM real numbers $\tau^{a\alpha}$ by $\tau^{a\alpha} \equiv \boldsymbol{\tau}(\mathbf{e}^a, \mathbf{f}^\alpha)$. Then for any $\boldsymbol{\lambda} \in T^*$ and $\boldsymbol{\mu} \in U^*$ we have

$$\begin{aligned} \boldsymbol{\tau}(\boldsymbol{\lambda}, \boldsymbol{\mu}) &= \boldsymbol{\tau}(\lambda_a \mathbf{e}^a, \mu_\alpha \mathbf{f}^\alpha) \\ &= \lambda_a \mu_\alpha \boldsymbol{\tau}(\mathbf{e}^a, \mathbf{f}^\alpha) \qquad \text{(on using the bilinearity of } \boldsymbol{\tau}) \\ &= \tau^{a\alpha} \lambda_a \mu_\alpha = \tau^{a\alpha} \mathbf{e}_{a\alpha}(\boldsymbol{\lambda}, \boldsymbol{\mu}). \end{aligned}$$

So for any $\boldsymbol{\tau} \in T \otimes U$, we have $\boldsymbol{\tau} = \tau^{a\alpha} \mathbf{e}_{a\alpha}$, showing that the set $\{\mathbf{e}_{a\alpha}\}$ spans $T \otimes U$. Moreover, $\{\mathbf{e}_{a\alpha}\}$ is an independent set, for if $x^{a\alpha} \mathbf{e}_{a\alpha} = \mathbf{0}$, then

$$0 = x^{a\alpha} \mathbf{e}_{a\alpha}(\mathbf{e}^b, \mathbf{f}^\beta) = x^{a\alpha} \delta^b_a \delta^\beta_\alpha = x^{b\beta}$$

for all b, β, on using equation (C.20).

Thus we have shown that the dimension of $T \otimes U$ is the product of the dimensions of T and U, and that in a natural way the bases $\{\mathbf{e}_a\}$ of T and $\{\mathbf{f}_\alpha\}$ of U induce a basis$\{\mathbf{e}_{a\alpha}\}$ of $T \otimes U$, the components $\tau^{a\alpha}$ of any $\boldsymbol{\tau} \in T \otimes U$ relative to this basis being given in terms of the dual bases of T^* and U^* by $\tau^{a\alpha} = \boldsymbol{\tau}(\mathbf{e}^a, \mathbf{f}^\alpha)$.

Let us now investigate how the components $\tau^{a\alpha}$ and the induced basis vectors $\mathbf{e}_{a\alpha}$ transform when new bases are introduced into T and U. Suppose that the bases of T and U are changed according to

$$\mathbf{e}_{a'} = X^b_{a'}\mathbf{e}_b, \qquad \mathbf{f}_{\alpha'} = Y^\beta_{a'}\mathbf{f}_\beta. \tag{C.21}$$

This induces a new basis $\{\mathbf{e}_{a'\alpha'}\}$ of $T \otimes U$, and for any $(\boldsymbol{\lambda}, \boldsymbol{\mu}) \in T^* \times U^*$,

$$\begin{aligned}\mathbf{e}_{a'\alpha'}(\boldsymbol{\lambda}, \boldsymbol{\mu}) &= \lambda_{a'}\mu_{\alpha'} = X^b_{a'}Y^\beta_{\alpha'}\lambda_b\mu_\beta \\ &= X^b_{a'}Y^\beta_{\alpha'}\mathbf{e}_{b\beta}(\boldsymbol{\lambda}, \boldsymbol{\mu}).\end{aligned}$$

So

$$\mathbf{e}_{a'\alpha'} = X^b_{a'}Y^\beta_{\alpha'}\mathbf{e}_{b\beta}. \tag{C.22}$$

Similarly, for components (see Exercise C.3.2),

$$\tau^{a'\alpha'} = X^{a'}_b Y^{\alpha'}_\beta \tau^{b\beta}. \tag{C.23}$$

A vector which is a member of the tensor product of two spaces (or more, see below) is called a *tensor.* The tensor product defined above is a product of spaces. It is possible to define a tensor which is the tensor product $\boldsymbol{\lambda} \otimes \boldsymbol{\mu}$ of individual tensors $\boldsymbol{\lambda}$ and $\boldsymbol{\mu}$ by setting

$$\boldsymbol{\lambda} \otimes \boldsymbol{\mu} \equiv \lambda^a \mu^\alpha \mathbf{e}_{a\alpha}, \tag{C.24}$$

where λ^a and μ^α are the components of $\boldsymbol{\lambda}$ and $\boldsymbol{\mu}$ relative to bases of T and U which induce the basis $\{\mathbf{e}_{a\alpha}\}$ of $T \otimes U$. Although this definition is given via bases, it is in fact basis-independent (see Exercise C.3.3). In particular we have

$$\mathbf{e}_a \otimes \mathbf{f}_\alpha = \mathbf{e}_{a\alpha}. \tag{C.25}$$

The tensor product $\boldsymbol{\lambda} \otimes \boldsymbol{\mu}$ belongs to $T \otimes U$, but not all tensors in $T \otimes U$ are of this form. Those that are are called *decomposable.*

Having established the basic idea of the tensor product of vector spaces we can extend it to three or more spaces. However, given three spaces T, U, and V, we can form their tensor product in two ways: $(T \otimes U) \otimes V$ or $T \otimes (U \otimes V)$. These two spaces clearly have the same dimension, and are in fact naturally isomorphic, in the sense that we can set up a basis-independent one-to-one correspondence between their members, just as we did with T and T^{**}. This is done by choosing bases $\{\mathbf{e}_a\}$, $\{\mathbf{f}_\alpha\}$, $\{\mathbf{g}_A\}$ in T, U, V respectively (three ranges, so three alphabets), letting $\tau^{a\alpha A}(\mathbf{e}_a \otimes \mathbf{f}_\alpha) \otimes \mathbf{g}_A$ in $(T \otimes U) \otimes V$ correspond to $\tau^{a\alpha A}\mathbf{e}_a \otimes (\mathbf{f}_\alpha \otimes \mathbf{g}_A)$ in $T \otimes (U \otimes V)$, and then showing that this correspondence is basis-independent. Because of the natural isomorphism one identifies these spaces and the notation $T \otimes U \otimes V$ is unambiguous.

An alternative way of defining $T \otimes U \otimes V$ is as the space of trilinear functions on $T^* \times U^* \times V^*$. This leads to a space which is naturally isomorphic to those of the preceding paragraph, and all three are identified. Other natural isomorphisms exist, for example between $T \otimes U$ and $U \otimes T$, or between $(T \otimes U)^*$ and $T^* \otimes U^*$, and whenever they exist, the spaces are identified.

Exercises C.3

1. Show that the functions $\mathbf{e}_{a\alpha} : T^* \times U^* \to \mathbb{R}$, defined by equation (C.19), are bilinear functionals.

2. Verify the transformation formula for components (equation (C.23)).

3. Prove that the definition of the tensor product $\boldsymbol{\lambda} \otimes \boldsymbol{\mu}$ of two vectors $\boldsymbol{\lambda}$ and $\boldsymbol{\mu}$ is basis independent.

C.4 The space T^r_s

We shall now restrict the discussion to tensor-product spaces obtained by taking repeated tensor products of just one space T and/or its dual T^*. We introduce the following notation:

$$\underbrace{T \otimes T \otimes \cdots \otimes T}_{r \text{ times}} \equiv T^r,$$

$$\underbrace{T^* \otimes T^* \otimes \cdots \otimes T^*}_{s \text{ times}} \equiv T_s,$$

$$T^r \otimes T_s \equiv T^r_s.$$

In particular, $T = T^1$ and $T^* = T_1$.

A member of T^r is a *contravariant tensor of rank* r, a member of T_s is a *covariant tensor of rank* s, while a member of T^r_s is a *a mixed tensor of rank* $(r+s)$. A member of T^r is also referred to as a *tensor of type* $(r,0)$, a member of T_s is as a *tensor of type* $(0,s)$, and a member of T^r_s as a *tensor of type* (r,s). Note that this nomenclature labels contravariant and covariant vectors as tensors of type $(1,0)$ and type $(0,1)$ respectively. Scalars may be included in the general scheme of things by regarding them as type $(0,0)$ tensors.

A basis $\{\mathbf{e}_a\}$ of T (of dimension N) induces a dual basis $\{\mathbf{e}^a\}$ of T^*, and these together yield a basis $\{\mathbf{e}^{b_1 \dots b_s}_{a_1 \dots a_r}\}$ of T^r_s. Each tensor $\boldsymbol{\tau} \in T^r_s$ has N^{r+s} unique components relative to the induced basis:

$$\boxed{\boldsymbol{\tau} = \tau^{a_1 \dots a_r}_{b_1 \dots b_s} \mathbf{e}^{b_1 \dots b_s}_{a_1 \dots a_r}.} \tag{C.26}$$

A change of basis of T induces a change of basis of T^r_s under which the components transform according to

$$\boxed{\tau^{a'_1 \dots a'_r}_{b'_1 \dots b'_s} = X^{a'_1}_{c_1} \cdots X^{a'_r}_{c_r} X^{d_1}_{b'_1} \cdots X^{d_s}_{b'_s} \tau^{c_1 \dots c_r}_{d_1 \dots d_s},} \tag{C.27}$$

where $[X^{a'}_{b}]$ is the matrix representing the change of basis of T and $[X^{d}_{b'}]$ is its inverse. We have thus reproduced the transformation formula (1.73), but here the matrices arise from a change of basis of T, rather than being Jacobian matrices from a change of coordinates.

C.5 From tensors to tensor fields

The main purpose of this appendix is to arrive at a more mathematically formal definition of a tensor field on a manifold and to give a more complete answer to the question posed in the title of Section 1.10.

The key concept required is that of the tangent space $T_P(M)$ at each point P of a manifold M. This is a vector space that is the analog of the tangent plane existing at each point P of a surface. However, a manifold and its tangent spaces are abstract mathematical objects and not as in the picture we have of a surface and its tangent planes as geometrical objects in sitting in three-dimensional Euclidean space: there is no higher dimensional-dimensional space in which they are embedded.

Once we have the vector space $T_P(M)$, we can introduce its dual $T^*_P(M)$ and use them to construct a tensor product space

$$(T^r_s)_P(M) \equiv T_P(M) \otimes \cdots \otimes T_P(M) \otimes T^*_P(M) \otimes \cdots \otimes T^*_P(M)$$

at each point P of M, as explained in the preceding section. A type (r, s) tensor field $\boldsymbol{\tau}$ can then be defined as an assignment of a member of $(T^r_s)_P(M)$ to each point P of M.

The formal definition of a manifold is given in the next section, and this is followed by a section explaining the concept of the tangent space at each point of a manifold. Finally, in Section C.8 we arrive at the formal definition of a tensor field and reconcile this with the less formal treatment of Section 1.8.

C.6 Manifolds

A differentiable manifold (or manifold for short) is a generalization of a surface in the sense that

(a) it has a dimension, N say, so that points in it may be labeled by N real coordinates $x^1, x^2, \ldots, x^N$;
(b) it can support a differentiable structure; i.e., the functions involved in changes of coordinates are differentiable.

There is, however, one important respect in which it differs from a surface, namely that it is a thing in itself, and we do not consider it embedded in some higher-dimensional Euclidean space. Our formal definition is arrived at by starting with a set M, and then giving it sufficient structure so that it becomes a manifold.

Let M be a set of points and ψ a one-to-one map from a subset U of M onto an open set in $\mathbb{R}^N$. ($\mathbb{R}^N$ is the set of N-tuples $(x^1, x^2, \ldots, x^N)$, where each x^a is real. An *open* set S of $\mathbb{R}^N$ is one with the property that each point of it may be surrounded by a ball B centered on the point in question, such that B lies entirely in S. The map ψ maps U *onto* S if for each $s \in S$ we have $\psi(u) = s$ for some $u \in U$.)[5] U is a *coordinate neighborhood*, ψ is a *coordinate function*, and the pair (U, ψ) together is a *chart*. The purpose of ψ is to attach coordinates to points in U, and if P is a point in U we shall call the chart (U, ψ) a *coordinate system* about P.

Now let $\{(U_\alpha, \psi_\alpha)\}$ be a collection of charts (α being a label which distinguishes different members of the collection) with the following properties:

(a) The collection $\{U_\alpha\}$ covers M, i.e., each point of M is a member of at least one U_α.

(b) ψ_α maps U_α into $\mathbb{R}^N$ with the same N for all α.

(c) For all α, β, $\psi_\alpha \circ \psi_\beta^{-1}$ and $\psi_\beta \circ \psi_\alpha^{-1}$ are differentiable functions from $\mathbb{R}^N \to \mathbb{R}^N$ wherever they are defined. (They are defined only if U_α and U_β intersect. The inverse maps ψ_α^{-1} and ψ_β^{-1} are defined because ψ_α and ψ_β are one-to-one.)

(d) This collection is maximal in the sense that if (U, ψ) is a chart, ψ mapping U onto an open set in $\mathbb{R}^N$ with $\psi \circ \psi_\alpha^{-1}$ and $\psi_\alpha \circ \psi^{-1}$ differentiable for all α for which they exist, then (U, ψ) belongs to the collection $\{(U_\alpha, \psi_\alpha)\}$.

A collection of charts having these properties is called an *atlas*, and M together with its atlas is an *N-dimensional differentiable manifold.*

Note that we do not claim that a manifold can be covered by a single coordinate neighborhood (though some can), and that is why we have a whole collection of charts. Property (c) tells us how to relate things in the overlap region of two coordinate neighborhoods.

Figure C.1 illustrates the situation which arises when two coordinate neighborhoods U_α and U_β intersect. The intersection is shaded, as are its images in $\mathbb{R}^N$ under ψ_α and ψ_β. The functions $\psi_\alpha \circ \psi_\beta^{-1}$ and $\psi_\beta \circ \psi_\alpha^{-1}$ are one-to-one differentiable functions which map one shaded region of $\mathbb{R}^N$ onto the other, as shown.

We should say something about the meaning of the word *differentiable.* Consider a function $f : S \to \mathbb{R}^L$, where S is an open set in $\mathbb{R}^K$. Then f is given by L component functions $f^1, \ldots, f^L$, each of which is a function of K variables, and f is said to be *differentiable of class C^r* if each f^a possesses continuous partial derivatives up to and including those of order r (r a positive integer). By a *differentiable function* we shall mean one of class C^r, where r is

[5]See, for example, Apostol, 1974, §§ 2.6, 3.2, 3.3 for details.

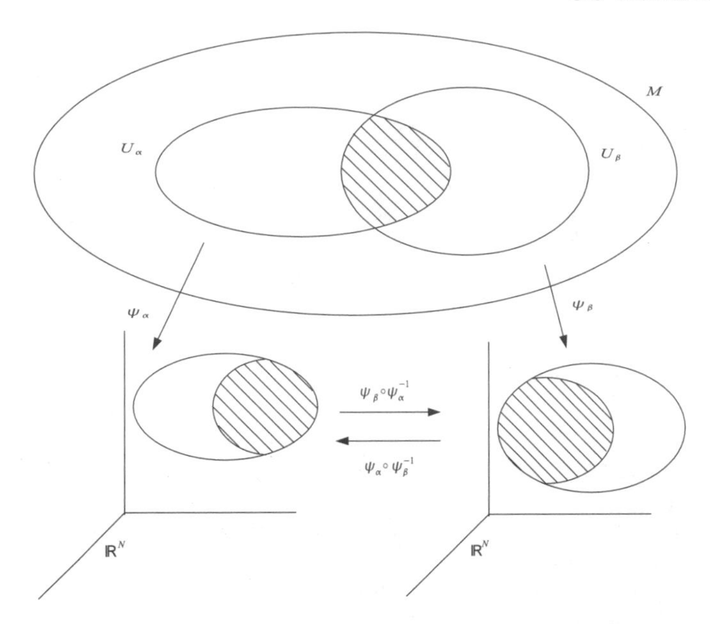

Fig. C.1. Overlapping coordinate neighborhoods.

sufficiently large to ensure that operations which depend on the continuity of partial derivatives (such as interchanging the order of partial differentiation) are valid, and that certain concepts are well-defined.

Let us consider some consequences of our definition. If (U, ψ) and (U', ψ') are charts with intersecting coordinate neighborhoods, then ψ assigns coordinates $(x^1, x^2, \ldots, x^N)$, say, to points in $U \cap U'$, while ψ' assigns coordinates $(x^{1'}, x^{2'}, \ldots, x^{N'})$, and the primed coordinates are given in terms of the unprimed coordinates by equations

$$x^{a'} = f^a(x^1, \ldots, x^N), \tag{C.28}$$

where $(f^1, \ldots, f^N) \equiv f = \psi' \circ \psi^{-1}$. ($A \cap B$ denotes the set of points common to the sets A and B.) The unprimed coordinates are similarly given in terms of the primed coordinates by equations

$$x^a = g^a(x^{1'}, \ldots, x^{N'}), \tag{C.29}$$

where $(g^1, \ldots, g^N) \equiv g = \psi \circ (\psi')^{-1}$. The function f and its inverse g are both one-to-one and differentiable, and it follows that the Jacobians $\left|\partial x^{a'} / \partial x^b\right|$ and $\left|\partial x^a / \partial x^{b'}\right|$ are non-zero.

Conversely, if we have a chart (U, ψ) and a system of equations of the form (C.28) with Jacobian $\left|\partial x^{a'} / \partial x^b\right|$ non-zero for values of x^a which are the coordinates of some point P in U, then it is possible to construct a coordinate system (U', ψ') about P whose coordinates are related to those of (U, ψ) by equations (C.28). See the exercise below.

Exercise C.6

1. In an N-dimensional manifold, (U, ψ) is a chart, and ψ maps U onto an open set V in $\mathbb{R}^N$. A differentiable function f maps a subset S of V into a set T in $\mathbb{R}^N$, and is such that the Jacobian of f is non-zero at $\psi(\mathrm{P}) \in S$. (P is a point of the manifold.) Use the inverse-function theorem[6] to show that there exists a chart (U', ψ') with $\mathrm{P} \in U'$, U' a subset of U, and $\psi' = f \circ \psi$ in U'.

C.7 The tangent space at each point of a manifold

The tangent space at any point of a manifold is the generalization of the tangent plane to a surface at any point of it. With a surface embedded in three-dimensional space we can readily realize the tangent plane as a plane in the embedding space. However, a manifold has no embedding space, so we must devise some implicit means of defining the tangent space in terms of the structure available to us. The way to do this is to make use of curves in the manifold. We first define a tangent vector to a curve at a point of it, and then the tangent space at that point as the vector space of tangent vectors to curves passing through the point.

In any coordinate neighborhood of an N-dimensional manifold M we can define a *curve* by means of N continuous functions $x^a(u)$, where u belongs to some real interval. As the parameter u varies we obtain the coordinates $x^a = x^a(u)$ of points on the curve. We shall confine the discussion to *regularly parameterized curves*, i.e., ones that may be parameterized in such a way that that at each point the derivative $\dot{x}^a(u)$ exist, and not all of them are zero.

Consider now a curve γ given by $x^a(u)$, and let P be a point on it. We can choose the parameterization so that $x^a(0) = (x^a)_P$, the coordinates of P. We define the *tangent vector* $\boldsymbol{\lambda}$ to γ at P to be the N-tuple $(\dot{x}^1(0), \dot{x}^2(0), \ldots, \dot{x}^N(0))$.[7] Thus a curve with a given parameterization yields a tangent vector, and conversely each N-tuple $\boldsymbol{\lambda} \equiv (\lambda^1, \ldots, \lambda^N)$ is the tangent vector to some curve through P, e.g., that given by

$$x^a(u) = \lambda^a u + (x^a)_P. \tag{C.30}$$

Since we have defined a tangent vector as an N-tuple, and under obvious definitions of addition and scalar multiplication the set of N-tuples is a vector

[6] See, for example, Apostol, 1974, § 13.3

[7] Our definition of a tangent vector makes use of N-tuples related to a coordinate system, and in this respect is somewhat unsatisfactory. However, we believe it has advantages of simplicity when compared with more sophisticated definitions, such as defining it as an equivalence class of curves, or as a directional derivative.

space, it follows that the set of all tangent vectors at P is a vector space. This space is the *tangent space* $T_P(M)$ of M at P. A basis of this space is $\{(\mathbf{e}_a)_P\}$, where $(\mathbf{e}_a)_P$ is the N-tuple with 1 in its ath position, and zeros elsewhere. We can then put

$$\boldsymbol{\lambda} = (\lambda^1, \ldots, \lambda^N) = \lambda^a (\mathbf{e}_a)_P,$$

and the λ^a of the N-tuple are the components of $\boldsymbol{\lambda}$ relative to the basis.

The basis vector $(\mathbf{e}_a)_P$ is in fact the tangent vector to the ath *coordinate curve* through P. That is, the curve obtained by keeping all the coordinates except the ath fixed. It is given by

$$x^b(u) = \delta^b_a u + (x^b)_P,$$

so that $x^b(0) = (x^b)_P$ as required; we also have that $\dot{x}^b(0) = \delta^b_a$, so the N-tuple $(\dot{x}^1(0), \dot{x}^2(0), \ldots, \dot{x}^N(0))$ has 1 in its ath position and zeros elsewhere. That is, the tangent vector to the ath coordinate curve is indeed $(\mathbf{e}_a)_p$.

The basis $\{(\mathbf{e}_a)_P\}$ whose members are tangent vectors to the coordinate curves through P is called the *natural basis* associated with the coordinate system. If we have another coordinate system about P with coordinates $x^{a'}$, then this gives rise to a new natural basis $\{(\mathbf{e}_{a'})_P\}$ of $T_P(M)$, and we can investigate the form of the transformation formula for vector components.

The curve γ through P will be given by $x^{a'}(u)$, say, in the new coordinate system, and the components of the tangent vector at P relative to the new natural basis are $\dot{x}^{a'}(0)$. But

$$\dot{x}^{a'}(0) = \left(\frac{\partial x^{a'}}{\partial x^b}\right)_P \dot{x}^b(0),$$

and since any vector $\boldsymbol{\lambda}$ in the tangent space is the tangent vector to some curve, we have in general that

$$\lambda^{a'} = (X^{a'}_b)_P \lambda^b, \qquad \text{(C.31)}$$

where $X^{a'}_b \equiv \dfrac{\partial x^{a'}}{\partial x^b}$.

If we similarly let $X^b_{a'} \equiv \dfrac{\partial x^b}{\partial x^{a'}}$, then, on using the chain rule,

$$X^a_{b'} X^{b'}_c = \frac{\partial x^a}{\partial x^{b'}} \frac{\partial x^{b'}}{\partial x^c} = \frac{\partial x^a}{\partial x^c} = \delta^a_c,$$

showing that the matrix $[X^a_{b'}]$ is the inverse of $[X^{a'}_b]$. It follows from the transformation formulae established in Section C.1 that the transformation formula for the basis elements is

$$(\mathbf{e}_{a'})_P = (X^b_{a'})_P (\mathbf{e}_b)_P.$$

To sum up, a change of coordinates about a point P of the manifold M induces a change of the natural basis of $T_P(M)$, and the matrices involved in the change of basis vectors and associated components are the Jacobian matrices of the coordinate transformation formulae evaluated at P.

Exercise C.7

1. Within a coordinate neighborhood a regularly parameterized smooth curve is given by $x^a(u)$, $a \le u \le b$. Show that if the parameter is changed to $u' = f(u)$, where f is a differentiable function whose derivative is nowhere zero for $a \le u \le b$, then u' is also a regular parameter. Show also that the tangent vector $(dx^a/du')_P$ is proportional to the tangent vector $(dx^a/du)_P$.

C.8 Tensor fields on a manifold

Having defined the tangent space $T_P(M)$, we may go on to define its dual $T_P^*(M)$, and hence build up spaces $(T_s^r)_P(M)$ of type (r, s) tensors at P. We then define a type (r, s) *tensor field* on M as an assignment to each point P of M a member of $(T_s^r)_P(M)$. We denote the set of all type (r, s) tensor fields on M by $T_s^r(M)$. In particular $T^1(M) = T(M)$ is the set of contravariant vector fields on M and $T_1(M) = T^*(M)$ is the set of covariant vector fields on M. A *scalar field* is simply a real-valued function on M.

In any coordinate neighborhood, scalar fields and components of vector and tensor fields may be regarded as functions of the coordinates. It follows from the preceding section that, under a change of coordinates, the components of a type (r, s) tensor field therefore transform according to equation (C.27), where

$$X_b^{a'} \equiv \frac{\partial x^{a'}}{\partial x^b} \quad \text{and} \quad X_{b'}^a \equiv \frac{\partial x^a}{\partial x^{b'}}.$$

So, through this more formal approach, we have obtained the defining property of a tensor field given in Section 1.8, justifying the less formal approach adopted in the main body of the text.

All the tensor fields considered in this text are assumed to be *differentiable*, i.e., their components are differentiable functions of the coordinates. In order that this concept be well defined, the differentiability class of the functions involved in the definition of M must be at least C^2. This condition is also sufficient to justify the change of order of partial differentiation of Jacobian matrix elements. That is, we can assert that in general

$$X_{bc}^{a'} = X_{cb}^{a'}, \quad \text{where} \quad X_{bc}^{a'} \equiv \frac{\partial^2 x^{a'}}{\partial x^b \partial x^c},$$

and this holds for all a, b and for all overlapping coordinate neighborhoods where we have coordinates $x^{a'}$ expressed in terms of coordinates x^a.

Solutions

Given here are solutions to the exercises and problems, or hints towards obtaining solutions, possible methods of approach, and comments. Some solutions are worked in more detail than others. If the exercise involves verifying a claim made in the text, for example, then little or no help is given, and where the reader is asked for a numerical result or a formula, then only the final answer is given. These are not 'model answers', but simply given as a check on answers or approaches. In all cases the reader should provide full details.

Chapter 1

Exercises

1.1.1 $\mathbf{e}_\rho = \cos\phi\,\mathbf{i} + \sin\phi\,\mathbf{j}$, $\mathbf{e}_\phi = -\rho\sin\phi\,\mathbf{i} + \rho\cos\phi\,\mathbf{j}$, $\mathbf{e}_z{=}\mathbf{k}$;
$\mathbf{e}^\rho = \cos\phi\,\mathbf{i} + \sin\phi\,\mathbf{j}$, $\mathbf{e}^\phi = -\rho^{-1}\sin\phi\,\mathbf{i} + \rho^{-1}\cos\phi\,\mathbf{j}$, $\mathbf{e}^z = \mathbf{k}$.

1.1.2 (a) Eliminate $\mathbf{j}$ and $\mathbf{k}$ from the system of equations given in Example 1.1.2 to get $\mathbf{i}$ in terms of $\mathbf{e}_r$, $\mathbf{e}_\theta$, and $\mathbf{e}_\phi$.
(b) and (c) Similar to (a), using the equations found in Exercise 1.1.1 for (b) and equations (1.12) for (c).

1.2.1 $\boldsymbol{\lambda}\cdot\mathbf{e}_j = \lambda_k\mathbf{e}^k\cdot\mathbf{e}_j = \lambda_k\delta^k_j = \lambda_j$.

1.2.2 Since $\{\mathbf{e}^j\}$ is a basis, we can write $\mathbf{e}_i = a_{ij}\mathbf{e}^j$ for some a_{ij}. Take the dot product with $\mathbf{e}_k$ to show that $a_{ij} = g_{ij}$. A similar argument gives $\mathbf{e}^i = g^{ij}\mathbf{e}_j$.

1.2.3 (a) $\lambda^i\lambda_i = \boldsymbol{\lambda}\cdot\boldsymbol{\lambda}$; (b) $\mu_i\lambda^i = \boldsymbol{\mu}\cdot\boldsymbol{\lambda}$; (c) 0.

1.2.4 They are diagonal matrices.

1.2.5 Straightforward substitution.

1.2.6 $\mu^i = -\frac{1}{2}v\delta_1^i - \frac{3}{2}u\delta_2^i + (v^2 + 3u^2 + 1)\delta_3^i$.

1.2.7 (a) 3; (b) 2; (c) N; (d) 4.

1.3.1 $ds^2 = (4v^2+2)du^2+(4u^2+2)dv^2+dw^2+8uv\,du\,dv+4v\,du\,dw+4u\,dv\,dw$.

1.3.2 It is one turn of a helix. Its length is $2\pi\sqrt{1+a^2}$.

1.3.3 This follows from equation (1.39).

1.4.1 $U_{i'}^k U_j^{i'} = \dfrac{\partial u^k}{\partial u^{i'}}\dfrac{\partial u^{i'}}{\partial u^j} = \dfrac{\partial u^k}{\partial u^j} = \delta_j^k$, and similarly for $U_i^{k'} U_{j'}^i = \delta_j^k$;
$\delta_j^k = \mathbf{e}^k \cdot \mathbf{e}_j = U_{i'}^k \mathbf{e}^{i'} \cdot U_j^{l'} \mathbf{e}_{l'} = U_{i'}^k U_j^{l'} \delta_l^i = U_{i'}^k U_j^{i'}$, and similarly for $\delta_j^k = U_i^{k'} U_{j'}^i$.

1.4.2 Straightforward substitutions.

1.4.3 $\hat{G} = \hat{U}^{\mathrm{T}} G \hat{U}$; $ds^2 = d\rho^2 + \rho^2 d\phi^2 + dz^2$.

1.5.1 $\mathbf{e}^i \cdot \boldsymbol{\tau}(\mathbf{e}_j) = \mathbf{e}^i \cdot \tau_j^k \mathbf{e}_k = \tau_j^k \delta_k^i = \tau_j^i$.
The primed version is $\tau_{m'}^{i'} = \mathbf{e}^{i'} \cdot \boldsymbol{\tau}(\mathbf{e}_{m'})$. Put $\mathbf{e}^{i'} = U_k^{i'} \mathbf{e}^k$, $\mathbf{e}_{m'} = U_{m'}^l \mathbf{e}_l$, and then simplify, using the linearity of $\boldsymbol{\tau}$.

1.6.1 You need to show that the expressions obtained for the basis vectors in Example 1.6.1 satisfy $\mathbf{e}^u \cdot \mathbf{e}_u = \mathbf{e}^v \cdot \mathbf{e}_v = 1$ and $\mathbf{e}^u \cdot \mathbf{e}_v = \mathbf{e}^v \cdot \mathbf{e}_u = 0$.

1.6.2 (a) $ds^2 = a^2(d\theta^2 + \sin^2\theta\, d\phi^2)$; (b) $ds^2 = a^2 d\phi^2 + dz^2$;
(c) $ds^2 = (2+4v^2)du^2 + 8uv\,du\,dv + (2+4u^2)dv^2$.

1.6.3 It is flat, because we can slit it along a generator and flatten it. Also, using coordinates w, z, where $w = az$ gives the Euclidean line element $ds^2 = dw^2 + dz^2$. An answer using the curvature tensor must await Chapter 3.

1.8.1 The matrix version of $\tau_{i'j'} = U_{i'}^k U_{j'}^l \tau_{kl}$ is $T' = \hat{U}^{\mathrm{T}} T U$. So if $T = [\tau_{kl}] = [\delta_{kl}] = I$, then for a transformation from spherical to cylindrical coordinates (as in Example 1.4.1)

$$T' = \hat{U}^{\mathrm{T}} U = \begin{bmatrix} \sin\theta & (\cos\theta)/r & 0 \\ 0 & 0 & 1 \\ \cos\theta & -(\sin\theta)/r & 0 \end{bmatrix} \begin{bmatrix} \sin\theta & r\cos\theta & 0 \\ 0 & 0 & 1 \\ \cos\theta & -r\sin\theta & 0 \end{bmatrix} \neq [\delta_{ij}].$$

1.8.2 If $\tau^{ab} = \tau^{ba}$, then $\tau^{c'd'} = X_a^{c'} X_b^{d'} \tau^{ab} = X_b^{d'} X_a^{c'} \tau^{ba} = \tau^{d'c'}$.

1.8.3 If $\sigma_{ab} = \sigma_{ba}$ and $\tau^{ab} = -\tau^{ba}$, then $\sigma_{ab}\tau^{ab} = -\sigma_{ba}\tau^{ba} = -\sigma_{ab}\tau^{ab}$, on relabeling dummy suffixes. This implies that $\sigma_{ab}\tau^{ab} = 0$.

1.8.4 $\tau^{ab} = \sigma^{ab} + \kappa^{ab}$, where $\sigma^{ab} \equiv \frac{1}{2}(\tau^{ab} + \tau^{ba})$ is symmetric and $\kappa^{ab} \equiv \frac{1}{2}(\tau^{ab} - \tau^{ba})$ is skew-symmetric, and similarly for τ_{ab}.

1.9.1 If g_{ab} is positive definite, then $g_{ab}(\lambda^a + x\mu^a)(\lambda^b + x\mu^b) \geq 0$ for all x. That is, $ax^2 + 2bx + c \geq 0$ for all x, where $a \equiv g_{ab}\mu^a\mu^b$, $b \equiv g_{ab}\lambda^a\mu^b$ and $c \equiv g_{ab}\lambda^a\lambda^b$. This means that the quadratic equation $ax^2 + 2bx + c = 0$ has no real roots, so $b^2 \leq ac$, from which the required result follows.

1.9.2 Using a new parameter $t' = f(t)$ with $a' < t' < b'$ (where $a' = f(a)$, $b' = f(b)$), we get a length $L' = \int_{b'}^{a'} \left| g_{ab}\frac{dx^a}{dt'}\frac{dx^b}{dt'} \right|^{1/2} dt'$. Now use $t' = f(t)$ as a substitution to evaluate the integral. This gives

$$L' = \int_b^a \left| g_{ab}\frac{dx^a}{dt'}\frac{dx^b}{dt'} \right|^{1/2} \frac{dt'}{dt}dt = \int_b^a \left| g_{ab}\frac{dx^a}{dt}\frac{dx^b}{dt} \right|^{1/2} dt = L.$$

(This assumes that t' increases with t; some changes are needed if it decreases.)

1.9.3 The lengths are (a) $c(1 - 2m/r)^{1/2}$, (b) $(1 - 2m/r)^{-1/2}$ and (c) 0.
As ν^μ is a null vector, the only angle defined is that between λ^μ and μ^μ, which is equal to $\pi/2$.
Only λ^μ and μ^μ form an orthogonal pair.

Problems

1.1 (a) $u = u_0$: an ellipsoid centered on the origin;
$v = v_0$: a squashed cone with its vertex at the origin and elliptical cross-sections;
$w = w_0$: a half-plane with its edge at the z-axis.

(b) $\mathbf{e}_u = a \sin v \cos w\, \mathbf{i} + b \sin v \sin w\, \mathbf{j} + c \cos v\, \mathbf{k}$;
$\mathbf{e}_v = au \cos v \cos w\, \mathbf{i} + bu \cos v \sin w\, \mathbf{j} - cu \sin v\, \mathbf{k}$;
$\mathbf{e}_w = -au \sin v \sin w\, \mathbf{i} + bu \sin v \cos w\, \mathbf{j}$;
$\mathbf{e}^u = a^{-1} \sin v \cos w\, \mathbf{i} + b^{-1} \sin v \sin w\, \mathbf{j} + c^{-1} \cos v\, \mathbf{k}$;
$\mathbf{e}^v = (au)^{-1} \cos v \cos w\, \mathbf{i} + (bu)^{-1} \cos v \sin w\, \mathbf{j} - (cu)^{-1} \sin v\, \mathbf{k}$;
$\mathbf{e}^w = -(au \sin v)^{-1} \sin w\, \mathbf{i} + (bu \sin v)^{-1} \cos w\, \mathbf{j}$.

(c) Check that $\mathbf{e}^u \cdot \mathbf{e}_u = \mathbf{e}^v \cdot \mathbf{e}_v = \mathbf{e}^w \cdot \mathbf{e}_w = 1$ and $\mathbf{e}^u \cdot \mathbf{e}_v = \mathbf{e}^u \cdot \mathbf{e}_w = \mathbf{e}^v \cdot \mathbf{e}_u = \mathbf{e}^v \cdot \mathbf{e}_w = \mathbf{e}^w \cdot \mathbf{e}_u = \mathbf{e}^w \cdot \mathbf{e}_v = 0$, using your answer to (b).

1.2 From the geometry of the figure, points on the torus have position vector $\mathbf{r} = \mathbf{a} + \mathbf{b}$, where $\mathbf{a} = a(\cos\phi\, \mathbf{i} + \sin\phi\, \mathbf{j})$ and $\mathbf{b} = b\cos\theta(\cos\phi\, \mathbf{i} + \sin\phi\, \mathbf{j}) + b\sin\theta\, \mathbf{k}$.

$$\mathbf{e}_\theta = -b\sin\theta\cos\phi\,\mathbf{i} - b\sin\theta\sin\phi\,\mathbf{j} + b\cos\theta\,\mathbf{k};$$
$$\mathbf{e}_\phi = -(a + b\cos\theta)\sin\phi\,\mathbf{i} + (a + b\cos\theta)\cos\phi\,\mathbf{j}.$$
$$g_{11} = b^2;\quad g_{12} = 0;\quad g_{22} = (a + b\cos\theta)^2.$$

1.3 You need to show that there exists a non-singular matrix $M \equiv [\mu^a_b]$, such that $ML = [1\ 0\ \ldots\ 0]^{\mathrm{T}}$, where $L \equiv [\lambda^b]$. This can be done by taking the first row of M to be $\|L\|^{-2} L^{\mathrm{T}}$, where $\|L\|^2 \equiv L^{\mathrm{T}}L$ (which is non-zero as $L \neq 0$) and then choosing the remaining rows to be independent and orthogonal to L^{T}.

1.4 Using the hint gives $\tau^{bc}\delta^a_1 + \tau^{ca}\delta^b_1 + \tau^{ab}\delta^c_1 = 0$. Taking $(a, b, c) = (1, 1, 1)$ gives $\tau^{11} = 0$; taking $(a, b, c) = (1, 1, i)$ $(i \neq 1)$ and using the symmetry of τ^{ab} gives $\tau^{1i} = \tau^{i1} = 0$ $(i \neq 1)$; taking $(a, b, c) = (1, i, j)$ $(i \neq 1 \neq j)$ gives $\tau^{ij} = 0$ $(i \neq 1 \neq j)$. This covers all components.

1.5 By invariance, $\tau_{ab}\lambda^a\lambda^b = \tau_{a'b'}\lambda^{a'}\lambda^{b'}$. Take $\lambda^a = \mu^a + \nu^a$, where μ^a, ν^a are arbitrary vectors. Then symmetry of τ_{ab} and invariance of $\tau_{ab}\mu^a\mu^b$, $\tau_{ab}\nu^a\nu^b$ yield $\tau_{ab}\mu^a\nu^b = \tau_{a'b'}\mu^{a'}\nu^{b'}$ for arbitrary μ^a, ν^a. Required result follows from quotient theorem.

1.6 Given property implies that $\tau_{abcd}(\alpha^a + \beta^a)\mu^b(\alpha^c + \beta^c)\mu^d = 0$ for all α^a, β^a, μ^a. On expanding and using given property of τ_{abcd}, this gives $\tau_{abcd}\alpha^a\mu^b\beta^c\mu^d = 0$ for all α^a, β^a, μ^a. Similarly you can show that $\tau_{abcd}\lambda^a\alpha^b\lambda^c\beta^d = 0$ for all λ^a, α^a, β^a. Now take $\tau_{abcd}(\alpha^a + \beta^a)(\gamma^b + \delta^b)(\alpha^c + \beta^c)(\gamma^d + \delta^d) = 0$ (from given property), expand and simplify (using the given property and the two results just obtained) to arrive at
$\tau_{abcd}\alpha^a\gamma^b\beta^c\delta^d + \tau_{abcd}\alpha^a\delta^b\beta^c\gamma^d + \tau_{abcd}\beta^a\gamma^b\alpha^c\delta^d + \tau_{abcd}\beta^a\delta^b\alpha^c\gamma^d = 0$
for all α^a, β^a, γ^a, δ^a. Relabeling gives
$(\tau_{abcd} + \tau_{adbc} + \tau_{cbad} + \tau_{cdab})\alpha^a\gamma^b\beta^c\delta^d = 0$ for arbitrary α^a, β^a, γ^a, δ^a, from which the required result follows.

1.7 (a) Expressions for $\mathbf{e}_{i'}$ are easily obtained from a diagram.

$$[X^{i'}_j] = \begin{bmatrix} \cos\theta & \sin\theta & 0 \\ -\sin\theta & \cos\theta & 0 \\ 0 & 0 & 1 \end{bmatrix};\ [X^i_{j'}] = \begin{bmatrix} \cos\theta & -\sin\theta & 0 \\ \sin\theta & \cos\theta & 0 \\ 0 & 0 & 1 \end{bmatrix}.$$

(b) $[L^i] = [0\ \ 15m\ \ 0]^{\mathrm{T}}$.

(c) $$[I^{i'}_{j'}] = \begin{bmatrix} m\sin^2\theta & m\sin\theta\cos\theta & 0 \\ m\sin\theta\cos\theta & m\cos^2\theta & 0 \\ 0 & 0 & m \end{bmatrix};$$

$$[\omega^{i'}] = \begin{bmatrix} 15\sin\theta \\ 15\cos\theta \\ 0 \end{bmatrix};\ [L^{i'}] = \begin{bmatrix} 15m\sin\theta \\ 15m\cos\theta \\ 0 \end{bmatrix}.$$

1.8 Check that $\Lambda(\phi)\Lambda(-\phi) = I$, where $\Lambda(\phi)$ is the displayed matrix. For the boost, $X^{\mu'}_{\nu} = \Lambda^{\mu'}_{\nu}$, so $[X^{\nu}_{\mu'}] = [\Lambda^{\mu'}_{\nu}]^{-1} = \Lambda(-\phi)$. The transformation formula $g_{\mu'\nu'} = X^{\alpha}_{\mu'}X^{\beta}_{\mu'}\eta_{\alpha\beta}$ translates to the matrix equation $G' = \Lambda(-\phi)^{\mathrm{T}}H\Lambda(-\phi)$, where $H = [\eta_{\alpha\beta}]$. It is straightforward to check that this gives $G' = H$.

1.9 If $X \equiv [X^{a'}_{b}]$, then $g_{a'b'} = X^{c}_{a'}X^{d}_{b'}g_{cd}$ translates to the matrix equation $G' = (X^{-1})^{\mathrm{T}}GX^{-1}$. So you get $G' = I$ if $X = P^{-1}$, where P is a diagonalizing matrix such that $P^{\mathrm{T}}GP = I$. The coordinate transformation $x^{a'} = s^{a}_{b}x^{b}$, where $[s^{a}_{b}] = P^{-1}$, gives $X = P^{-1}$, as required.

1.10 Take $t^{\mu} = \delta^{\mu}_{0}$, $n^{\mu} = \delta^{\mu}_{0} + \delta^{\mu}_{1}$ and $s^{\mu} = \delta^{\mu}_{1}$, as suggested.
(a) Orthogonality condition $\eta_{\mu\nu}\lambda^{\mu}t^{\nu} = 0$ implies that $\lambda^{0} = 0$, which means that λ^{μ} is spacelike.
(b) Orthogonality condition $\eta_{\mu\nu}\lambda^{\mu}n^{\nu} = 0$ implies that $\lambda^{0} = \lambda^{1}$. Hence $\eta_{\mu\nu}\lambda^{\mu}\lambda^{\nu} = -(\lambda^{2})^{2} - (\lambda^{3})^{2}$. If $\lambda^{2} = \lambda^{3} = 0$, then λ^{μ} is null, otherwise it is spacelike.
(c) The given vectors t^{μ}, n^{μ} are orthogonal to s^{μ}, as is the spacelike vector $p^{\mu} \equiv \delta^{\mu}_{2}$.

1.11 $\mathbf{e}_{t'} = \gamma\,\mathbf{e}_{t} + \gamma v\,\mathbf{e}_{x}$; $\mathbf{e}_{x'} = (\gamma v/c^{2})\,\mathbf{e}_{t} + \gamma\,\mathbf{e}_{x}$; $\mathbf{e}_{y'} = \mathbf{e}_{y}$; $\mathbf{e}_{z'} = \mathbf{e}_{z}$.
Actual directions depend on units and scales used. If these are such that a null vector is inclined at 45° to $\mathbf{e}_{t}$ and $\mathbf{e}_{x}$ (as is usual), then $\mathbf{e}_{t'}$ is inclined at an angle $\arctan(|v|/c)$ to $\mathbf{e}_{t}$, and $\mathbf{e}_{x'}$ is inclined at the same angle to $\mathbf{e}_{x}$. Inclinations are such that for $v > 0$ the angle between $\mathbf{e}_{t'}$ and $\mathbf{e}_{x'}$ is less than a right angle (so the axes close up), while for $v < 0$ it is greater (so the axes open out).

Chapter 2

Exercises

2.0.1 They do not deviate; ant cannot decide.

2.1.1 Use $\frac{du^{i}}{ds} = \frac{du^{i}}{dt}\frac{dt}{ds}$ and $\frac{d^{2}u^{i}}{ds^{2}} = \frac{d^{2}u^{i}}{dt^{2}}\left(\frac{dt}{ds}\right)^{2} + \frac{du^{i}}{dt}\frac{d^{2}t}{ds^{2}}$. Solve $\frac{d^{2}t}{ds^{2}} = 0$, which gives $h(s) = 0$.

2.1.2 (a) Follows from defining equation (1.80).
(b) $\pm 2L\dot{L} = \dot{g}_{ab}\dot{x}^{a}\dot{x}^{b} + 2g_{ab}\ddot{x}^{a}\dot{x}^{b}$, on using symmetry of g_{ab}.
(c) $\pm 2L\dot{L} = \partial_{c}g_{ab}\dot{x}^{c}\dot{x}^{a}\dot{x}^{b} - 2g_{ab}\Gamma^{a}_{cd}\dot{x}^{c}\dot{x}^{d}\dot{x}^{b}$.
(d) $\pm 2L\dot{L} = \partial_{c}g_{ab}\dot{x}^{c}\dot{x}^{a}\dot{x}^{b} - g_{ab}g^{ae}(\partial_{c}g_{ed} + \partial_{d}g_{ce} - \partial_{e}g_{cd})\dot{x}^{c}\dot{x}^{d}\dot{x}^{b}$.
(e) Simplify, using $g_{ab}g^{ae} = \delta^{e}_{b}$.

2.1.3 Note that $L^2 = \pm g_{ab}\frac{dx^a}{du}\frac{dx^b}{du} = \pm g_{ab}\frac{dx^a}{ds}\frac{dx^b}{ds}(\frac{ds}{du})^2 = (\frac{ds}{du})^2$, so $L = \text{const}$ implies that $\frac{du}{ds} = \text{const} = A$, say. So $u = As + B$.

2.1.4 Show that $u' = f(u)$ transforms equation (2.12) to $\frac{d^2x^a}{du'^2} + \Gamma^a_{bc}\frac{dx^b}{du'}\frac{dx^c}{du'} = h(u)\frac{dx^a}{du'}$, where $h(u) = -\frac{d^2u'}{du^2}(\frac{du'}{du})^{-2}$. So if u' is also affine, then $h(u) = 0$, which gives $d^2u'/du^2 = 0$ and $u' = Au + B$. (See Ex. 2.1.1 for similar argument.)

2.1.5 Comparing $ds^2 = a^2(d\theta^2 + \sin^2\theta\, d\phi^2)$ with $ds^2 = g_{AB}du^A du^B$ gives $g_{11} = a^2$, $g_{12} = g_{21} = 0$ and $g_{22} = a^2\sin^2\theta$. For the Lagrangian $L \equiv \frac{1}{2}a^2(\dot\theta^2 + \sin^2\theta\,\dot\phi^2)$, Euler–Lagrange equations yield the geodesic equations $\ddot\theta - \sin\theta\cos\theta\,\dot\phi^2 = 0$, $\ddot\phi + 2\cot\theta\,\dot\theta\dot\phi = 0$, from which you can pick out the non-zero connection coefficients.

2.1.6 In the notation of Example 2.1.1, a line of longitude can be parameterized by $u^A = (s/a)\delta^A_1 + \phi_0\delta^A_2$ ($\phi_0 = \text{const}$). So $\dot u^A = (1/a)\delta^A_1$ and $\ddot u^A = 0$. So for the geodesic equation to be satisfied, we need $(1/a^2)\Gamma^A_{11} = 0$, which is true by Exercise 2.1.5.

2.1.7 Show that equations (2.20) are satisfied by $t = u$, $r = r_0$, $\theta = \theta_0$, $\phi = \phi_0$.

2.2.1 Substitution from (2.26) into (2.24) verifies that the differential equations are satisfied and putting $t = 0$ in (2.26) verifies that the initial conditions (2.25) are satisfied.

2.2.2 Reversed if $\omega = \frac{1}{2}$, so $\theta_0 = \pi/3$ or $60°$.

2.2.3 Example 2.1.1 shows that the equator ($\theta_0 = \pi/2$) is a geodesic and has tangent vector $\mu^A \equiv a^{-1}\delta^A_2$. For transport round the equator, $\lambda^A = a^{-1}\cos\alpha\,\delta^A_1 + a^{-1}\sin\alpha\,\delta^A_2$. A straightforward calculation gives $\frac{\pi}{2} - \alpha$ for the angle between μ^A and λ^A.

2.2.4 Differentiation of $g_{c'd'} = X^a_{c'}X^b_{d'}g_{ab}$ gives $\partial_{e'}g_{c'd'} = X^a_{e'c'}X^b_{d'}g_{ab} + X^a_{c'}X^b_{e'd'}g_{ab} + X^a_{c'}X^b_{d'}X^f_{e'}\partial_f g_{ab}$, from which you should deduce that $\Gamma_{d'e'c'} = X^b_{d'}X^f_{e'}X^a_{c'}\Gamma_{bfa} + X^a_{e'c'}X^b_{d'}g_{ab}$. Required result follows by contraction with $g^{d'h'} = X^{d'}_i X^{h'}_j g^{ij}$.

2.2.5 Note that differentiation of $X^d_{b'}X^{a'}_d = \delta^a_b$ with respect to $x^{c'}$ gives $X^d_{c'b'}X^{a'}_d + X^d_{b'}X^f_{c'}X^{a'}_{fd} = 0$.

2.2.6 Use (2.39) and the fact that $\dot\lambda^{a'} = \frac{d}{dt}(X^{a'}_f\lambda^f) = X^{a'}_f\dot\lambda^f + X^{a'}_{ef}\dot x^e\lambda^f$.

2.3.1 See solution to Exercise 2.2.6.

2.3.2 Take $\tau_{ab} = \lambda_a \mu_b$. Then $D\tau_{ab}/dt = (D\lambda_a/dt)\mu_b + \lambda_a(D\mu_b/dt) =$
$(\dot{\lambda}_a - \Gamma^c_{ad}\lambda_c \dot{x}^d)\mu_b + \lambda_a(\dot{\mu}_b - \Gamma^c_{bd}\mu_c \dot{x}^d) =$
$(\dot{\lambda}_a \mu_b + \lambda_a \dot{\mu}_b) - \Gamma^c_{ad}\lambda_c \mu_b \dot{x}^d - \Gamma^c_{bd}\lambda_a \mu_c \dot{x}^d = \dot{\tau}_{ab} - \Gamma^c_{ad}\tau_{cb}\dot{x}^d - \Gamma^c_{bd}\tau_{ac}\dot{x}^d$.
Formula for $D\tau^a_b/dt$ is obtained similarly by taking $\tau^a_b = \lambda^a \mu_b$.

2.3.3 Follows from equation (2.45) with $\lambda^a = \dot{x}^a$.

2.3.4 Follows from the fact that $D(g_{ab}\dot{x}^a \dot{x}^b)/du = 0$, because $Dg_{ab}/du = 0$ and (for an affinely parameterized geodesic) $D\dot{x}^a/du = 0$.

2.4.1 As noted in the text, $(X^{a'}_d)_{\mathrm{O}} = \delta^a_d$, so
$(g_{ab})_{\mathrm{O}} = (g_{c'd'} X^{c'}_a X^{d'}_b)_{\mathrm{O}} = (g_{c'd'} \delta^c_a \delta^d_b)_{\mathrm{O}} = (g_{a'b'})_{\mathrm{O}}$.

2.5.1 Timelike. No, because it is not in free-fall.

2.5.2 If $f^\mu = 0$, then $Dp^\mu/d\tau \equiv dp^\mu/d\tau + \Gamma^\mu_{\nu\sigma} p^\nu \dot{x}^\sigma = 0$, where $p^\mu = m(dx^\mu/d\tau)$. Required result follows, as m is constant.

2.7.1 Suppose that $g^{\mu\nu} = \eta^{\mu\nu} + f^{\mu\nu}$, where the $f^{\mu\nu}$ are small. Then $\delta^\mu_\nu = g^{\mu\sigma} g_{\sigma\nu} = (\eta^{\mu\sigma} + f^{\mu\sigma})(\eta_{\sigma\nu} + h_{\sigma\nu})$ gives (to first order) $f^{\mu\sigma}\eta_{\sigma\nu} = -\eta^{\mu\sigma} h_{\sigma\nu}$. Required approximation is given by contracting with $\eta^{\nu\rho}$.
Note that $\Gamma_{\rho\nu\sigma} = \frac{1}{2}(\partial_\nu h_{\sigma\rho} + \partial_\sigma h_{\nu\rho} - \partial_\rho h_{\nu\sigma})$, then contract with $g^{\mu\rho} = \eta^{\mu\rho} - h^{\mu\rho}$ and discard second-order terms to get required approximation.

2.7.2 Equation (2.81) says that $d^2x^i/dt^2 = -\partial V/\partial x^i$ $(i = 1, 2, 3)$, which are the three component equations of $\mathbf{a} = -\nabla V$. So $\mathbf{F} = m\mathbf{a} = -m\nabla V$.

2.9.1 Quick way to get (2.86) is to differentiate (2.85) to get $dT = dt$,
$dX = dx \cos\omega t - dy \sin\omega t - \omega(x \sin\omega t + y \cos\omega t)dt$,
$dY = dx \sin\omega t + dy \cos\omega t + \omega(x \cos\omega t - y \sin\omega t)dt$, $dZ = dz$, and then substitute into (2.84). Why does this work?
Either invert $[g_{\mu\nu}]$ to get $[g^{\mu\nu}]$, or check that $[g_{\mu\nu}][g^{\nu\sigma}] = I$, using matrix methods.

2.9.2 (a) Euler–Lagrange equations give geodesic equations in form $g_{\nu\sigma}\ddot{x}^\sigma + \Gamma_{\nu\alpha\beta}\dot{x}^\alpha \dot{x}^\beta = 0$. Use $g^{\mu\nu}$ to raise ν and then pick out $\Gamma^\mu_{\alpha\beta}$.
(b) Straightforward, but tedious. (c) Straightforward.

2.9.3 Substitute $x = \rho\cos\phi$, $y = \rho\sin\phi$ in (2.87) and take appropriate combinations of the middle pair of equations.

2.9.4 Note that $h_{0k} = (\omega/c)(y\delta^1_k - x\delta^2_k)$, so $[\partial_j h_{0k}] = \begin{bmatrix} 0 & -\omega/c & 0 \\ \omega/c & 0 & 0 \\ 0 & 0 & 0 \end{bmatrix}$.

Problems

2.1 Euler–Lagrange equations yield $(1+2v^2)\ddot{u} + 2uv\ddot{v} + 4v\dot{u}\dot{v} = 0$, $2uv\ddot{u} + (1+2u^2)\ddot{v} + 4u\dot{u}\dot{v} = 0$. These are satisfied by $u = u_0$, $v = As + B$, (u_0, A, B constant), and also by $v = v_0$, $u = Cs + D$, (v_0, C, D constant), showing that all parametric curves are geodesics. (Note that $g_{AB}\dot{u}^A\dot{u}^B = 1$ implies that $2(1+2u_0^2)A^2 = 1$ and $2(1+2v_0^2)C^2 = 1$.)

2.2 Geodesic equations are $\ddot{\rho} - \rho\dot{\phi}^2 = 0$, $\rho^2\ddot{\phi} + 2\rho\dot{\rho}\dot{\phi}$. These are satisfied by $\phi = \phi_0$, $\rho = As + B$ (using arc-length s as parameter) and you can take $A = 1$, $B = 0$. (Why?) Equations of parallel transport are $\dot{\lambda}^1 - \rho\lambda^2\dot{\phi} = 0$, $\dot{\lambda}^2 + \rho^{-1}\lambda^1\dot{\phi} + \rho^{-1}\lambda^2\dot{\rho} = 0$ (got by picking out Γ^A_{BC} from the geodesic equations), which reduce to $\dot{\lambda}^1 = 0$, $\dot{\lambda}^2 + \rho^{-1}\lambda^2\dot{\rho} = 0$ for the ray. These are satisfied by $\lambda^1 = \lambda^1_0$, $\lambda^2 = (\rho_0/\rho)\lambda^2_0$. Straightforward to check claims regarding length and angle.

2.3 For the line element, put $\theta = \theta_0$ in (1.40). For the second geodesic equation, use the fact that ϕ is a cyclic coordinate. Rest of problem is straightforward, following instructions given.

2.4 Apply (2.56) to $\lambda_{a;b} - \lambda_{b;a}$.

2.5 Gamma-terms cancel to leave $B_{abc} = A_{ab;c} + A_{bc;a} + A_{ca;b}$.

2.6 Use (1.40).

2.7 See Example 2.1.2, which takes you through a similar exercise.

2.8 The observer is in a position similar to the observer at rest in the rotating system K' of Section 2.9, and could assert that there exists a gravitational-type force (i.e., the centrifugal force) to balance the pull of the Sun on the Earth.

Chapter 3

Exercises

3.1.1 In such a system, $u^\mu = c\delta^\mu_0$ at P.

3.1.2 In the observer's rest frame, his 4-velocity is $U^\mu = (c, \mathbf{0})$ and the 4-momentum of the particle is $p^\mu = (E/c, \mathbf{p})$, so $p_\mu U^\mu = E$, which is the energy he assigns to the particle. Since $p_\mu U^\mu$ is invariant, it gives the assigned energy in any coordinate system.

3.1.3 All terms have the dimensions of pressure: $ML^{-1}T^{-2}$.

3.1.4 $u^\nu u_\nu = g_{\nu\sigma}u^\nu u^\sigma = c^2$ implies that $g_{\nu\sigma}(u^\nu{}_{;\mu}u^\sigma + u^\nu u^\sigma{}_{;\mu}) = 0$, on differentiating and noting that $g_{\nu\sigma;\mu} = 0$. This simplifies to give $2u^\nu{}_{;\mu}u_\nu = 0$.

3.2.1 (a) Equation (3.12) can be written as $\lambda_{a;bc} - \lambda_{a;cb} = -R_{adbc}\lambda^d$, because $R^d{}_{abc}\lambda_d = R_{dabc}\lambda^d = -R_{adbc}\lambda^d$. Required result follows on raising a.
(b)
$(\lambda^a\mu^b)_{;cd} = (\lambda^a{}_{;c}\mu^b + \lambda^a\mu^b{}_{;c})_{;d} = \lambda^a{}_{;cd}\mu^b + \lambda^a{}_{;c}\mu^b{}_{;d} + \lambda^a{}_{;d}\mu^b{}_{;c} + \lambda^a\mu^b{}_{;cd}$.
So $\tau^{ab}{}_{;cd} - \tau^{ab}{}_{;dc} = (\lambda^a\mu^b)_{;cd} - (\lambda^a\mu^b)_{;dc} = (\lambda^a{}_{;cd} - \lambda^a{}_{;dc})\mu^b + \lambda^a(\mu^b{}_{;cd} - \mu^b{}_{;dc}) = -R^a{}_{ecd}\lambda^e\mu^b - R^b{}_{ecd}\lambda^a\mu^e = -R^a{}_{ecd}\tau^{eb} - R^b{}_{ecd}\tau^{ae}$.
(c) $\tau^{ab}_{c\,;de} - \tau^{ab}_{c\,;ed} = -R^a{}_{fde}\tau^{fb}_c - R^b{}_{fde}\tau^{af}_c + R^f{}_{cde}\tau^{ab}_f$.

3.2.2 Straightforward using the defining equation (3.13) and the fact that $\Gamma^d_{ab} = \Gamma^d_{ba}$.

3.2.3 $R_{abcd} = g_{ae}R^e{}_{bcd} = g_{ae}(\partial_c\Gamma^e_{bd} - \partial_d\Gamma^e_{bc} + \Gamma^f_{bd}\Gamma^e_{fc} - \Gamma^f_{bc}\Gamma^e_{fd}) =$
$\partial_c(g_{ae}\Gamma^e_{bd}) - (\partial_c g_{ae})\Gamma^e_{bd} - \partial_d(g_{ae}\Gamma^e_{bc}) + (\partial_d g_{ae})\Gamma^e_{bc} + \Gamma^f_{bd}\Gamma_{afc} - \Gamma^f_{bc}\Gamma_{afd} =$
$\partial_c\Gamma_{abd} - \partial_d\Gamma_{abc} - g^{ef}(\Gamma_{fbd}\partial_c g_{ae} - \Gamma_{fbc}\partial_d g_{ae} - \Gamma_{ebd}\Gamma_{afc} + \Gamma_{ebc}\Gamma_{afd})$.
Required result follows on putting
$\partial_c\Gamma_{abd} = \frac{1}{2}\partial_c(\partial_b g_{ad} + \partial_d g_{ba} - \partial_a g_{bd})$, with a similar expression for $\partial_d\Gamma_{abc}$, and $\partial_c g_{ae} = \Gamma_{aec} + \Gamma_{eac}$, with a similar expression for $\partial_d g_{ae}$.

3.2.4 $0 = R^a{}_{bca} + R^a{}_{cab} + R^a{}_{abc} = R_{bc} - R^a{}_{cba} + 0 = R_{bc} - R_{cb}$.

3.3.1 $\oint d(\xi^c\xi^d) = 0$ implies that $\oint(\xi^c d\xi^d + \xi^d d\xi^c) = 0$, so
$f^{cd} \equiv \oint \xi^c d\xi^d = \oint \xi^c d\xi^d - \frac{1}{2}\oint(\xi^c d\xi^d + \xi^d d\xi^c) = \frac{1}{2}\oint(\xi^c d\xi^d - \xi^d d\xi^c)$.

3.3.2 For parallel transport along OQ and RS, which are geodesics, the length of λ^A and the angle it makes with the geodesic are both constant. Use these facts to verify the expressions given for λ^A_{Q} and λ^A_{S}. Along QR, use (2.26) with $\alpha = 0$, θ_0 replaced by $\theta_0 + \varepsilon$ and $t = 2\varepsilon$. Along SO, use (2.26) with $\alpha = -2\varepsilon\cos(\theta_0 + \varepsilon)$, θ_0 replaced by $\theta_0 - \varepsilon$ and $t = -2\varepsilon$, and then simplify the resulting expressions using $\cos(\theta_0 - \varepsilon) - \cos(\theta_0 + \varepsilon) = 2\sin\theta_0\sin\varepsilon$.

3.3.3 To second order, $\cos(4\varepsilon\sin\theta_0\sin\varepsilon) = 1$, giving $\Delta\lambda^1 = 0$. Because $\sin(4\varepsilon\sin\theta_0\sin\varepsilon) = 4\varepsilon^2\sin\theta_0$ to second order, the zeroth-order approximation $a\sin\theta_0$ is sufficient for the denominator $a\sin(\theta_0 - \varepsilon)$, giving $\Delta\lambda^2 = 4\varepsilon^2/a$.

3.4.1 Straightforward, but tedious, following instructions given in text.

3.5.1 Since $\delta^\mu_\mu = 4$, contraction gives $-R = \kappa T$. Put $R = -\kappa T$ in field equation (3.38).

3.7.1 Straightforward, but tedious, using given expressions for $R_{\mu\nu}$ and $\Gamma^\mu_{\nu\sigma}$.

3.7.2 $g_{00} = 1 + k/r$, $g_{11} = -1 + k(x^1)^2/(r+k)r^2$ (with similar expressions for g_{22} and g_{33}), $g_{23} = kx^2x^3/(r+k)r^2$ (with similar expressions for g_{31} and g_{12}), $g_{0i} = 0$ $(i = 1,2,3)$, where $r = \sqrt{(x^1)^2 + (x^2)^2 + (x^3)^2}$ and $k = -2MG/c^2$.

Problems

3.1 For a non-zero R_{ABCD}, $A \neq B$ and $C \neq D$. Only possibilities are R_{1212}, $R_{1221} = -R_{1212}$, $R_{2112} = -R_{1212}$ and $R_{2121} = R_{1212}$. Only non-zero connection coefficients are $\Gamma^1_{22} = \sin\theta\cos\theta$, $\Gamma^2_{12} = \Gamma^2_{21} = \cot\theta$, with $\Gamma_{122} = -a^2\sin\theta\cos\theta$, $\Gamma_{212} = \Gamma_{221} = a^2\sin\theta\cos\theta$. So $R_{1212} = \frac{1}{2}(\partial_2\partial_1 g_{21} - \partial_2\partial_2 g_{11} + \partial_1\partial_2 g_{12} - \partial_1\partial_1 g_{22}) - (\Gamma^F_{11}\Gamma_{F22} - \Gamma^F_{12}\Gamma_{F21}) = -\frac{1}{2}\partial^2(a^2\sin^2\theta)/\partial\theta^2 + (\cot\theta)(a^2\sin\theta\cos\theta) = a^2\sin^2\theta$.
For the Ricci tensor and curvature scalar, use $R_{AB} = g^{CD}R_{CABD} = g^{11}R_{1AB1} + g^{22}R_{2AB2}$ and $R = g^{AB}R_{AB} = g^{11}R_{11} + g^{22}R_{22}$, where $g^{11} = 1/a^2$ and $g^{22} = 1/a^2\sin^2\theta$.

3.2 Raising a and contracting with d gives $R_{bc} = (2-N)S_{bc} - Sg_{bc}$, where $S \equiv g^{ad}S_{ad}$, from which you can deduce (by raising b and contracting with c) that $R = 2(1-N)S$. So, if $N > 2$, you get (by eliminating S) $S_{bc} = \frac{1}{2(N-2)(N-1)}Rg_{bc} - \frac{1}{N-2}R_{bc}$, showing that $S_{bc} = S_{cb}$.
Raising a and contracting with c in $R_{abcd;e} + R_{abde;c} + R_{abec;d} = 0$ (Bianchi identity) gives
$(N-3)(S_{bd;e} - S_{be;d}) + g_{bd}(S_{;e} - S^c{}_{e;c}) - g_{be}(S_{;d} - S^c{}_{d;c}) = 0$, from which you can deduce (by raising b and contracting with d) that $2(N-2)(S_{;e} - S^c{}_{e;c}) = 0$. So if $N > 2$, then $S_{;e} - S^c{}_{e;c} = 0$ and $(N-3)(S_{bd;e} - S_{be;d}) = 0$. Hence for $N > 3$, $S_{bd;e} = S_{be;d}$.

3.3 $T^{\mu\nu}{}_{;\mu} = 0$ gives $(\rho u^\mu)_{;\mu}u^\nu + \rho u^\mu u^\nu{}_{;\mu} = 0$, and required result follows on showing that $(\rho u^\mu)_{;\mu} = 0$, for then $u^\nu{}_{;\mu}u^\mu = 0$, which is the geodesic equation. Contraction of equation above with u_ν gives $c^2(\rho u^\mu)_{;\mu} = -\rho(u^\nu{}_{;\mu}u_\nu u^\mu)$. But $u^\nu{}_{;\mu}u_\nu = 0$ (from Exercise 3.1.4), so $(\rho u^\mu)_{;\mu} = 0$, as required.

3.4 Check that $E^{\mu\nu}{}_{;\mu} = -\mu_0^{-1}[F^{\rho\mu}{}_{;\mu}F_\rho{}^\nu + F^{\rho\mu}F_\rho{}^\nu{}_{;\mu} - \frac{1}{2}g^{\mu\nu}(F_{\rho\sigma;\mu}F^{\rho\sigma})] = -\mu_0^{-1}[F^{\rho\mu}{}_{;\mu}F_\rho{}^\nu - \frac{1}{2}g^{\nu\beta}F^{\rho\mu}(F_{\beta\rho;\mu} + F_{\mu\beta;\rho} + F_{\rho\mu;\beta})]$.
So $E^{\mu\nu}{}_{;\mu} = F^\nu{}_\rho j^\rho$, on using Maxwell's equations (A.55) and (A.56) adapted to curved spacetime.

Differentiation gives $T^{\mu\nu}{}_{;\mu} = (\mu u^\mu)_{;\mu} u^\nu + \mu u^\mu u^\nu_{;\mu} + F^\nu{}_\rho j^\rho$. Continuity equation gives $(\mu u^\mu)_{;\mu} = 0$ and equation of motion gives $\mu u^\nu_{;\mu} u^\mu = -F^\nu{}_\rho j^\rho$ (see equation(A.60)), hence $T^{\mu\nu}{}_{;\mu} = 0$.

3.5 $\lambda^a_{(\text{via A})} = \lambda^a_{\text{O}} + (\Gamma^a_{bc}\lambda^b)_{\text{O}}(\xi^c + \eta^c) + (\partial_d \Gamma^a_{bc} + \Gamma^a_{ec}\Gamma^e_{bd})_{\text{O}}\lambda^b_{\text{O}}\xi^d\eta^c$, with a similar expression for $\lambda^a_{(\text{via B})}$.

3.6 Newtonian gravitational theory attributes energy to the gravitational field and recognizes mass as its source; special relativity equates energy with mass. Any theory that attempts to merge these ideas will have the gravitational field acting as its own source, which is a situation that leads to non-linear field equations.
Principle of superposition does not hold, because it applies to solutions of linear equations only.

3.7 Useful intermediate results are:
$1 - \frac{2GM}{rc^2} = \left(1 - \frac{GM}{2\rho c^2}\right)^2\left(1 + \frac{GM}{2\rho c^2}\right)^{-2}; \quad dr = \left(1 + \frac{GM}{2\rho c^2}\right)\left(1 - \frac{GM}{2\rho c^2}\right) d\rho.$

Chapter 4

Exercises

4.1.1 Use $\int\sqrt{\frac{x+a}{x+b}}\,dx = \sqrt{(x+a)(x+b)} + (a-b)\ln(\sqrt{x+a} + \sqrt{x+b}) + c.$

4.2.1 Use the expansions $(1+x)^n = 1 + nx + n(n-1)x^2/2! + \cdots$ and $\ln(1+x) = x - x^2/2 + \cdots$, which are valid for $|x| < 1$.

4.3.1 Take $r_R = \infty$ in (4.17) to get $\Delta\nu/\nu_E \approx -GM/c^2 r_E \approx -0.741 \times 10^{-3}$.

4.4.1 Equations are $-(1-2m/r)^{-1}\ddot{r} - (mc^2/r^2)\dot{t}^2 + (1 - 2m/r)^{-2}(m/r^2)\dot{r}^2 + r(\dot{\theta}^2 + \sin^2\theta\dot{\phi}^2) = 0$ and $-r^2\ddot{\theta} - 2r\dot{r}\dot{\theta} + r^2\sin\theta\cos\theta\dot{\phi}^2 = 0$, which is clearly satisfied by $\theta = \pi/2$.
Equation (4.21) follows on putting $\dot{\theta} = 0$ and $\sin\theta = 1$.

4.4.2 Straightforward, following instructions in text.

4.4.3 For radially traveling light, $dr/dt = \pm c(1 - 2m/r)$, which gives the coordinate time for the round-trip to be $\Delta t = (2/c)\int_{r_2}^{r_1}[r/(r-2m)]\,dr = (2/c)(r_1 - r_2 + 2m\ln[(r_1-2m)/(r_2-2m)])$.
Observer at r_0 sees start and finish separated by same coordinate time Δt, which gives a measured time $\Delta\tau = \sqrt{1 - 2GM/r_0}\,\Delta t$.

4.5.1 Expression for $d\phi/d\bar{u}$ gives

$$\Delta\phi = \frac{\varepsilon}{2}\int_{\bar{u}_1}^{\bar{u}_2}\frac{\bar{u}-1}{\sqrt{\beta^2-(\bar{u}-1)^2}}\,d\bar{u} + (1+\tfrac{3}{2}\varepsilon)\int_{\bar{u}_1}^{\bar{u}_2}\frac{d\bar{u}}{\sqrt{\beta^2-(\bar{u}-1)^2}}.$$

First integral can be evaluated by substituting $v^2 = \beta^2 - (\bar{u}-1)^2$ and second integral by substituting $\bar{u} - 1 = \beta\sin\psi$. This leads to (4.44). Note that $\beta^2 = (\bar{u}_1 - 1)^2 = (\bar{u}_2 - 1)^2$, so on feeding in the limits, there is no contribution from the first integral.

4.7.1 $\alpha \approx 3\pi m_\odot/r = 3\pi M_\odot G/c^2 r \approx 9.3 \times 10^{-8}\,\text{rad}$.

4.8.1 Put $ct = v - r - 2m\ln(r/2m - 1)$ to get $c\,dt = dv - (1 - 2m/r)^{-1}dr$.

4.8.2 Substitution leads to $\tau(r_0, r) = (2r_0^{3/2}/c\sqrt{2m})\int_\alpha^{\pi/2}\sin^2\psi\,d\psi$, where $\alpha \equiv \arcsin\sqrt{r/r_0}$. Evaluate this using $\sin^2\psi = \frac{1}{2}(1-\cos 2\psi)$.

4.8.3 Getting $\tau(4m, 2m)$ is straightforward, on noting that $\arcsin(1/\sqrt{2}) = \pi/4$. For $\tau(4m, r)$ we have

$$\frac{8m^{3/2}}{c\sqrt{2m}}\left(\frac{\pi}{2} - \arcsin\sqrt{\frac{r}{4m}} + \sqrt{\frac{r}{4m}}\sqrt{1-\frac{r}{4m}}\right) \to \frac{2\sqrt{2}m\pi}{c}$$

as $r \to 0$.

4.8.4 Because she cannot accelerate beyond the speed of light, Alice's world line lies inside the forward "light-cones" of null geodesics that emanate from each point on her world line. We see from Fig. 4.13 that, once she has fallen beyond $r = 2m$, all possible world lines end up at $r = 0$. Even if she could travel at the speed of light, she could not escape.

4.9.1 The differentials are:

$$du = \frac{1}{4m}(r/2m-1)^{1/2}e^{r/4m}\left(\frac{\cosh(ct/4m)}{1-2m/r}\,dr + c\sinh(ct/4m)\,dt\right),$$

$$dv = \frac{1}{4m}(r/2m-1)^{1/2}e^{r/4m}\left(\frac{\sinh(ct/4m)}{1-2m/r}\,dr + c\cosh(ct/4m)\,dt\right),$$

and $\cosh^2 x - \sinh^2 x = 1$.

4.10.1 Straightforward.

4.10.2 $[J] = ML^2T^{-1}$, $[M] = M$, and $[c] = LT^{-1}$, so $[J/Mc] = ML^2T^{-1}M^{-1}L^{-1}T = L$.

4.10.3 Check that $g_{\mu\nu}g^{\nu\sigma} = \delta_\mu^\sigma$.

4.10.4 Straightforward and almost obvious.

4.10.5 Separation of the first equation into its real and imaginary parts gives: $x = (r\cos\Phi - a\sin\Phi)\sin\theta$, $\quad y = (r\sin\Phi + a\cos\Phi)\sin\theta$, where $\Phi = \phi + a\int_\infty^r \frac{dr}{\Delta}$. Result follows by showing that
$(r\cos\Phi - a\sin\Phi) = (r^2 + a^2)^{1/2}\cos(\Phi + \beta)$,
$(r\sin\Phi + a\cos\Phi) = (r^2 + a^2)^{1/2}\sin(\Phi + \beta)$, where $\beta = \arctan(a/r)$.

4.10.6 The event horizons coincide, each tending towards the surface given by $r = m$. The infinite-redshift surfaces remain distinct, tending towards the two surfaces given by $r = m(1 \pm \sin\theta)$. All three touch at the North and South Poles, where $\theta = 0$ and $\theta = \pi$.

Problems

4.1 The "equatorial plane" has line element

$ds^2 = (1 - 2m/r)^{-1}dr^2 + r^2 d\phi^2$, which gives

$\Gamma^1_{11} = -m/r(r - 2m)$, $\Gamma^1_{22} = -(r - 2m)$, $\Gamma^2_{12} = \Gamma^2_{21} = 1/r$,

as the only non-zero connection coefficients (where $x^1 \equiv r$ and $x^2 \equiv \phi$). A routine calculation gives ${R^1}_{212} = -m/r \neq 0$, so the "equatorial plane" is not flat.

4.2 From Exercise 3.1.2, the energy is $p_\mu U^\mu = g_{\nu\mu}p^\nu U^\mu$. For the observer, $U^\mu = (dt/d\tau)\delta^\mu_0 = (1 - 2m/r_0)^{-1/2}\delta^\mu_0$, since for him $d\tau^2 = (1 - 2m/r_0)dt^2$.

So $E = (1 - 2m/r_0)^{-1/2} g_{\nu 0}p^\nu = c^2(1 - 2m/r_0)^{1/2}p^0$.

But $p^0 = \mu\dot{t} = \mu k/(1 - 2m/r_0)$, from equation(4.22), so $E = \mu k c^2(1 - 2m/r_0)^{-1/2}$.

4.3 Clearly, $E \to \mu c^2 k$ as $r_0 \to \infty$, so this is the energy of the particle according to the "observer at infinity". However, for this observer, $E^2 = p^2c^2 + \mu^2c^4$, where p is the magnitude of the particle's 3-velocity.

So $\mu^2c^4k^2 = p^2c^2 + \mu^2c^4$, which implies that $k^2 \geq 1$, as $p^2c^2 \geq 0$.

4.4 From the discussion of photons in Section 4.4, we have $(d\phi/dt)^2 = \dot{\phi}^2/\dot{t}^2 = mc^2/r^3 = c^2/27m^2$, which gives $\Delta t = 6\pi m\sqrt{3}/c$ as the change in t for an orbit at $r = 3m$. So the observer at $r = 3m$ measures a proper time period of
$\Delta\tau = \sqrt{1 - 2m/r}\,\Delta t = \Delta t/\sqrt{3} = 6\pi m/c$.

As noted in equation (4.13), the coordinate time difference between two events at the same point in space remains constant when propagated along null geodesics to a spatially fixed observer. Hence the observer at infinity measures a proper time period of $6\pi m\sqrt{3}/c$, since $\Delta t = \Delta\tau$ at infinity.

4.5 Differentiation of equation (4.39) leads to $d^2u/d\phi^2 + u = 3mu^2$, which converts to $d^2r/d\phi^2 - 2r^{-1}(dr/d\phi)^2 = r - 3m$, on putting $u = 1/r$. If we now perturb this by putting $r = 3m + \eta$, where η is small relative to $3m$, and work to first order in η and its derivatives, we get $d^2\eta/d\phi^2 - \eta = 0$. This has exponential (rather than trigonometric) solutions, indicating that η does not remain small. Hence the orbit is unstable.

4.6 The particle's energy per unit mass is $\tilde{E} = \frac{1}{2}(\dot{r}^2 + r^2\dot{\phi}^2) - GM/r$ and its angular momentum per unit mass is $h = r^2\dot{\phi}$. Dividing the equation for $\tilde{E}$ by $\dot{\phi}^2$ and then putting $\dot{\phi}^2 = h^2/r^4$ gives $\tilde{E}r^4/h^2 = \frac{1}{2}\left((dr/d\phi)^2 + r^2\right) - GMr^3/h^2$. The required result follows from putting $r = u^{-1}$, rearranging, and setting $E = 2\tilde{E}/h^2$.

4.7 $2GM_\oplus/c^2 \approx 8.89 \times 10^{-3}\,\mathrm{m}$.

4.8 $\rho_\odot/\rho_\oplus = \left(M_\odot/M_\odot^3\right) / \left(M_\oplus/M_\oplus^3\right) = M_\oplus^2/M_\odot^2 \ll 1$.

4.9 $dR = (1 + m/2\rho)^2 d\rho, \quad d\tau = (2\rho - m)(2\rho + m)^{-1}dt$.

4.10 (a) Equation (4.33) (with $GM = mc^2$) gives 5835 s for the orbital period. From Section 4.7, the geodesic effect for a circular orbit is $3\pi m/r$ radians per orbit. This leads to an annual deviation of about 6659 milliseconds.

(b) Imagine a finite-length gyroscope pointing towards a star on the celestial equator. The inner (or *back*) end of the gyroscope's axis will suffer a greater dragging than the outer (or *front*) end, and half a revolution later the front of the gyroscope will suffer a greater drag than the back (as a sketch should make clear). The drag will also vary at different parts of an equatorial circular orbit. This angular deviation, in the opposite sense to the rotation of the Earth, would be measurable; but in a polar circular orbit it is constant, and in the same sense as the rotation.

(c) Same as in (a); zero.

Chapter 5

Exercises

5.1.1 $R_{\mu\nu} \equiv R^{\alpha}{}_{\mu\nu\alpha} = \Gamma^{\alpha}_{\mu\alpha,\nu} - \Gamma^{\alpha}_{\mu\nu,\alpha} + \Gamma^{\beta}_{\mu\alpha}\Gamma^{\alpha}_{\beta\nu} - \Gamma^{\beta}_{\mu\nu}\Gamma^{\alpha}_{\beta\alpha}$,

where $\Gamma^{\mu}_{\nu\sigma} = \frac{1}{2}\left(h^{\mu}_{\sigma,\nu} + h^{\mu}_{\nu,\sigma} - h_{\nu\sigma}{}^{,\mu}\right)$.

Terms like $\Gamma^{\beta}_{\mu\alpha}\Gamma^{\alpha}_{\beta\nu}$ yield products of $h_{\mu\nu}$ (or their derivatives) and are neglected. Hence

$R_{\mu\nu} = \Gamma^{\alpha}_{\mu\alpha,\nu} - \Gamma^{\alpha}_{\mu\nu,\alpha} = \frac{1}{2}(h_{,\mu\nu} - h^{\alpha}_{\nu,\mu\alpha} - h^{\alpha}_{\mu,\nu\alpha} + h_{\mu\nu,\alpha}{}^{\alpha})$,

$R = \eta^{\mu\nu}R_{\mu\nu} = \frac{1}{2}\left(h_{,\mu}{}^{\mu} - h^{\alpha}_{\mu,\alpha}{}^{\mu} - h^{\alpha\mu}{}_{,\alpha\mu} + h^{\mu,\alpha}_{\mu}{}_{\alpha}\right) = h_{,\alpha}{}^{\alpha} - h^{\alpha\beta}{}_{,\alpha\beta}$.

5.1.2 $\bar{h} = \eta^{\mu\nu}(h_{\mu\nu} - \frac{1}{2}h\eta_{\mu\nu}) = h - 2h = -h$.

Hence, $h_{\mu\nu} = \bar{h}_{\mu\nu} + \frac{1}{2}h\eta_{\mu\nu} = \bar{h}_{\mu\nu} - \frac{1}{2}\bar{h}\eta_{\mu\nu}$.

5.1.3 $g^{\mu'\nu'} = \eta^{\mu'\nu'} - h^{\mu'\nu'} = X^{\mu'}_{\alpha}X^{\nu'}_{\beta}g^{\alpha\beta} = (\delta^{\mu}_{\alpha} + \xi^{\mu}{}_{,\alpha})(\delta^{\nu}_{\beta} + \xi^{\nu}{}_{,\beta})(\eta^{\alpha\beta} - h^{\alpha\beta})$.

Multiplying out, neglecting products of small quantities, and rearranging lead to $h^{\mu'\nu'} = h^{\mu\nu} - \xi^{\mu,\nu} - \xi^{\nu,\mu}$.

Contracting with $\eta_{\mu\nu}$ then gives $h' = h - 2\xi^{\mu}{}_{,\mu}$.

Also, $\bar{h}^{\mu'\nu'} \equiv h^{\mu'\nu'} - \frac{1}{2}h'\eta^{\mu\nu}$

$= (h^{\mu\nu} - \xi^{\mu,\nu} - \xi^{\nu,\mu}) - \frac{1}{2}\eta^{\mu\nu}(h - 2\xi^{\alpha}{}_{,\alpha}) = \bar{h}^{\mu\nu} - \xi^{\mu,\nu} - \xi^{\nu,\mu} + \eta^{\mu\nu}\xi^{\alpha}{}_{,\alpha}$.

To show that $X^{\mu}_{\nu'} = \delta^{\mu}_{\nu} - \xi^{\mu}{}_{,\nu}$, check that $X^{\rho'}_{\mu}X^{\mu}_{\nu'} = \delta^{\rho}_{\nu}$.

5.1.4 Equation (5.19) shows that $\kappa T^{\mu\nu}$ is a small quantity of the same order of magnitude as $h_{\mu\nu}$. Moreover, the proper conclusion from (5.19) and (5.20) is that $\kappa T^{\mu\nu}{}_{;\nu} = 0$. This does not conflict with $\kappa T^{\mu\nu}{}_{,\nu} = 0$, as the difference between the two expressions consists of products of the form connection coefficient times $\kappa T^{\mu\nu}$, which are negligible second-order quantities.

5.2.1 Equation (5.23) gives $A^{\mu 0}k_0 + A^{\mu j}k_j = 0$, so $A^{00}k_0 + A^{0j}k_j = 0$ (1) and $A^{i0}k_0 + A^{ij}k_j = 0$ (2). Because k_μ is null, $k_0 \neq 0$, so we can divide by it. Hence (2) allows us to express A^{i0} in terms of A^{ij}, and (1) allows us to express A^{00} in terms of A^{0j} and hence in terms of A^{ij}. (Note that $A^{\mu\nu}$ is symmetric.)

5.2.2 $\Gamma^{\mu}_{00} = \frac{1}{2}\eta^{\mu\nu}(2h_{\nu 0,0} - h_{00,\nu}) = 0$, as $h_{0\mu} = \bar{h}_{0\mu} = 0$ in the TT gauge.

Also, $\Gamma^{\mu}_{0\nu} = \frac{1}{2}\eta^{\mu\alpha}(0 + h_{\nu\alpha,0} + 0) = \frac{1}{2}h^{\mu}_{\nu,0}$.

5.2.3 To first order in $h_{\mu\nu}$, $\delta_{ij}\zeta^i\zeta^j = \delta_{ij}(\xi^i + \frac{1}{2}h^i_k\xi^k)(\xi^j + \frac{1}{2}h^j_l\xi^l)$

$= \delta_{ij}\xi^i\xi^j + \frac{1}{2}\delta_{ij}(h^i_k\xi^k\xi^j + h^j_l\xi^i\xi^l) = \delta_{ij}\xi^i\xi^j + \delta_{ij}h^i_k\xi^k\xi^j$

$= \delta_{ij}\xi^i\xi^j - \eta_{ij}h^i_k\xi^k\xi^j = \delta_{ij}\xi^i\xi^j - h_{jk}\xi^j\xi^k$

$= (\delta_{ij} - h_{ij})\xi^i\xi^j = \bar{g}_{ij}\xi^i\xi^j = d^2$.

5.2.4 For (5.33), we have $\zeta^i = \xi^i + \frac{1}{2}\eta_{k\mu}h^{i\mu}\xi^k = \xi^i - \frac{1}{2}\delta_{kj}h^{ij}\xi^k$.
Substituting for $h^{ij} = \bar{h}^{ij}$ gives $\zeta^i = \xi^i - \frac{1}{2}\alpha\delta_{kj}e_1^{ij}\cos k(x_0 - x_3)\,\xi^k$, which yields $\zeta^1 = \xi^1 - \frac{1}{2}\alpha\cos k(x_0 - x_3)\,\xi^1$,
$\zeta^2 = \xi^2 + \frac{1}{2}\alpha\cos k(x_0 - x_3)\,\xi^2$ and $\zeta^3 = \xi^3 = 0$, as required.
A similar chain of substitutions verifies (5.34).

5.2.5 The effect is to substitute $|\alpha|\cos(k(x_0 - x_3) + \theta)$ for $\alpha\cos k(x_0 - x_3)$, wherever it occurs. This gives a phase-shift, which rotates all the figures in the table through θ.

5.3.1 Set $x^j = \pm(0, a\cos\omega t, a\sin\omega t)$, so that the dumbbell rotates about the x^1 axis. For the plane-wave approximation, we get
$$\left[\bar{h}^{ij}\right] \approx \frac{8GMa^2\omega^2}{c^4 r}\Re\left[e^{ij}\exp\frac{2i\omega}{c}(x^0 - x^3)\right],$$
where $\left[e^{ij}\right] = \begin{bmatrix} 0 & 0 & 0 \\ 0 & 1 & -i \\ 0 & -i & -1 \end{bmatrix}$, in place of (5.43).
On transforming to the TT gauge, we have $A^{1'1'} = \frac{1}{2}(A^{11} - A^{22})$ and $A^{1'2'} = A^{12}$, which gives $\left[e^{ij}\right] = \begin{bmatrix} -\frac{1}{2} & 0 & 0 \\ 0 & \frac{1}{2} & 0 \\ 0 & 0 & 0 \end{bmatrix} = -\frac{1}{2}\left[e_1^{ij}\right]$, so the wave is linearly polarized.

5.3.2 (a) Use (5.44) with $I = \text{mass} \times \text{length}^2/12$.
(b) If r is the radius of the orbit, then $GM_\odot^2/(2r)^2 = M_\odot r\omega^2$.
So $r^3 = GM_\odot/4\omega^2$, giving $I = 2M_\odot(GM_\odot/4\omega^2)^{2/3}$ for the moment of inertia. Now use (5.44) with this expression for I and a value for ω calculated from the given period.

Problems

5.1 With $\dot{x}^\mu = c\delta_0^\mu$, (3.41) becomes $D^2\xi^\mu/d\tau^2 + c^2 R^\mu{}_{0\nu0}\xi^\nu = 0$, where (using the result of Exercise 5.2.2)
$R^\mu{}_{0\nu0} = \Gamma^\mu_{00,\nu} - \Gamma^\mu_{0\nu,0} + \Gamma^\rho_{00}\Gamma^\mu_{\rho\nu} - \Gamma^\rho_{0\nu}\Gamma^\mu_{\rho0} = -\frac{1}{2}h^\mu_{\nu,00} - \frac{1}{4}h^\rho_{\nu,0}h^\mu_{\rho,0}$.
So the equation of geodesic deviation is
$D^2\xi^\mu/d\tau^2 - c^2(\frac{1}{2}h^\mu_{\nu,00} + \frac{1}{4}h^\rho_{\nu,0}h^\mu_{\rho,0})\xi^\nu = 0$. (*)
But $\xi^\mu = \text{const}$ gives
$D\xi^\mu/d\tau = 0 + \Gamma^\mu_{\nu\sigma}\xi^\nu\dot{x}^\sigma = \frac{1}{2}ch^\mu_{\nu,0}\xi^\nu$, which gives
$D^2\xi^\mu/d\tau^2 = \frac{1}{2}c^2h^\mu_{\nu,00}\xi^\nu + \frac{1}{2}c\Gamma^\mu_{\rho\lambda}h^\rho_{\nu,0}\xi^\nu\dot{x}^\lambda =$
$\frac{1}{2}c^2h^\mu_{\nu,00}\xi^\nu + \frac{1}{4}c^2h^\mu_{\rho,0}h^\rho_{\nu,0}\xi^\nu$, showing that (*) is satisfied.

5.2 Taking $A^{\mu\nu} = \alpha(e_1^{\mu\nu} + ie_2^{\mu\nu})$ and putting $\theta = k(x^0 - x^3)$ gives

$\zeta^i = \left(\xi^1 - \frac{1}{2}\alpha(\xi^1 \cos\theta - \xi^2 \sin\theta), \xi^2 + \frac{1}{2}\alpha(\xi^1 \sin\theta + \xi^2 \cos\theta), 0\right)$
as the counterpart to (5.33) or (5.34). For clarity, put $\zeta^i = (x, y, z)$ and $\xi^1 = a$, $\xi^2 = b$, so that the relative separation is given by (x, y, z) with $x = a - \frac{1}{2}\alpha(a\cos\theta - b\sin\theta)$, $y = b + \frac{1}{2}\alpha(a\sin\theta + b\cos\theta)$ and $z = 0$. These yield $(x-a)^2 + (y-b)^2 = \frac{1}{4}\alpha^2(a^2+b^2)$, which is the equation of a circle.

5.3 Equation (5.36) gives $\bar{h}^{\mu\nu} = -(4G/c^4 r)Mc^2\delta^\mu_0\delta^\nu_0 = -(4m/r)\delta^\mu_0\delta^\nu_0$. So $\bar{h}_{\mu\nu} = -(4m/r)\delta^0_\mu\delta^0_\nu$ and $\bar{h} = -4m/r$. Hence $g_{\mu\nu}$
$= \eta_{\mu\nu} + \bar{h}_{\mu\nu} - \frac{1}{2}\bar{h}\eta_{\mu\nu} = (1 + 2m/r)\eta_{\mu\nu} - (4m/r)\delta^0_\mu\delta^0_\nu$, which gives
$c^2 d\tau^2 = (1 - 2m/r)c^2 dt^2 - (1 + 2m/r)(dx^2 + dy^2 + dz^2)$.
For large r, $1/r \approx 1/\rho$, so the above gives
$c^2 d\tau^2 = (1 - 2m/\rho)c^2 dt^2 - (1 + 2m/\rho)(d\rho^2 + \rho^2 d\theta^2 + \rho^2 \sin^2\theta d\phi^2)$,
which is in agreement with (4.67) when we use the approximations $(1 - m/2\rho)^2(1 + m/2\rho)^{-2} \approx 1 - 2m/\rho$ and $(1 + m/2\rho)^4 \approx 1 + 2m/\rho$, valid for large ρ.

5.4 Take the positions of the particles at time t to be
$x^j = \pm(a\cos\omega t, a\sin\omega t, 0)$ and $x^j = \pm(-a\sin\omega t, a\cos\omega t, 0)$.
Then in place of the matrix in (5.42) we get

$$\begin{bmatrix} \cos^2\omega t & \cos\omega t \sin\omega t & 0 \\ \cos\omega t \sin\omega t & \sin^2\omega t & 0 \\ 0 & 0 & 0 \end{bmatrix}_{\text{ret}} + \begin{bmatrix} \sin^2\omega t & -\sin\omega t\cos\omega t & 0 \\ -\sin\omega t\cos\omega t & \cos^2\omega t & 0 \\ 0 & 0 & 0 \end{bmatrix}_{\text{ret}} = \begin{bmatrix} 1 & 0 & 0 \\ 0 & 1 & 0 \\ 0 & 0 & 0 \end{bmatrix}_{\text{ret}}.$$

We therefore get a static metric tensor and no radiation.

Chapter 6

Exercises

6.0.1 $z \equiv \Delta\lambda/\lambda_0 = \lambda/\lambda_0 - 1 = \gamma(1 + v/c) - 1$, from (A.41). To first order in v/c, this gives $z = v/c$.

6.2.1 Straightforward, but tedious, using the connection coefficients (6.4) and the given expression for $R_{\mu\nu}$.

6.2.2 The derivative of (6.8) is $2\dot{R}\ddot{R} = (8\pi G/3)(\dot{\rho}R^2 + 2\rho R\dot{R})$. Eliminate $\ddot{R}$ from this and (6.6).

6.2.3 With $u^\mu = \delta^\mu_0$, LHS of (6.10) $= (\rho + p)u^\nu{}_{;0} = (\rho + p)\Gamma^\nu_{00} = 0$.
As p depends only on t, $p_{,\mu} = \dot{p}\delta^0_\mu$, so
RHS of (6.10) $= (g^{\mu\nu} - \delta^\mu_0\delta^\nu_0)\dot{p}\delta^0_\mu = (g^{0\nu} - \delta^\nu_0)\dot{p} = 0$.

6.3.1 Straightforward checks.

6.4.1 A quick way of getting the spatial geodesic equations is to put $\dot{t} = R' = 0$ in the last three of equations (2.20) and note that dots now represent differentiation with respect to s. The resulting equations are satisfied when $\dot{\theta} = \dot{\phi} = 0$ provided $\ddot{r} + kr(1 - kr^2)^{-1}\dot{r}^2 = 0$, whose solution gives r in terms of s.

6.4.2 The null geodesic equations are given by equations (2.20), where dots denote differentiation with respect to an affine parameter u, say. These are satisfied when $\dot{\theta} = \dot{\phi} = 0$ provided $\ddot{t} + RR'(1 - kr^2)^{-1}\dot{r}^2 = 0$ and $\ddot{r} + 2R'R^{-1}\dot{t}\dot{r} + kr(1 - kr^2)^{-1}\dot{r}^2 = 0$. This pair can be solved simultaneously to give t and r in terms of u.

6.4.3 Let $Q = L_G(t_R)/c$, so that $Q = \Delta t + \frac{1}{2}H_0(\Delta t)^2 + \cdots$ (*). Assume an expansion of Δt in powers of Q: $\Delta t = a_0 + a_1 Q + a_2 Q^2 + \cdots$, then substitute in (*) to deduce that $a_0 = 0$, $a_1 = 1$, and $a_2 = -\frac{1}{2}H_0$, as required.

6.5.1 Straightforward, using pointers in text.

6.5.2 Equation (6.48) can be rearranged to give $z = (2c/(2c - v))^2 - 1$, which shows that, as $v \to c$, $z \to \infty$. So there is no upper bound on the observable redshift.

6.6.1 Put $T^{\mu\nu} = 0$ in (6.50) and contract with $g_{\mu\nu}$ to get $R - 2R + 4\Lambda = 0$, which gives $R = 4\Lambda$. Then put $R = 4\Lambda$ back in the field equation.

6.6.2 Use equation (6.14), the fact that $A^2 = 8\pi G\rho R^3/3$, and the defining equations $H = \dot{R}/R$ and $\rho_c = 3H^2/8\pi G$.

6.7.1 Straightforward exercise on scaling.

Problems

6.1 The non-zero components of the Ricci tensor are $R_{11} = -2k/(1 - kr^2)$, $R_{22} = -2kr^2$, $R_{33} = -2kr^2 \sin^2\theta$. (They can be got from (6.5) by setting $R(t) = 1$.) So $R = g^{ij}R_{ij} = -2k - 2k - 2k = -6k$.

6.2 From (6.14), $R(\dot{R}^2 + k) = A^2$. Differentiating this leads to $\dot{R}^2 + k + 2R\ddot{R} = 0$, which gives (on using present-day values) $\dot{R}_0^2 + k - 2q_0\dot{R}_0^2 = 0$ (*). Substituting $\dot{R}_0^2 = A^2/R_0 - k$ and solving for $A^2/2$ gives the first expression.
The second expression can be obtained from the first by noting that (*) yields $H_0^2 R_0^2 = k/(2q_0 - 1)$.

From (6.19), $R_0 = \frac{1}{2}A^2(1 - \cos\psi_0)$, so (as $k = 1$)
$1 - \cos\psi_0 = (2q_0 - 1)/q_0$. Hence, $\cos\psi_0 = (1 - q_0)/q_0$.
A similar argument (using (6.20)) gives $\cosh\psi_0 = (1 - q_0)/q_0$ for the open model.
Finally, for the closed model (with $k = 1$), the age t_0 is obtained from
$t_0 = \frac{1}{2}A^2(\psi_0 - \sin\psi_0) = (q_0/H_0)(2q_0 - 1)^{-3/2}(\psi_0 - \sin\psi_0)$.
With $q_0 \approx 1$, $\psi_0 \approx \pi/2$, so $t_0 \approx (\pi/2 - 1)/H_0$.
The age for the open model can be found in a similar way.

6.3 (a) Just under 36×10^9 light yrs. (b) Just under 96×10^9 yrs and just under 144×10^9 light yrs. (c) Extremely large ($v = 2c$ gives $z = \infty$) and $z = 3$.

6.4 Use equation (6.47).

6.5 To three significant figures, the answers are:
(a) 2.36 hrs; (b) 1.58 km.

6.6 (a) Use equation (6.39).
(b) 9.7×10^9 yrs, compared with 10.9×10^9 yrs.

6.7 12.4×10^9 yrs.

6.8 2969 K.

Appendix A

Exercises

A.1.1 The first and third of the expressions involving A, B and C give $B^2C^2c^4 = (c^2 + A^2v^2)(A^2 - 1)$, while the second expression gives $B^2C^2c^4 = A^4v^2$. Subtraction then leads to $A^2(c^2 - v^2) = c^2$, from which we get $A = (1 - v^2/c^2)^{1/2}$. The expressions for B and C then follow easily. (The signs taken for the square roots should be justified.)

A.1.2 The inverse matrix is $\begin{bmatrix} \gamma & v\gamma/c & 0 & 0 \\ v\gamma/c & \gamma & 0 & 0 \\ 0 & 0 & 1 & 0 \\ 0 & 0 & 0 & 1 \end{bmatrix}$.

The velocity of K relative to K' is $-v$.

A.1.3 If v/c is negligible, then $\gamma = 1$, and the Galilean equations follow.

A.2.1 Use the addition formulae $\cosh(\psi + \phi) = \cosh\psi\cosh\phi + \sinh\psi\sinh\phi$ and $\sinh(\psi + \phi) = \sinh\psi\cosh\phi + \cosh\psi\sinh\phi$.

A.2.2 The result follows from the identity
$(1 - v/c)(1 - w/c) = (1 + vw/c^2) - (v + w)/c$. For $v < c$ and $w < c$ makes the LHS positive, showing that $v + w < c(1 + vw/c^2)$.

A.5.1 Take λ^μ to be the position vector of a point P in spacetime, so that its coordinates are given by $ct = \lambda^0$, $x = \lambda^1$, $y = \lambda^2$, $z = \lambda^3$. Then
$\eta_{\mu\nu}\lambda^\mu\lambda^\nu > 0 \iff x^2 + y^2 + z^2 < c^2t^2 \iff$ P inside the cone,
$\eta_{\mu\nu}\lambda^\mu\lambda^\nu = 0 \iff x^2 + y^2 + z^2 = c^2t^2 \iff$ P on the cone,
$\eta_{\mu\nu}\lambda^\mu\lambda^\nu < 0 \iff x^2 + y^2 + z^2 > c^2t^2 \iff$ P outside the cone.

A.5.2 Substitute $t = \gamma(t' + xv'^2/c^2)$, $x = \gamma(x' + vt')$, $y = y'$, $z = z'$ into (A.20).

A.6.1 (a) $u^\mu = (c, 0, 0, 0)$, (b) $u^\mu = (\gamma c, \gamma \mathbf{v})$.
No, because $d\tau = 0$ for neighboring points on a photon's path.

A.6.2 Use the fact that $u_\mu u^\mu = c^2$ implies (on differentiating) that $u_\mu du^\mu/d\tau = 0$.
From (A.24) and (A.30), $u_\mu f^\mu = \gamma^2(\mathbf{F}\cdot\mathbf{v} - \mathbf{v}\cdot\mathbf{F}) = 0$.

A.6.3 Use $\cos^2\phi + \sin^2\phi = 1$ to get
$\gamma^2 m^2 v^2 = (\nu^2 - 2\nu\bar{\nu}\cos\theta + \bar{\nu}^2)h^2/c^2$ and equate this with
$\gamma^2 m^2 v^2 = m^2(\gamma^2 - 1)c^2 = (h\nu/c + mc - h\bar{\nu}/c)^2 - m^2c^2$.

A.7.1 $\tan\theta' = \dfrac{k^{2'}}{k^{1'}} = \dfrac{k^2}{-(\gamma v/c)k^0 + \gamma k^1} = \cdots = \dfrac{\tan\theta}{\gamma(1 - (v/c)\sec\theta)}$.

A.8.1 $A^\mu{}_{,\mu} = \partial A^0/\partial(ct) + \partial A^1/\partial x + \partial A^2/\partial y + \partial A^3/\partial z$
$= \nabla\cdot\mathbf{A} + c^{-2}\partial\phi/\partial t = \nabla\cdot\mathbf{A} + \mu_0\varepsilon_0\partial\phi/\partial t$.

A.8.2 $F_{01} = A_{0,1} - A_{1,0} = \partial(\phi/c)/\partial x - \partial A_1/\partial(ct)$
$= (\partial\phi/\partial x - \partial A_1/\partial t)/c = (\partial\phi/\partial x + \partial A^1/\partial t)/c = -E^1/c$,
and similarly for the other components.

A.8.3 Use $F^\mu{}_\nu = \eta^{\mu\sigma}F_{\sigma\nu}$ and $F^{\mu\nu} = F^\mu{}_\sigma\eta^{\sigma\nu}$.

A.8.4 It is straightforward to show that $\mu = 0$ in (A.55) gives (A.44) and that $\mu = 1, 2, 3$ gives (A.46). Note that (A.56) comprises just four distinct non-trivial equations, which can be obtained by taking (μ, ν, σ) equal to $(1, 2, 3)$, $(0, 2, 3)$, $(0, 1, 3)$ and $(0, 1, 2)$ in turn. In this way, (A.43) and (A.45) can be derived.

A.8.5 Use (A.26) and the fact that $dt/d\tau = \gamma$.

Problems

A.1 $\Delta\tau/\Delta t = 3/3.000\,000\,015 = \sqrt{1 - v^2/c^2} \;\Rightarrow\; v/c \approx 10^{-4}$.
This gives $v \approx 3 \times 10^4\,\mathrm{m\,s^{-1}}$, or about 67,500 mph.

A.3 $\lambda = \lambda_0/\sqrt{3} \approx 365.35\,\mathrm{nm}$.

A.4 The matrix version of $F^{\mu'\nu'} = F^{\alpha\beta}\Lambda^{\mu'}_{\alpha}\Lambda^{\nu'}_{\beta}$ is $F' = \Lambda F \Lambda^{\mathrm{T}}$. Use this matrix equation with F and F' given by (A.63) and Λ by (A.13).

A.5 With $0 \le v < c$, we get a graph that increases with v, having the value one when $v = 0$ and tending to infinity as $v \to c$.

A.6 Let K be the lab frame and K' the rocket frame. Because the switch is fixed in the lab frame, $\Delta x = 0$, so
$\Delta t' = \gamma\,\Delta t \;\Rightarrow\; \gamma = 5/3 \;\Rightarrow\; v/c = 4/5$.
Hence $\Delta x' = -\gamma v\,\Delta t = -(4/3)c\,\Delta t = -4c = -12 \times 10^8\,\mathrm{m}$ and
$v = 4c/5 = 2.4 \times 10^8\,\mathrm{m\,s^{-1}}$.

A.7 Use $j^{\mu'} = \Lambda^{\mu'}_{\nu} j^{\nu}$, where $j^{\mu} = (\rho_0, 0, 0, 0)$ and $\left[\Lambda^{\mu'}_{\nu}\right]$ is the matrix of equation (A.13).

A.8 Kinetic energy $= m\gamma c^2 - mc^2 = mc^2(\sqrt{4/3} - 1) \approx 9.75 \times 10^{17}$ joules.
Momentum $= \mathbf{p} = \gamma m\mathbf{v} = \gamma m(-c/2, 0, 0)$, which has magnitude
$1.21 \times 10^{10}\,\mathrm{kg\,m\,s^{-1}}$.

A.9 Use the hint to show that $u^{\mu}_{\mathrm{source}}k_{\mu} = 2\pi/\lambda_0$ and $u^{\mu}k_{\mu} = 2\pi/\lambda$.

A.10 Time dilation gives a laboratory half-life of
$\Delta t = \gamma\,\Delta\tau = 12.55 \times 10^{-8}$ sec, in which time a pion travels
$0.99c \times 12.55 \times 10^{-8} = 37.3\,\mathrm{m}$, in agreement with the observation in the laboratory.
In the pions' rest frame, we have the contracted distance
$37.3/\gamma = 5.26\,\mathrm{m}$, which is covered by the lab in a time
$5.26/0.99c = 1.77 \times 10^{-8}\,\mathrm{s}$, in agreement with the half-life of a stationary pion.

A.11 Work in the rest frame of the material, where $u^{\mu} = (c, 0, 0, 0)$. For $\mu = 0$, the equation reduces to $0 = 0$, but for $\mu = i$ it gives
$j^i = \sigma c F^{i0}$. That is, $\mathbf{J} = \sigma\mathbf{E}$, which is Ohm's law.

A.12 Use a non-rotating frame, with origin at the center of the Earth. For an eastbound flight, we have $v_E = v_\oplus + v_A$, while for a westbound flight, we have $v_W = v_\oplus - v_A$, where $v_\oplus$ is the speed due to the

Earth's rotation, and v_A is the speed of the aircraft relative to the Earth. Note that $\Delta\tau_\oplus = \sqrt{1 - v_\oplus^2/c^2}\,\Delta t$,
$\Delta\tau_E = \sqrt{1 - v_E^2/c^2}\,\Delta t$, and $\Delta\tau_W = \sqrt{1 - v_W^2/c^2}\,\Delta t$.

Appendix B

Exercises

B.2.1 Working to first order in ε gives

$$\begin{aligned}\delta s_L &= \{1 + \varepsilon g_{AB}\dot{x}^A\dot{n}^B + \varepsilon g_{AB}\dot{n}^A\dot{x}^B + \varepsilon\partial_D g_{AB} n^D\dot{x}^A\dot{x}^B\}^{1/2}\delta s \\ &= \{1 + 2\varepsilon g_{AB}\dot{x}^A\dot{n}^B + \varepsilon\partial_D g_{AB} n^D\dot{x}^A\dot{x}^B\}^{1/2}\delta s \\ &= \{1 + \varepsilon g_{AB}\dot{x}^A\dot{n}^B + \tfrac{1}{2}\varepsilon\partial_D g_{AB} n^D\dot{x}^A\dot{x}^B\}\delta s.\end{aligned}$$

B.2.2

$$\begin{aligned}t_B Dn^B/ds &= g_{AB}\dot{x}^A\left(\dot{n}^B + \Gamma^B_{CD} n^C\dot{x}^D\right) \\ &= g_{AB}\dot{x}^A\dot{n}^B + \Gamma_{ACD} n^C\dot{x}^A\dot{x}^D.\end{aligned}$$

Substitute $\Gamma_{ACD} = \frac{1}{2}(\partial_C g_{AD} + \partial_D g_{CA} - \partial_A g_{CD})$ and cancel two terms to get (B.9).

B.2.3 Straightforward, using $t_A t^A = 1$, $t_A Dt^A/ds = 0$, and $t_A n^A = 0$.

Appendix C

Exercises

C.1.1 From (C.6), $\mathbf{e}_b = X^{c'}_b \mathbf{e}_{c'}$ and substitution in (C.5) gives $\mathbf{e}_{a'} = X^b_{a'} X^{c'}_b \mathbf{e}_{c'}$. By uniqueness of components we have $X^b_{a'} X^{c'}_b = \delta^c_a$.

C.2.1 Let f, g be linear functionals. Then for all $\alpha, \beta \in \mathbb{R}$ and $\mathbf{u}, \mathbf{v} \in T$,
$(f + g)(\alpha\mathbf{u} + \beta\mathbf{v})$

$$\begin{aligned}&= f(\alpha\mathbf{u} + \beta\mathbf{v}) + g(\alpha\mathbf{u} + \beta\mathbf{v}) && \text{(by definition of the sum)} \\ &= \alpha f(\mathbf{u}) + \beta f(\mathbf{v}) + \alpha g(\mathbf{u}) + \beta g(\mathbf{v}) && \text{(since } f \text{ and } g \text{ are linear)} \\ &= \alpha(f(\mathbf{u}) + g(\mathbf{u})) + \beta(f(\mathbf{v}) + g(\mathbf{v})) \\ &= \alpha(f + g)(\mathbf{u}) + \beta(f + g)(\mathbf{v}),\end{aligned}$$

showing that the sum $f + g$ is a linear functional.

C.2.2 $\boldsymbol{\mu} = \mu_b\mathbf{e}^b = \mu_b X^b_{a'}\mathbf{e}^{a'}$ (from (C.17)) $\Rightarrow \mu_{a'} = X^b_{a'}\mu_b$.
Similarly for the second of equations (C.18).

C.2.3 The identification of T^{**} with T means that (as a vector in T^{**}) $\boldsymbol{\lambda} = \lambda^a\mathbf{f}_a$, so $\boldsymbol{\lambda}(\boldsymbol{\mu}) = \lambda^a\mathbf{f}_a(\mu_b\mathbf{e}^b) = \lambda^a\mu_b\mathbf{f}_a(\mathbf{e}^b) = \lambda^a\mu_b\delta^b_a = \lambda^a\mu_a$.

C.3.1 For $\sigma, \tau \in \mathbb{R}$ and $\mathbf{u}, \mathbf{v} \in T^*$, the components of $\sigma\mathbf{u} + \tau\mathbf{v}$ relative to $\{\mathbf{e}^a\}$ are $\sigma u_a + \tau v_a$, so $\mathbf{e}_{a\alpha}(\sigma\mathbf{u} + \tau\mathbf{v}, \mathbf{w}) = (\sigma u_a + \tau v_a)w_\alpha = \sigma(u_a w_\alpha) + \tau(v_a w_\alpha) = \sigma\mathbf{e}_{a\alpha}(\mathbf{u}, \mathbf{w}) + \tau\mathbf{e}_{a\alpha}(\mathbf{v}, \mathbf{w})$.
Similarly we can show that
$\mathbf{e}_{a\alpha}(\mathbf{v}, \gamma\mathbf{w} + \delta\mathbf{x}) = \gamma\mathbf{e}_{a\alpha}(\mathbf{v}, \mathbf{w}) + \delta\mathbf{e}_{a\alpha}(\mathbf{v}, \mathbf{x})$.

C.3.2 $\tau^{a'\alpha'} = \boldsymbol{\tau}(\mathbf{e}^{a'}, \mathbf{f}^{\alpha'})$ (by definition)

$$= \boldsymbol{\tau}(X_b^{a'}\mathbf{e}^b, Y_\beta^{\alpha'}\mathbf{f}^\beta) = X_b^{a'}Y_\beta^{\alpha'}\boldsymbol{\tau}(\mathbf{e}^b, \mathbf{f}^\beta) \quad \text{(using bilinearity)}$$

$$= X_b^{a'}Y_\beta^{\alpha'}\tau^{b\beta}.$$

C.3.3 Using primed bases, we would make the definition $\boldsymbol{\lambda} \otimes \boldsymbol{\mu} = \lambda^{b'}\mu^{\beta'}\mathbf{e}_{b'\beta'}$. But $\lambda^{b'}\mu^{\beta'}\mathbf{e}_{b'\beta'} = (X_a^{b'}\lambda^a)(Y_\alpha^{\beta'}\mu^\alpha)(X_{b'}^c Y_{\beta'}^\gamma \mathbf{e}_{c\gamma}) = \lambda^a\mu^\alpha\delta_a^c\delta_\alpha^\gamma\mathbf{e}_{c\gamma} = \lambda^a\mu^\alpha\mathbf{e}_{a\alpha}$, showing agreement with the definition using unprimed bases.

C.6.1 Let $x_0 = \psi(\mathrm{P})$. The inverse-function theorem asserts that there exist open sets $X \subset S$ and $Y = f(X) \subset T$ with $x_0 \in X$, such that the restriction of f to X has a differentiable inverse $g : Y \to X$. Then for the chart (U', ψ'), take $U' = \psi^{-1}(g(Y)) = \psi^{-1}(X) \subset U$ and $\psi' = f \circ \psi$, with ψ restricted to U'.

C.7.1 Because the derivative is nowhere zero, f is either a strictly increasing, or a strictly decreasing function, and has a differentiable inverse g, say. Suppose (for convenience) that it is increasing, so that $f(a) \le u' \le f(b)$. Then g is also increasing and $du/du' = g'(u') > 0$. The chain rule gives $dx^a/du' = (dx^a/du)(du/du')$, showing that $(dx^a/du')_P \neq 0$ at all points P. So u' is also a regular parameter and the tangent vector $(dx^a/du')_P$ is proportional to $(dx^a/du)_P$.
The strictly decreasing case is covered by a similar argument.

References

Anderson, J.D., Esposito, P.B., Martin, W., and Thornton, C. L. (1975) "Experimental test of general relativity using time-delay data from *Mariner 6* and *Mariner 7*," *Astrophys. J.*, **200,** 221–33.

Anderson, J.L. (1967) *Principles of Relativity Physics*, Academic Press, New York.

Apostol, T.M. (1974) *Mathematical Analysis*, 2nd ed., Addison–Wesley, Reading, Mass.

Birkhoff, G. and Mac Lane, S. (1977) *A Survey of Modern Algebra*, 4th ed., Macmillan, New York. (Now available as an AKP Classic, A.K. Peters, Wellesley, MA.)

Biswas, T. (1994) "Special Relativistic Newtonian Gravity," *Found. Phys.*, **24,** 4, 513–524.

Bondi, H. (1960) *Cosmology*, 2nd ed., Cambridge University Press, Cambridge.

Bondi, H. and Gold, T. (1948) "The steady state theory of the expanding universe," *Mon. Not. R. Astr. Soc.*, **108,** 252–70.

Boyer, R.H. and Lindquist, R.W. (1967) "Maximal analytic extension of the Kerr Metric," *J. Math. Phys.*, **8,** 265–81.

Carmeli, M., Fickler, S.I., and Witten, L., eds. (1970) *Relativity*, Plenum, New York.

Charlier, C.V.I. (1922) "How an infinite world may be built up," *Ark. Mat. Astron. Fys.*, **16,** No. 22.

Cousins, F.W. (1955) "A mystery of ancient China: The riddle of the south-seeking chariot," *Meccano Magazine*, **XL,** 498–9.

Damour, T. and Deruelle, N. (1985) "General relativistic celestial mechanics of binary systems. 1. The post-Newtonian motion," *Ann. Inst. H. Poincaré*, **43,** 107–32.

Damour, T. and Deruelle, N. (1986) "General relativistic celestial mechanics of binary systems. 2. The post-Newtonian timing formula," *Ann. Inst. H. Poincaré*, **44,** 263–92.

Duncombe, R.L. (1956) "Relativity effects for the three inner planets," *Astronom. J.*, **61,** 174–5.

Feenberg, E. (1959) "Doppler effect and time dilatation," *Am. J. Phys.*, **27,** 190.

Friedmann, A. (1922) "Über die Krümmung des Raumes," *Z. Phys.*, **10,** 377–86.

Goldstein, H., Poole, C., and Safko, J. (2002) *Classical Mechanics*, 3rd ed., Addison–Wesley, Reading, Mass.

Guth, A.H. (1981) "Inflationary Universe, a possible solution to the horizon and flatness problems," *Phys. Rev.*, **D23,** 347–56.

Hafele, J.C. and Keating, R.E. (1972) "Around-the-world atomic clocks: Observed relativistic time gains," *Science*, **177,** 168–70.

Halmos, P. R. (1974) *Finite-Dimensional Vector Spaces*, Springer-Verlag, New York.

Hawking, S.W. and Ellis, G.F.R. (1973) *The Large Scale Structure of Space-Time*, Cambridge University Press, Cambridge.

Hoffmann, B. (1972) *Albert Einstein: Creator and Rebel*, New American Library, New York.

Horgan, J. (1987) "Big-Bang Bashers," *Scientific American,* **257,** September 1987, 22–24.

Hoyle, F. (1948) "A new model for the expanding universe," *Mon. Not. R. Astr. Soc.*, **108,** 372–82.

Hulse, R.A. and Taylor, J.H. (1975) "Discovery of a pulsar in a binary system," *Astrophys. J.*, **195,** L51–3.

Kerr, R.P. (1963) "Gravitational field of a spinning mass as an example of algebraically special metrics," *Phys. Rev. Lett.*, **11,** 237–38.

Kilmister, C.W. (1973) *General Theory of Relativity*, Pergamon, Oxford.

Landau, L.D. and Lifshitz, E.M. (1980) *The Classical Theory of Fields*, 4th ed., Butterworth–Heinemann, Oxford.

Landau, L.D. and Lifshitz, E.M. (1987) *Fluid Mechanics*, 2nd ed, Butterworth–Heinemann, Oxford.

Logan, J.D. (1987) *Applied Mathematics: a Contemporary Approach*, Wiley, New York.

Mach, E. (1893) *The Science of Mechanics*, translation of the 9th German ed. by T.J. McCormack, Open Court Publishing Company, LaSalle.

Misner, C.W., Thorne, K.S., and Wheeler, J.A. (1973) *Gravitation*, Freeman, San Francisco.

Møller, C. (1972) *The Theory of Relativity*, 2nd ed., Oxford University Press, Oxford.

Munem, M.A. and Foulis, D.J. (1984) *Calculus*, Worth Publishers, New York.

Narlikar, J.V. and Padmanabhan, T. (1991) "Inflation for Astronomers", *Ann. Rev. Astron. Astrophys.*, **29,** 325–62.

Needham, J. and Ling, W. (1965) *Science and Civilisation in China*, Vol. 4, Part 2 (Mechanical Engineering), Cambridge University Press, Cambridge.

Ohanian, H.C. (1989) *Physics*, 2nd ed., Norton, New York.

Ohanian, H.C. and Ruffini, R. (1994) *Gravitation and Spacetime*, 2nd ed., Norton, New York.

Penzias, A.A. and Wilson, R.W. (1965) "A measurement of excess antenna temperature at 4080 Mc/s," *Astrophys. J.*, **142,** 419–21.

Pound, R.V. and Rebka, G.A. (1960) "Apparent weight of photons," *Phys. Rev. Lett.*, **4,** 337–41.

Riley, J.M. (1973) "A measurement of the gravitational deflection of radio waves by the Sun during 1972 October," *Mon. Not. R. Astr. Soc.*, **161,** 11P–14P.

Rindler, W. (1982) *Introduction to Special Relativity*, Oxford University Press, Oxford.

Robertson, H.P. (1935) "Kinematics and world structure," *Astrophys. J.*, **82,** 248–301.

Robertson, H.P. (1936) "Kinematics and world structure," *Astrophys. J.*, **83,** 257–71.

Schouten, J.A. (1954) *Ricci-Calculus*, 2nd ed., Springer, Berlin.

Shapiro, I.I. (1964) "Fourth test of general relativity," *Phys. Rev. Lett.*, **13,** 789–91.

Shapiro, I.I. (1968) "Fourth test of general relativity: Preliminary results," *Phys. Rev. Lett.*, **20,** 1265–9.

Shapiro, I.I., Ash, M.E., Campbell, D.B., Dyce, R.B., Ingalls, R.P., Jurgens, R.F., and Pettingill, G.H. (1971) "Fourth test of general relativity: New radar results," *Phys. Rev. Lett.*, **26,** 1132–5.

Shapiro, S.L. and Teukolsky, S.A. (1983) *Black Holes, White Dwarfs, and Neutron Stars: the Physics of Compact Objects*, Wiley, New York.

Stuckey, W.M. (1992) "Can galaxies exist within our particle horizon with Hubble recessional velocities greater than c?," *Am. J. Phys.* **60(2),** 142–46.

Symon, K.R (1971) *Mechanics*, 3rd ed., Benjamin Cummings, San Francisco.

Thirring, H. and Lense, J. (1918) "Über den Einfluss der Eigenrotation der Zentralkörper auf die Bewegung der Planeten und Monde nach der Einsteinschen Gravitationstheorie," *Phys. Zeit.*, **19**, 156–63.

Thorne, K.S. (1974) "The search for black holes," *Scientific American*, **231,** No. 6, 32–43.

Walker, A.G. (1936) "On Milne's theory of world structure," *Proc. London Math. Soc.*, **42,** 90–127.

Weinberg, S. (1972) *Gravitation and Cosmology: Principles and Applications of the General Theory of Relativity*, Wiley, New York.

Weisberg, J.M. and Taylor, J.H. (1984) "Observations of post-Newtonian timing effects in the binary pulsar PSR 1913+16," *Phys. Rev. Lett.*, **52,** 1348–50.

Winget, D.E., Hansen, C.J., Liebert, J., van Horn, H.M., Fontaine, G., Nather, R.E., Kepler, S.O., and Lamb, D.Q. (1987) "An independent method for determining the age of the universe," *Astrophys. J.*, **315,** L77–L81.

Young, P.J., Westphal, J.A., Kristian, J., Wilson, C.P., and Landauer, F.P. (1978) "Evidence for a supermassive object in the nucleus of the galaxy M87 from SIT and CCD area photometry," *Astrophys. J.*, **221,** 721-30.

Index